T0318727

CLOUD COMPUTING IN OCEAN AND ATMOSPHERIC SCIENCES

CLOUD COMPUTING IN OCEAN AND ATMOSPHERIC SCIENCES

Edited by

TIFFANY C. VANCE
Alaska Fisheries Science Center, NOAA Fisheries, Seattle, WA, USA

NAZILA MERATI
Merati and Associates, Seattle, WA, USA

CHAOWEI YANG
George Mason University, Fairfax, VA, USA

MAY YUAN
Geospatial Information Sciences, School of Economic, Political, and Policy Sciences, University of Texas at Dallas, Richardson, TX, USA

Amsterdam • Boston • Heidelberg • London
New York • Oxford • Paris • San Diego
San Francisco • Singapore • Sydney • Tokyo
Academic Press is an imprint of Elsevier

Academic Press is an imprint of Elsevier
125 London Wall, London EC2Y 5AS, UK
525 B Street, Suite 1800, San Diego, CA 92101-4495, USA
50 Hampshire Street, 5th Floor, Cambridge, MA 02139, USA
The Boulevard, Langford Lane, Kidlington, Oxford OX5 1GB, UK

Copyright © 2016 Elsevier Inc. All rights reserved. Tiffany C. Vance's editorial and chapter
contributions to the Work are the work of a U.S. Government employee.

No part of this publication may be reproduced or transmitted in any form or by any
means, electronic or mechanical, including photocopying, recording, or any information
storage and retrieval system, without permission in writing from the publisher. Details on
how to seek permission, further information about the Publisher's permissions policies and
our arrangements with organizations such as the Copyright Clearance Center and the
Copyright Licensing Agency, can be found at our website: www.elsevier.com/permissions.

This book and the individual contributions contained in it are protected under copyright
by the Publisher (other than as may be noted herein).

Notices
Knowledge and best practice in this field are constantly changing. As new research and
experience broaden our understanding, changes in research methods, professional practices,
or medical treatment may become necessary.

Practitioners and researchers must always rely on their own experience and knowledge in
evaluating and using any information, methods, compounds, or experiments described
herein. In using such information or methods they should be mindful of their own safety
and the safety of others, including parties for whom they have a professional responsibility.

To the fullest extent of the law, neither the Publisher nor the authors, contributors, or
editors, assume any liability for any injury and/or damage to persons or property as a
matter of products liability, negligence or otherwise, or from any use or operation of any
methods, products, instructions, or ideas contained in the material herein.

British Library Cataloguing-in-Publication Data
A catalogue record for this book is available from the British Library

Library of Congress Cataloging-in-Publication Data
A catalog record for this book is available from the Library of Congress

ISBN: 978-0-12-803192-6

For information on all Academic Press publications
visit our website at https://www.elsevier.com/

  Working together
to grow libraries in
ELSEVIER   Book Aid   developing countries
             International
www.elsevier.com • www.bookaid.org

Publisher: Janco Candice
Acquisition Editor: Louisa Hutchins
Editorial Project Manager: Rowena Prasad
Production Project Manager: Paul Prasad Chandramohan
Designer: Mark Rogers

Typeset by TNQ Books and Journals

In memory of Doug Nebert, whose gentle guidance and steadfast support was critical to many of the projects described in this book.

CONTENTS

LIST OF CONTRIBUTORS

A. Arribas
Met Office Informatics Lab, Exeter, UK

K. Butler
Esri, Redlands, CA, USA

H. Caumont
Terradue Srl, Rome, Italy

G. Cervone
Pennsylvania State University, University Park, PA, USA

B. Combal
IOC-UNESCO, Paris, France

R. Correa
European Centre for Medium-Range Weather Forecasts, Reading, UK

P. Dhingra
Microsoft Corporation, Seattle, WA, USA

R. Fatland
University of Washington, Seattle, WA, USA

D. Gannon
School of Informatics and Computing, Indiana University, Bloomington, IN, USA

R. Hogben
Met Office Informatics Lab, Exeter, UK

Q. Huang
University of Wisconsin–Madison, Madison, WI, USA

C.N. James
Embry-Riddle Aeronautical University, Prescott, AZ, USA

Y. Jiang
George Mason University, Fairfax, VA, USA

J. Li
University of Denver, Denver, CO, USA

W. Li
School of Geographical Sciences and Urban Planning, Arizona State University, Tempe, AZ, USA

K. Liu
George Mason University, Fairfax, VA, USA

P. MacCready
University of Washington, Seattle, WA, USA

B. McKenna
RPS ASA, Wakefield, RI, USA

R. Mendelssohn
NOAA/NMFS/SWFSC, Santa Cruz, CA, USA

N. Merati
Merati and Associates, Seattle, WA, USA

A. Merten
NOAA, National Ocean Service, Seattle, WA, USA

R. Middleham
Met Office Informatics Lab, Exeter, UK

N. Oscar
Oregon State University, Corvallis, OR, USA

T. Powell
Met Office Informatics Lab, Exeter, UK

R. Prudden
Met Office Informatics Lab, Exeter, UK

M. Ramamurthy
University Corporation for Atmospheric Research, Boulder, CO, USA

B. Raoult
European Centre for Medium-Range Weather Forecasts, Reading, UK; University of Reading, Reading, UK

N. Robinson
Met Office Informatics Lab, Exeter, UK

M. Saunby
Met Office Informatics Lab, Exeter, UK

J.L. Schnase
NASA Goddard Space Flight Center, Greenbelt, MD, USA

H. Shao
School of Geographical Sciences and Urban Planning, Arizona State University, Tempe, AZ, USA

K. Sheets
NOAA, National Weather Service, Bohemia, NY, USA

B. Simons
NOAA/NMFS/SWFSC, Santa Cruz, CA, USA

A. Sinha
Esri Inc., Redlands, CA, USA

S. Stanley
Met Office Informatics Lab, Exeter, UK

K. Tolle
Microsoft Research, Seattle, WA, USA

J. Tomlinson
Met Office Informatics Lab, Exeter, UK

T.C. Vance
Alaska Fisheries Science Center, NOAA Fisheries, Seattle, WA, USA

S. Wang
School of Geographical Sciences and Urban Planning, Arizona State University, Tempe, AZ, USA

J. Weber
University Corporation for Atmospheric Research, Boulder, CO, USA

R.S. Wigton
Bin Software Co., Bellevue, WA, USA

R. Wright
NOAA, National Ocean Service, Silver Spring, MD, USA

S. Wu
School of Geographical Sciences and Urban Planning, Arizona State University, Tempe, AZ, USA

J. Xia
George Mason University, Fairfax, VA, USA

C. Yang
George Mason University, Fairfax, VA, USA

M. Yuan
Geospatial Information Sciences, School of Economic, Political, and Policy Sciences, University of Texas at Dallas, Richardson, TX, USA

X. Zhou
School of Geographical Sciences and Urban Planning, Arizona State University, Tempe, AZ, USA

AUTHOR BIOGRAPHIES

Alberto Arribas, Science Fellow at Met Office (United Kingdom) and Head of Informatics Lab.

The Informatics Lab combines scientists, software engineers, and designers to make environmental science and data useful. We achieve this through innovation and experimentation, moving rapidly from concepts to working prototypes.

In the past, Alberto has led the development of monthly-to-seasonal forecasting systems, co-authored over 40 scientific papers, been a lecturer and committee member for organizations such as World Meteorological Organization or the US National Academy of Sciences and has been Associate Editor for the Quarterly Journal of the Royal Meteorological Society.

Kevin A. Butler is a member of the Geoprocessing and Analysis team at Esri working primarily with the spatial statistics and multidimensional data tools. He holds a Bachelor of Science degree in computer science from the University of Akron, and a doctorate in geography from Kent State University. Prior to joining ESRI, he was a senior lecturer and manager of GIScience research at the University of Akron, where he taught courses in spatial statistics, geographic information system (GIS) programming, and database design.

Hervé Caumont Products & Solutions Program Manager at Terradue (http://www.terradue.com) is in charge of developing and maintaining the company's business relationships across international projects and institutions. This goes through the coordination of R&D activities co-funded by several European Commission projects, and the management of corporate programs for business development, product line innovation, and solutions marketing. At the heart of this expertise, a set of flagship environmental systems designed for researchers with data-intensive requirements, and active contributions to the Open Geospatial Consortium (http://opengeospatial.org), the Global Earth Observations System of Systems (http://earthobservations.org), and the Helix Nebula European Partnership for Cloud Computing in Science (http://www.helix-nebula.eu).

Guido Cervone is associate director of the Institute for CyberScience, director of the laboratory for Geoinformatics and Earth Observation, and associate professor of geoinformatics in the Department of Geography and Institute for CyberScience at The Pennsylvania State University. In addition, he is affiliate scientist with the Research Application Laboratory (RAL) at the National Center of Atmospheric Research (NCAR) in Boulder, Colorado, and research fellow with the National Center for Supercomputing Applications (NCSA) at the University of Illinois at Urbana-Champaign, Illinois. He sits on the advisory committee of the United Nations Environmental Program (UNEP), Division of Early Warning and Assessment (DEWA). He received the Ph.D. in Computational Science and Informatics in 2005. His fields of expertise are geoinformatics, machine learning, and remote sensing. His research focuses on the development and application of computational algorithms for the analysis of spatiotemporal remote sensing, numerical modeling, and social media "Big Data." The problem domains of his research are related to environmental hazards and renewable energy. His research has been funded by Office of Naval Research (ONR), US Department of Transportation (USDOT), National Geospatial-Intelligence Agency (NGA), National Aeronautics and Space Administration (NASA), Italian Ministry of Research and Education, Draper Labs, and StormCenter Communications. In 2013, he received the "Medaglia di Rappresentanza" from the President of the Italian Republic for his work related to the Fukushima crisis.

He does not own a cell phone. He has sailed over 4000 offshore miles.

Bruno Combal studied atmospheric physics and has a Ph.D. on radiative transfer modeling. After 8 years of research on the assessment of vegetation biophysical parameters from space observations, he joined the European Commission Joint Research Center (JRC) in which he developed several satellite image-processing chains, and a computer system to process Eumet-Cast data in near real time (eStation). Since December 2012, he has worked for the Intergovernmental Oceanographic Commission (IOC) of United Nations Educational, Scientific and Cultural Organization (UNESCO) in Paris, as a scientific data and scientific computing expert in the Ocean Observations and Services section.

Ricardo Correa, European Center for Medium-Range Weather Forecasts (ECMWF). Ricardo has been working at ECMWF since 1997 in a number of different analyst roles ranging from the design and deployment of a wide area Multiprotocol Label Switching (MPLS) private network for meteorological data to projects such as Distributed European Infrastructure for Supercomputing Applications (DEISA) for establishing a supercomputer grid coupling the distributed resources of 11 National Super-computing Services across Europe. Currently, he leads the Network Applications Team and has a special interest in Cloud Computing, High-performance Computing, and distributed software design.

Prashant Dhingra is a Principal Program Manager with Microsoft where he works with data scientists and engineers to build a portfolio of Machine Learning models. He works to identify gaps and feature requirement for Azure Machine Learning (ML) and related technology and to ensure models are built efficiently, performance and accuracy are good, and they have a good return on investment. He is working with National Flood Interoperability Experiment (NFIE) to build a flood-forecasting solution.

Rob Fatland is the University of Washington Director of Cloud and Data Solutions. From a background in geophysics and a career built on computer technology, he works on environmental data science and real-world relevance of scientific results; from carbon cycle coupling to marine microbial ecology to predictive modeling that can enable us to restore health to coastal oceans.

Dennis Gannon is a computer scientist and researcher working on the application of cloud computing in science. His blog is at http://esciencegr oup.com. From 2008 until he retired in 2014, he was with Microsoft Research (MSR) and MSR Connections as the Director of Cloud Research Strategy. In this role, he helped provide access to cloud computing resources to over 300 projects in the research and education community. Gannon is a professor emeritus of Computer Science at Indiana University and the former science director of the Indiana Pervasive Technology Labs. His interests include large-scale cyber infrastructure, programming systems and tools, distributed and parallel computing, data analysis, and machine learning. He has published more than 200 refereed articles and three co-edited books.

Richard Hogben is a computer programmer and communications expert. His qualifications include a degree in physics, a diploma in Spanish, and a certificate in programming FORTRAN. Prior to joining the Met Office, he taught science to teenagers in Zimbabwe and did statistical analysis for a government agency in London. In recent years, he has worked on the development and support of the Met Office's web applications. He is now using his creative skills in the Informatics Lab.

Qunying Huang received her Ph.D. in Earth System and Geoinformation Science from George Mason University in 2011. She is currently an Assistant Professor in the Department of Geography at University of Wisconsin–Madison. Her fields of expertise are geographic information science (GIScience), cyberinfrastucture, Big Data mining, large-scale environmental modeling and simulation. She is very interested in applying different computing models, such as cluster, grid, graphics processing unit (GPU), citizen computing, and especially cloud computing, to address contemporary computing challenges in GIScience. Most recently, she is leveraging and mining social media data for various applications, such as emergency response, disaster coordination, and human mobility.

Curtis James is Professor of Meteorology and Department Chair of Applied Aviation Sciences at Embry–Riddle Aeronautical University (ERAU) in Prescott, Arizona. He has taught courses in beginning meteorology, aviation weather, thunderstorms, satellite and radar imagery interpretation, atmospheric physics, mountain meteorology, tropical meteorology, and weather forecasting for over 16 years. He has also served as Director of ERAU's Undergraduate Research Institute and as faculty representative to the University's Board of Trustees. He participates in ERAU's Study Abroad program, offering alternating summer programs each year in Switzerland and Brazil.

He earned a Ph.D. in Atmospheric Sciences from the University of Washington (2004) and participated in the Mesoscale Alpine Program (MAP, 1999), an international field research project in the European Alps. His research specialties include radar, mesoscale, and mountain meteorology. He earned his B.S. degree in Atmospheric Science from the University of Arizona (1995), during which time he gained operational experience as a student intern with the National Weather Service Forecast Office in Tucson, Arizona (1993–1995).

Yongyao Jiang is a Ph.D. student in Earth Systems and GeoInformation Sciences, at Department of Geography and GeoInformation Science and National Science Foundation (NSF) Spatiotemporal Innovation Center, George Mason University, Fairfax, Virginia. Prior to Mason, he earned his M.S. degree (2014) in GIScience from Clark University, Worcester, Massachusetts, and B.E. degree (2012) in remote sensing from Wuhan University, Wuhan, China. He has received the First Prize in the Robert Raskin CyberGIS student competition, Association of American Geographers. His research interests range from geospatial cyberinfrastructure, to data mining, and spatial data quality.

Jing Li received her M.S. degree in earth system science, and Ph.D. In Earth System and Geoinformation Science from George Mason University, Fairfax, Virginia, in 2009 and 2012, respectively. She is currently an Assistant Professor with the Department of Geography and the Environment, University of Denver, Denver, Colorado. Her research interests include spatiotemporal data modeling, geovisualization, and geocomputation.

Wenwen Li is an assistant professor in GIScience at Arizona State University. She obtained her B.S. degree in Computer Science from Beijing Normal University (Beijing, China); M.S. degree in Signal and Information

Processing from Chinese Academy of Sciences (Beijing, China), and her Ph.D. in Earth System and Geoinformation Science from George Mason University (Fairfax, Virginia). Her research interest is in cyberinfrastructure, semantic web, and space–time data mining.

Kai Liu is currently a graduate student in the Department of Geography and GeoInformation Sciences (GGS) in the College of Science at George Mason University. Previously, he was a visiting scholar at the Center of Intelligent Spatial Computing for Water/Energy Science (CISC), and worked for 4 years at Heilongjiang Bureau of Surveying and mapping in China. His previous education was at Wuhan University, China, B.A. degree in Geographic Information Science. His research focuses on geospatial semantics, geospatial metadata management, spatiotemporal cloud computing, and citizen science.

Parker MacCready is a Professor in the School of Oceanography at the University of Washington (UW), Seattle. He specializes in the physics of coastal and estuarine waters, often developing realistic computer simulations, and is the lead of the UW Coastal Modeling Group. The forecast models developed by his group have been applied to important problems such as ocean acidification, harmful algal blooms, hypoxia, and regional effects of global climate change. He received a B.A. degree in Architecture from Yale University in 1982, an M.S. degree in Engineering Science from California Institute of Technology in 1986, and a Ph.D. in Oceanography from UW in 1991. He has written nearly 50 research papers.

Brian McKenna is a Senior Programmer at RPS Group/Applied Science Associates (RPS/ASA). He is an atmospheric scientist and Information Technology (IT) Specialist. He has atmospheric modeling expertise in development and implementation of primitive models and advanced statistical models. His IT experience covers a broad range of data delivery and storage techniques and systems administration for high-performance computing (HPC) environments. Brian's interests include enhancing model performance and scalability with tighter integration from IT best practices and innovations. He has a B.S. degree in Meteorology from Pennsylvania State University and an M.S. degree in Atmospheric Sciences from the University of Albany.

Roy Mendelsohn is a Supervisory Operations Research Analyst at National Oceanic and Atmospheric Administration (NOAA)/National Marine Fisheries Service (NMFS)/Southwest Fisheries Science Center (SWFSC)/Environmental Research Division (ERD). He leads a group at ERD that serves a wide assortment of data (presently about 120TB) through a variety of web services and web pages. He has been actively involved in serving data since 1998, helped write NOAA's Global Earth Observation—Integrated Data Environment (GEO-IDE) framework and as well as the original Integrated Ocean Observing Systems Data Management and Communication (IOOS DMAC) Plan. He has been involved in projects related to data sharing in IOOS, Ocean Observatories Initiative Cyberinfrastructure (OOICI), and the Federal GeoCloud Project among others and has served on NOAA's Data Management and Integration Team since its inception. In his spare time, he does large-scale statistical modeling of climate change in the ocean.

Nazila Merati is an innovator successful at marketing and executing uses of technology in science. She focuses on peer data sharing for scientific data, integrating social media information for science research, and model validation. Nazila has more than 20 years of experience in marine data discovery and integration, geospatial data modeling and visualization, data stewardship including metadata development and curation, cloud computing, and social media analytics and strategy.

Amy Merten is the Chief of the Spatial Data Branch, NOAA's Assessment and Restoration Division, Office of Response and Restoration (OR&R) in Seattle, Washington. Amy developed the original concept for an online mapping/data visualization tool known as "ERMA" (Environmental Response Mapping Application). Amy oversees the data management and visualization activities for the Deepwater Horizon natural resource damage assessment case. Dr. Merten is the current Chair of the Arctic Council's Emergency Prevention, Preparedness and Response Work Group. Dr. Merten received her doctorate (2005) and Masters degree (1999) in Marine, Estuarine, and Environmental Sciences with a specialization in Environmental Chemistry from the University of Maryland; and a Bachelor of Arts (1992) from the University of Colorado, Boulder in Environmental, Organismic and Population Biology.

Ross Middleham is a member of the Met Office Informatics Lab. Creative design is what I do. I live and breathe design, taking inspiration from everything around me. I like to surround myself with designs, objects, and things that inspire me. Having these things can help to create that spark when you need it. I particularly love all things retro—1970s oranges and 1980s neons always catch my eye.

I work as Design Lead across the Met Office, collaborating with other organizations, agencies, and universities on a wide range of creative projects. I recently developed an event called 'Design Storm' as a way of helping to bring together industry creatives and undergraduates to inspire, collaborate, and innovate.

Nels Oscar studies graphics, data visualization, and how to make sense of it at Oregon State University, where he is currently pursuing a Ph.D. in Computer Science. He has worked on projects ranging from the visualization of volumetric ocean state forecasts to topic-specific sentiment analysis on Twitter. He spends a significant chunk of his time figuring out new and creative ways to re-purpose web browsers.

Thomas Powell is a member of the Met Office Informatics Lab. For me the Informatics Lab presents an exciting opportunity to work closer with the Met Office's world-leading scientists. I am really hoping to gain an insight into some of the clever stuff they do and help add some magical, cutting-edge technology fairy dust to better convey what's really going on.

Prior to joining the Lab, I have been primarily working in middleware with Java in the Met Office's Data Services team. I have a real appetite to learn and as such have dabbled in various front- and back-end technologies, something I am really looking forward to expanding upon while working in the Lab.

Outside of work, my main passion is sports, especially rugby! I play for my local team and enjoy the social side of rugby as much as the playing side. I have some exciting things going on this year; I have just got married, in August, to my long-term girlfriend Nikki. We are currently working on extending our house, and we have just become the proud owners of a new Labrador puppy "Harry."

Rachel Prudden is a member of the Met Office Informatics Lab. After studying Math at Southampton University, I joined the Met Office as a Visual Weather developer in 2012. Since then, I have been involved in various projects related to data visualization, mainly working in Python and JavaScript. I have always been curious about the scientific side of meteorology, and I would like to see the Lab start to bridge the gap between science and technology.

Mohan Ramamurthy is the Director of the Unidata program at the University Corporation for Atmospheric Research (UCAR) in Boulder, Colorado. He joined UCAR after spending nearly 17 years on the faculty in the Department of Atmospheric Sciences at the University of Illinois at Urbana–Champaign. Dr. Ramamurthy has bachelor's and master's degrees in Physics and Ph.D. in Meteorology. Over the past three decades, Mohan Ramamurthy has conducted research on a range of topics in mesoscale meteorology, numerical weather prediction, information technology, data services, and computer-mediated education, publishing over 50 peer-reviewed papers on those topics.

Dr. Ramamurthy pioneered the use of the then-emergent World Wide Web (and its precursor, Gopher) in the early 1990s for the dissemination of weather and climate information and multimedia educational modules, and was involved in the development of collaborative visualization tools for geoscience education. Dr. Ramamurthy is a Fellow of the American Meteorological Society.

As the Director of Unidata, Dr. Ramamurthy oversees a National Science Foundation-sponsored program and a cornerstone data facility that provides data services, tools, and cyberinfrastructure leadership to universities and the broader geoscience community.

Baudouin Raoult, ECMWF. Baudouin has been working for ECMWF since 1989, and has been involved in the design and implementation of ECMWF's Meteorological Archival and Retrieval System (MARS), ECMWF's data manipulation and visualization software (Metview), as well as ECMWF's data portals and web-based interactive charts, among other activities. He has been involved in several European Union-funded projects and is member of the World Meteorological Organization's Expert Team on the World Meteorological Organization (WMO) Information System Centers. Baudouin is currently principal software architect and strategist at ECMWF.

Niall Robinson is a member of the Met Office Informatics Lab. Niall has been researching atmospheric science for 8 years. He lived in the rainforest for three months, studying the chemical make-up of atmospheric aerosols for his Ph.D. He has been involved in experiments in the field and from research aircraft, from central London to the Rocky Mountains. He moved to the Met Office Hadley Center two years ago, where he studied the modeling of climate dynamics and multiyear forecasting. Recently, he's taken a slightly different challenge as a member of the newly formed Met Office Informatics Lab, where he sits on the boundary between science, technology, and design.

Michael Saunby develops software for postprocessing and exchange of monthly-to-decadal forecasts. His areas of expertise include scientific software development and project management. Michael is presently developing cloud-computing services for processing and sharing monthly-to-decadal forecasts.

Michael has been developing meteorological software since 1987, first at Reading University's Department of Meteorology, briefly at the ECMWF, and since 1996 at the Met Office. In April 2012, Michael helped organize and deliver the International Space Apps Challenge hackathon. He continues to design and deliver collaborative innovation events at the Met Office and across the United Kingdom.

John Schnase is a senior computer scientist and the climate informatics functional area lead in NASA's Goddard Space Flight Center's Office of Computational and Information Sciences and Technology. He is a graduate of Texas A&M University. His work focuses on the development of advanced information systems to support Earth science. Dr. Schnase is a Fellow of the American Association for the Advancement of Science (AAAS), a member of the Executive Committee of the Computing Accreditation Commission (CAC) of the Accreditation Board for Engineering and Technology (ABET), a former member of the President's Council of Advisors on Science and Technology (PCAST) Panel on Biodiversity and Ecosystems, and currently co-Chairs the Ecosystems Societal Benefit Area of the Office of Science and Technology Policy (OSTP) National Observation Assessment.

Hu Shao is currently a Ph.D. student in GIScience at Arizona State University. He obtained both his B.S. degree in Geographic Information Systems and M.S. degree in Cartography and Geographic Information Systems from Peking University (Beijing, China). His research interests are in Cyberinfrastructure, Geographic Data Retrieval, and Social Media Data Mining.

Kari Sheets is a Program and Management Analyst at the National Oceanic and Atmospheric Administration's National Weather Service. Prior to rejoining the National Weather Service, Kari was a Physical Scientist with NOAA's National Ocean Service Office of Response and Restoration (OR&R) where she was the lead for the Environmental Response Management Application (ERMA®) New England and Atlantic regions and ERMA's migration to a cloud-computing infrastructure. Ms. Sheets holds a Bachelor of Science in Atmospheric Science from the University of Louisiana at Monroe and a Masters of Engineering in Geographic Information Systems (GIS) from the University of Colorado at Denver. Kari spent the first 11 years of her career at the National Weather Service (NWS) working on numerical weather prediction guidance, GIS development to support gridded forecasting and guidance production, and overall NWS GIS collaboration and projects. Currently, Ms. Sheets leads the Geographic Information Systems Project of the National Weather Service's Integrated Dissemination Program.

Bob Simons is an IT Specialist at the NOAA/NMFS/SWFSC/ Environmental Research Division. Bob is the creator of ERDDAP, a data server which is used by over 50 organizations around the world. Bob has participated in data service activities with IOOS, OOICI, Open Network Computing (ONC), and NOAA's Data Management and Integration Team, among others.

Amit Sinha specializes in GIS, cloud computing and Big Data applications, and has deep interests in spatially querying and mining information from very large datasets in climate and other domains. He also has expertise in the use of machine-learning algorithms to build predictive models, and seeks innovative techniques to integrate them with cluster-computing tools such as Apache Hadoop and Apache Spark. He has authored, and helped develop desktop- and cloud-based geospatial software applications that are used worldwide. He is currently employed as a Senior GIS Software Engineer at Esri, Inc.

Simon Stanley works on long-range forecasting applications development. Simon's activities focus on developing science for user-relevant predictions. His current work includes an analysis of predictability of United Kingdom seasonal precipitation—using output from the high-resolution seasonal prediction system *GloSea,* and the potential for applications to hydrological predictions. He is also investigating observed correlations in United Kingdom regional temperature and precipitation. Simon joined the Met Office Hadley Center in October 2012 after graduating with a B.Sc. degree in Mathematics from Nottingham Trent University.

Kristin M. Tolle is the Director of the Data Science Initiative in Microsoft Research Outreach, Redmond, Washington.

Since joining Microsoft in 2000, Dr. Tolle has acquired numerous patents and worked for several product teams including the Natural Language Group, Visual Studio, and the Microsoft Office Excel Team. Since joining Microsoft Research's outreach program in 2006, she has run several major initiatives from biomedical computing and environmental science to more traditional computer and information science programs around natural user interactions and data curation. She was also directed the development of the Microsoft Translator Hub and the Environmental Science Services Toolkit.

She is also one of the editors and authors of one of the earliest books on data science, *The Fourth Paradigm: Data Intensive Scientific Discovery*. Her current focus is developing an outreach program to engage with academics on data science in general and more specifically around using data to create meaningful and useful user experiences across devices and platforms.

Prior to joining Microsoft, Tolle was an Oak Ridge Science and Engineering Research Fellow for the National Library of Medicine and a Research Associate at the University of Arizona Artificial Intelligence Lab managing the group on medical information retrieval and natural language processing. She earned her Ph.D. in Management of Information Systems with a minor in Computational Linguistics.

Dr. Tolle's present research interests include global public health as related to climate change, mobile computing to enable field scientists and inform the public, sensors used to gather ecological and environmental data, and integration and interoperability of large heterogeneous environmental data sources. She collaborates with several major research groups in Microsoft Research including eScience, computational science laboratory, computational ecology and environmental science, and the sensing and energy research group.

Jacob Tomlinson is an engineer with experience in software development and operational system administration. He uses these skills to ensure the Met Office Informatics Lab is building prototypes on the cutting edge of technology.

Tiffany C. Vance is a geographer working for the National Oceanic and Atmospheric Administration (NOAA). She received her Ph.D. in geography and ecosystem informatics from Oregon State University. Her research addresses the application of multidimensional GIS to both scientific and historical research, with an emphasis on the use and diffusion of techniques for representing three- and four-dimensional data. Ongoing projects include developing cloud-based applications for particle tracking and data discovery, supporting enterprise GIS adoption at NOAA, developing histories of environmental variables affecting larval pollock recruitment and survival in Shelikof Strait, Alaska, and the use of GIS and visualizations in the history of recent arctic science. She was a participant in the first US Geological Survey (USGS)-initiated GeoCloud Sandbox to explore the use of the cloud for geospatial applications.

Sizhe Wang is a Masters student in GIScience at Arizona State University. He obtained his bachelor degree majoring in GIScience in China University of Geosciences (Wuhan, China). His current research interests focus on cyberinfrastructure, spatial data discovery and retrieval, spatial data visualization, and spatiotemporal data analysis.

Jeff Weber is a Scientific Project Manager at the Unidata Program Center, a division of the University Corporation for Atmospheric Research in Boulder, Colorado. Jeff has created case studies, maintained the Internet Data Distribution system, worked on visualization tools, managed cloud implementation, and many other activities to support the Unidata community since 1998. Jeff received the National Center for Atmospheric Research (NCAR) award for Outstanding Accomplishment in Education and Outreach in 2006, and continues to reach out to the community.

Mr. Weber earned his B.S and M.S. degrees from the University of Colorado (1984, 1999) with a focus on Arctic Climate and Remote Sensing. Jeff spent the 1997–1998 field seasons on the Greenland Ice Sheet collecting data and installing towers to support the Program for Regional Climate Assessment (PARCA) sponsored by NASA.

Jeff continues to stay active in his community, supporting science as the NCAR science wizard, and continuing outreach to many of the Boulder area schools. Jeff is married with three children, and they all enjoy the outdoor activities that are available in the Boulder area.

Scott Wigton is a co-founder and Managing Director at Bin Software. Bin's software products fuel scientific insight and discovery through data-intensive visualization, simulation, and modeling using the emerging generation of affordable virtual reality (VR), atmospheric research (AR), and holographic hardware. Prior to founding Bin, Mr. Wigton was an engineer and product leader at Microsoft for two decades, where he held a range of technical roles. He served as General Project Manager (GPM) for the company's Virtual Earth/Bing Maps geospatial platform in the run-up to the release of the Bing search engine. Among other key roles, he led product engineering for Bing's local search relevance effort, held leadership roles in the company's Technical Computing and HPC-for-cloud efforts, and served as a Director of Engineering for early high-scale social content efforts. His software patents fall mainly in the storage systems area. Mr. Wigton received his B.S. degree in Chemical Engineering from the University of Virginia in 1984, with an emphasis in biochemical systems and thesis focus in the computational modeling of the James River estuary in Virginia. Mr. Wigton also holds an M.F.A. degree from the University of Arizona, where he held a teaching appointment in the Department of Rhetoric and Composition.

Robb Wright is a geographer working for NOAA. He has an M.A. degree in Geography and GIS from the University of Maryland and a B.A. degree in Geography from Virginia Polytechnic Institute and State University. He has worked on the Environmentally Sensitivity Index Data Viewer and other tools to make data discoverable and viewable online.

Sheng Wu is a lecturer in the School of Computer and Information Science at Southwest University (Chongqing, China). He obtained his M.S. degree in Computer Science from Southwest University and Ph.D. in Cartography and Geography Information System at the Institute of Remote Sensing and Digital Earth, Chinese Academy of Sciences (Beijing, China). He is now a visiting professor at Arizona State University. Sheng's research interest is in cyberinfrastructure, distributed spatiotemporal services, and semantic web.

Jizhe Xia earned his Ph.D. from George Mason University in August 2015, and he is working as a postdoctoral researcher at a cloud-computing company. His research interests include high-performance computing, web service quality, and cyberinfrastructure.

Chaowei Phil Yang received his Ph.D. from Peking University in 2000 and was recruited as a tenure track Assistant Professor of Geographic Information Science in 2003 by George Mason University. He was promoted as Associate Professor with tenure in 2009 and granted Full Professorship in 2014.

His research focuses on utilizing spatiotemporal principles to optimize computing infrastructure to support science discoveries and engineering development. He is leading GIScience computing by proposing several research frontiers including distributed geographic information processing, geospatial cyberinfrastructure, and spatial computing. These research directions are further consolidated through his research, publications, and workforce training activities. For example, he has been funded as Principal Investigator (PI) by multiple resources such as National Science Foundation (NSF) and NASA with over $5 M expenditures. He has also participated in several large projects total over $20 M. He has published over 100 papers, edited three books, and eight special issues for international journals. He is writing two books and editing two special issues. His publications have been among the top five cited and read papers of International Journal of Digital Evidence (IJDE) and Computers, Environment and Urban Systems (CEUS). His Proceedings of the National Academy of Sciences (PNAS) spatial computing definition paper was captured by Nobel Intent Blog in 2011. The spatial computing direction was widely accepted by the computer science community in 2013.

May Yuan is Ashbel Smith Professor of Geospatial Information Science at University of Texas at Dallas. May Yuan studies temporal GIS and its applications to geographic dynamics. She is a member of the Mapping Science Committee at the National Research Council (2009–2014), Associate Editor of the International Journal of Geographical Information Science, member of the editorial boards of Annals of American Association of Geographers and Cartography and Geographical Information Science, and a member of the academic committee of the United States Geospatial Intelligence Foundation.

Xiran Zhou is a Ph.D. student at Arizona State University. He obtained his B.S. degree in Geoscience from Ningbo University (Ningbo, China); and M.S. degree in Surveying Engineering from Wuhan University (Wuhan, China). His research interests are remote sensing data classification, cyberinfrastructure, and machine learning.

FOREWORD

Human society has always been dependent on and at the mercy of the forces of wind and sea. Recorded observations of the tide were performed by the early Greeks, whereas direct measurements of the air began in the Renaissance. The rather ancient fields of oceanic and atmospheric sciences may offer the greatest successes, and the greatest challenges, to the comparatively recent technology of Cloud computing. Success will be found because the Cloud approach is ideally suited to analyzing the enormous data volumes resulting from the evolution of sensors and numerical models: instead of attempting to deliver copies of data to all users in their own facilities, the Cloud brings the users to the data to compute in place on scalable, rentable infrastructure. This advantage is magnified when data from multiple sources are brought together to better address today's pressing multidisciplinary science and policy issues; indeed, the very fact that disparate data about the Earth are naturally related to each other by concepts of location and time provides a unifying framework that will help drive success. The Cloud also permits low-risk experimentation in developing customized products for end-users such as decision-makers, emergency responders, businesses, and citizens who may not have the expertise to directly work with the source data. However, notable challenges exist. The enormous computing power required to generate operational forecasts of complex physical problems occurring on scales from seconds to years, and from centimeters to thousands of kilometers, will likely continue to require dedicated, on-premises computing resources. There are technical issues involved in getting data into the Cloud, or into the specific Cloud that the user may prefer. Existing standards and tools for data access and manipulation are mostly focused on the older approach of transferring data to the user's facility, and may need adaptation. The pay-as-you-go cost model is a hurdle for some procurements. Policy issues of attribution, authoritativeness, traceability, and the respective roles of the government and private sector remain to be solved. Nevertheless, these challenges are surmountable, and it is likely that the new paradigm of Cloud computing will find tremendous success in the fields of oceanic and atmospheric sciences. The papers in this volume illustrate how we are now beginning to take advantage of this opportunity and to resolve some of the difficulties.

Jeff de La Beaujardière, Ph.D.
Data Management Architect
National Oceanic and Atmospheric Administration

ACKNOWLEDGMENTS

We wish to thank all of the contributors to this book for all the work they have done both on the projects described and in crafting their chapters. We would also like to thank the reviewers for the chapters. Without their willingness to review, often on an absurdly tight timeline, and the thoughtful, helpful, and demanding yet fair comments they sent, this process would have been much harder for the editors (and the authors).

The reviewers are:

Lori Armstrong	Esri
Janet Duffy-Anderson	NOAA/National Marine Fisheries Service
Jennifer Ferdinand	NOAA/National Marine Fisheries Service
Yingjie Hu	Department of Geography, University of California Santa Barbara
Thomas Huang	System Architect at JPL (NASA)
Scott Jacobs	NOAA/National Weather Service
Zhenlong Li	Department of Geography University of South Carolina
Ann Matarese	NOAA/National Marine Fisheries Service
Linda Mangum	University of Maine, Orono
Don Murray	NOAA/ESRL/PSD and Colorado University—CIRES
Ivonne Ortiz	University of Washington—Joint Institute for the Study of the Atmosphere and Ocean
Jon Rogers	University of Dundee
Jack Settlemaier	NOAA/National Weather Service
Stephan Smith	NOAA/National Weather Service
Malcolm Spaulding	Professor Emeritus Department of Ocean Engineering University of Rhode Island
Min Sun	Department of Geography and Geoinformation Science, George Mason University
Stan Thomas	Department of Computer Science Wake Forest University
Vera Trainer	NOAA/National Marine Fisheries Service
Kevin Tyle	Department of Atmospheric Sciences, University at Albany—SUNY
David Wong	Department of Geography and Geoinformation Science, George Mason University

and other reviewers who wish to remain anonymous.

Candice Janco was our original editor at Elsevier and she got the whole process started with support and unbridled enthusiasm. Shoshana Goldberg deftly shepherded the middle of the process. Rowena Prasad has been the Editorial Project Manager and she has answered all of our questions patiently, provided invaluable advice, and has been incredibly understanding of the challenges of wrangling this many authors and chapters. Paul Prasad Chandramohan has guided the production process and ensured that the final result is something we can all be proud of.

This research is contribution EcoFOCI-0855 to NOAA's Ecosystems and Fisheries-Oceanography Coordinated Investigations.

INTRODUCTION

When we first were approached to put together this book, we knew that our colleagues and peers were using the cloud to do great things. It was not until we saw the paper topics emerge that we discovered the wide array of frameworks and applications that existed within the disciplines of oceanic and atmospheric sciences.

Distributed computing and resource sharing toward developing models and sharing scientific results are not new concepts to science. Grid computing and virtual environments have been used to bring researchers together in one place to collaborate and compute. Today one can do similar tasks by committing code to remote repositories, storing and sharing files via cloud storage systems, and communicating in workgroups via one shared platform.

High-performance computing is no longer solely the realm of the computer scientist, but something that we take for granted when we store our music and photos or use software that exists solely in the cloud to manage client relationships. We harvest open data sources that governments make public, and we can connect to map services to create maps without having to buy expensive software. The cloud serves our data and software and is used to manage our daily work lives, and, for the most part, we have no idea that we are using such services. For the first time, the evolution of cloud computing for science is developing at the same rate as consumer-based cloud applications and it is changing the way science develops applications.

This book provides an overview and introduction to the use of cloud computing in the atmospheric and oceanographic sciences. Rather than being an introduction to the infrastructure of cloud computing, the authors focus on scientific applications and provide examples showing capabilities most needed in the domain sciences. The book is divided into three sections—the first gives a broad picture of cloud computing's use in atmospheric and ocean sciences. The first chapter provides a primer on cloud computing as a reference for the rest of the book. Kevin Butler and Nazila Merati's chapter on how analysis patterns shows provides a language for describing the use of the cloud in scientific research and provides examples of a variety of applications. Scott Wigton's paper explains how workflows are critical to cloud computing. Mohan Ramamurthy details the transition to cloud-based cyberinfrastructure at Unidata and how this transition fits into Unidata's wider mission. Bruno Combal and Hervé Caumont illustrate the ways in which cloud services can be used to analyse climate model outputs for studying climate change in the oceans and how these analyses can be shared. Niall Robinson and the team at the United Kingdom Met Office Informatics Lab show how they are using the cloud both in their day-to-day work for communication and collaboration and also for the development of visualizations of Met Office weather predictions. Curtis James and Jeff Weber detail the use of the cloud for teaching and specifically the creation of a cloud-based version of the Advanced Weather Interactive Processing System II (AWIPSII) weather forecasting system. Baudouin Raoult and Ricardo Correa describe ways to make massive datasets generated by the European Center for Medium Range Weather Forecasting (ECMWF) available via a public or commercial cloud.

The second section focuses on how cloud computing has changed the face of cyberinfrastructure and how greater computing power, algorithm development, and predictive analytics to detect behaviors and help guide decision making, have moved from sophisticated command centers to cloud-based solutions. Wen Wen Li and others examine the ways in which they have created a cyber infrastructure to support a variety of data management, analysis, and visualization tasks. Jiang et al. describe a portal hosted in the cloud that enables researchers to discover and share resources about the Polar Regions. John Schnase describes the creation of a climate analytics service at National Aeronautics and Space Administration (NASA) to move analyses closer to the massive datasets that are now being generated. Prashant Dhinghra et al. explain how the cloud can be used to create a platform to support better modeling and prediction of flooding and the ways these improved analyses can save lives. Amit Sinha shows how Big Data tools such as Hadoop and geographic information systems (GIS) can help analyze large datasets.

Applications of cloud-based computing are featured in the third section of the book. The projects range from how regional models run in the cloud can be used to monitor harmful algal blooms and ocean acidification to developing data platforms hosted in the cloud that give a common operating picture to first responders to natural hazards. The case studies not only describe a research problem and how they came to use cloud computing as a solution, but also give the reader a realistic assessment of some of the drawbacks of implementing cloud computing. Rob Fatland and others describe LiveOcean, a tool originally developed to assist with efforts to mitigate the effects of ocean acidification which also provides a model for a modular scientific data management system. Qunying Huang and Guide Cervone provide a case study that shows how data analytics can be used to analyze social media data to help with crisis relief. Brian McKenna details how deploying a meteorological/ocean forecasting system in the cloud to decrease the times needed to run the models and to make maintenance of the models easier. Li, Liu and Huang write of creating a version of the NASA ModelE climate model that includes a web portal to set model parameters, cloud instances of the model, and a data repository. Kari Sheets and others describe the challenges moving the Environmental Resource Management Application (ERMA) environmental response tool to the cloud. Roy Mendelssohn and Bob Simons provide a cautionary tale of some of the more subtle cost–benefit considerations when moving a large data service to the cloud.

The book concludes with a brief essay by May Yuan on the road ahead.

This is an exciting time for the world of cloud computing and how scientists access data, serve their data and models, and innovate the ways they communicate, analyze, and consume services using the cloud platform. We hope that the papers in this volume both educate readers about the tenets and applications of cloud computing in ocean and atmospheric sciences and inspire them to explore how cloud technologies can help further their research goals.

CHAPTER 1

A Primer on Cloud Computing

T.C. Vance

Alaska Fisheries Science Center, NOAA Fisheries, Seattle, WA, USA

Cloud computing is more prevalent in your personal and professional life than you may realize. If you have checked out an eBook from your local library, opened Office 365 to create documents and spreadsheets, used Google maps to find a location, stored or shared photographs using Flickr, or sent messages with Gmail, you have used cloud computing. In your research, if you have stored files and shared data with collaborators on an article via Dropbox, participated in a seminar via GoToMeeting, utilized ArcGIS Online to create and share maps, delineated a watershed using a geoprocessing service, visualized data using Tableau Public, or worked with a colleague whose climate model is hosted on Amazon Web Services (AWS), you have used some aspect of cloud computing. All of these tools depend upon data storage and computing resources found in "the cloud." These resources allow you access to virtually unlimited amounts of storage (though more storage will cost more), rapidly scalable computing (so you get your email as quickly at 3 pm as at 3 am), and available almost anywhere you have connectivity to the Internet (in your office, aboard a ship, or at a research field station). You do not have to buy and install software, specify the computer resource you need, know the size of the files you want to read in your email, or otherwise manage your use of these tools and services. You simply use them.

The most widely accepted definition of cloud computing comes from the US National Institute of Standards and Technology (NIST). NIST states:

> *Cloud computing is a model for enabling ubiquitous, convenient, on-demand network access to a shared pool of configurable computing resources (e.g., networks, servers, storage, applications, and services) that can be rapidly provisioned and released with minimal management effort or service provider interaction.*
>
> **Mell and Grance (2011)**

For oceanic and atmospheric scientists, cloud computing can provide a powerful and flexible platform for modeling, analysis, and data storage. High-performance resources can be acquired, and released, as needed

Cloud Computing in Ocean and Atmospheric Sciences
ISBN 978-0-12-803192-6
http://dx.doi.org/10.1016/B978-0-12-803192-6.00001-3

without the necessity of creating and supporting infrastructure. The resource can support any of a number of operating systems and analysis software. Teams can collaborate and use the same resources from geographically diverse locations. Large datasets can be stored in the cloud and accessed from any location with Internet connectivity. With the rise of analytical capabilities on the cloud, researchers can analyze large datasets using remote resources as needed without having to purchase or install software.

In the United States, the Federal Geospatial Data Committee's (FGDC) GeoCloud Sandbox has created geospatial cloud services that can be used by a variety of government agencies. The Sandbox supports virtual server instances that can be configured and shared. It provides a place to create and deploy web services and to evaluate costs and performance. It also provides an example of gaining security accreditation that can be used by agencies seeking accreditation for their own cloud operations. GeoCloud Projects have included a United States Census Bureau Topologically Integrated Geographic Encoding and Referencing (TIGER)/Line server to provide base layers for spatial analyses, warehousing, and dissemination of data for The National Map, and analysis of education data using Environmental Systems Research Institute (Esri) ArcServer (FGDC, 2014).

In Europe, the Helix Nebula project is working toward the development of a Science Cloud hosted on public/private commercial cloud resources. The supported projects range from biomedical research to computing and data management for the European Laboratory for Particle Physics (CERN)'s Large Hadron Collider. Another project is supporting the needs of the Group of Earth Observation to reduce risks from geohazards. An upcoming project will look at using the Helix Nebula to support the UNESCO Ocean and Coastal Information Supersite (Helix Nebula, 2014).

Individual US projects that have made good use of cloud computing include running regional numerical weather models on a cloud platform (Molthan et al., 2015), the deployment of the Ocean Biogeographic Information System (OBIS) and its 31.3 million observations using open source tools in a cloud environment (Fujioka et al., 2012), and LarvaMap, a cloud-based tool for running particle tracking models in support of studies of larval distributions and the early life history of commercially important fish species (see Fig. 1.1) (Vance et al., 2015). CloudCast is a project from the University of Massachusetts–Amherst to use cloud computing to create short-term weather forecasts tuned to the needs of individual customers (Krishnappa et al., 2013). Humphrey et al. (2012) used a cloud resource to calibrate models of watersheds with an eye on creating real-time interactive models of watersheds.

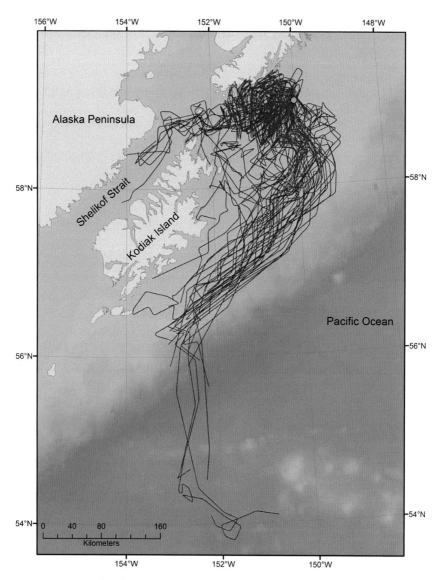

Figure 1.1 Example of the output from a cloud-based particle-tracking application called LarvaMap. Virtual fish larvae are released from a location (green dot) and their paths are calculated using a model running on the Amazon cloud. The red lines show the paths followed by the larvae over 60 days. *See Vance, T.C., Sontag, S., Wilcox, K., 2015. Cloudy with a chance of fish: ArcServer and cloud based fisheries oceanography applications. In: Wright, D., et al. (Eds.), Ocean Solutions, Earth Solutions. Esri Press, Redlands, California.*

Challenges to fully utilizing cloud resources can result from insufficient network connectivity and bandwidth. For government agencies and others, security concerns may limit the use of cloud resources until security accreditation is in place. The pay-as-you-go approach used by cloud providers may be challenging in a world of fixed yearly budgets and tight control of spending by line item or project. But, even with these challenges, a number of innovative cloud projects have been developed by oceanic and atmospheric scientists.

THE CHARACTERISTICS OF CLOUD COMPUTING

Cloud computing is defined by a number of characteristics. Users have the ability to perform for themselves many tasks previously limited to IT support personnel. They can request more storage or computing power via web interfaces and receive the new allocations automatically. Cloud resources can be used via a number of devices—sometimes referred to as thick and thin clients—from high-end workstations to mobile phones. You can read your email as easily on your phone as you can on your desktop computer. The cloud provider can make large amounts of storage and computing power available as a part of a pool of resources without each user having to purchase a fixed capacity in advance. The capacity is rented to the user for as long as it is needed and then released when the need is over. These increases or decreases in capacity can be obtained rapidly and automatically. Users pay for what they use, not for the full capacity of the system.

NIST defines a number of essential characteristics of cloud computing. These characteristics differentiate cloud computing from both desktop and mobile computing and define the ways a scientist interacts with a cloud computing resource and the capabilities of the resource itself. They are:

On-demand self-service.

A consumer can unilaterally provision computing capabilities, such as server time and network storage, as needed automatically without requiring human interaction with each service provider.

Broad network access.

Capabilities are available over the network and accessed through standard mechanisms that promote use by heterogeneous thin or thick client platforms (e.g., mobile phones, tablets, laptops, and workstations).

Resource pooling.

The provider's computing resources are pooled to serve multiple consumers using a multi-tenant model, with different physical and virtual resources dynamically assigned and reassigned according to consumer demand. There is a sense of location independence in that the customer generally has no control or knowledge over the exact location of the provided resources but may be able to specify location at a higher level of abstraction (e.g., country, state, or datacenter). Examples of resources include storage, processing, memory, and network bandwidth.

Rapid elasticity.

Capabilities can be elastically provisioned and released, in some cases automatically, to scale rapidly outward and inward commensurate with demand. To the consumer, the capabilities available for provisioning often appear to be unlimited and can be appropriated in any quantity at any time.

Measured service.

Cloud systems automatically control and optimize resource use by leveraging a metering capability at some level of abstraction appropriate to the type of service (e.g., storage, processing, bandwidth, and active user accounts). Resource usage can be monitored, controlled, and reported, providing transparency for both the provider and consumer of the utilized service.

Mell and Grance (2011)

These characteristics allow scientists to procure (and pay for) only the resources they need at a given time. The fact that acquiring these capabilities is self-service means they can be rapidly acquired at any time. In turn, they can be released when they are no longer needed. Network access means the resources are available anywhere there is a robust enough network connection. Huge computing or data storage capacities can be available, but no single user, program, or organization has to support the resources. The ability to rapidly acquire resources supports disaster response, analyzing real time data for satellites or experiments, and any other projects that require quick access to data or bursts of computing resources. Disaster response could be as simple as ensuring that the plans and documents needed for a response are stored in the cloud to allow access even if local infrastructure has been destroyed or as complex as the Environmental Response Management Application (ERMA) system (see Fig. 1.2) described elsewhere in this book.

The fact that resource use is measured and charged can be both an advantage and a challenge. Because the costs are well-defined, they can be presented to funding agencies clearly. The fact that the costs are dependent upon usage means a successful project may end up costing more than originally projected

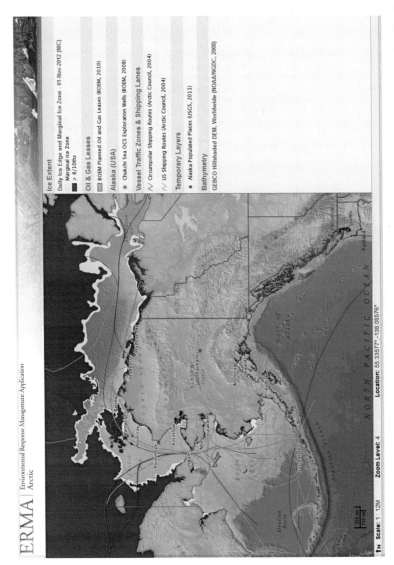

Figure 1.2 Example of a disaster response tool hosted in the cloud. NOAA's **Environmental Response Management Application®** (ERMA) is a web-based Geographic Information System (GIS) tool that assists both emergency responders and environmental resource managers in dealing with incidents that may adversely impact the environment. ERMA integrates and synthesizes various real-time and static datasets into a single interactive map, and provides fast visualization of the situation and improves communication and coordination among responders and environmental stakeholders (NOAA, 2014).

as more researchers take advantage of computational resources or more members of the public download data or use a web service.

SERVICE MODELS FOR CLOUD COMPUTING

Service models place cloud computing resources in a number of general categories by describing the ways in which cloud services are provided and the capabilities of the service. These model categories provide an easy way to refer to the services.

Software as a Service (SaaS) refers to software and applications that are hosted on a cloud resource. Examples include subscription services such as Microsoft's Office 365, the Adobe Creative Cloud for creating digital images, Esri's ArcGIS Pro, which provides professional level GIS applications via a web interface to a cloud-computing resource, and email via Google's Gmail and other cloud-based services. A prime aspect of SaaS is the fact that the user does not have to know anything about the underlying hardware, does not need to install or manage the software, and can use the software via a variety of devices (see Fig. 1.3).

Platform as a Service (PaaS) describes a situation in which the cloud provider provides an operating system, software libraries, and storage. The user can then install applications on the platform and run them. This is the level of service provided by Amazon Web Services (AWS), Microsoft Windows Azure, Google's App Engine, and other large cloud-infrastructure providers. An example of using PaaS would be a modeler who wants to run a large, CPU-intensive ocean or atmospheric circulation model. He or she would choose the cloud provider, specify the number and speed of the CPU cores, the amount of storage, the operating system, and any necessary compilers and libraries needed. Once the platform is created, often referred to as provisioning or spinning up, the user would install the model and run it. The user can store the parameters for the platform—operating system, size, libraries, storage, etc.—and start and stop the platform as needed. The sizes of the resources can be adjusted as needed. The user only pays for the time the platform is running.

Infrastructure as a Service (IaaS) describes a bare-bones situation in which computing and storage resources are provided, but the user must install operating systems, programs, applications, libraries, and any other needed elements. Google's Compute Engine, Microsoft's Azure, Rackspace Open Cloud, and Amazon Web Services (AWS) are examples of IaaS. Services such as Azure and AWS can be either platform or infrastructure depending on the capabilities chosen. IaaS might be used by a modeling consortium that wanted to provide computing resources for a comparison

Figure 1.3 Output from LiveOcean showing salinity values for the Strait of Juan de Fuca and the Washington coast in the northwestern United States. Darker colors show higher-salinity waters.

of models but in which each model ran on a different operating system, used different libraries, or the modelers wanted to tightly control all aspects of the model and its infrastructure.

NIST's original definition described three service models, but at least two more models have been developed since 2011. The original service models are:

Software as a Service (SaaS).

The capability provided to the consumer is to use the provider's applications running on a cloud infrastructure. The applications are accessible from various client devices through either a thin client interface, such as a web browser (e.g., web-based email), or a program interface. The consumer does not manage or control the underlying cloud infrastructure including network, servers, operating systems, storage, or even individual application capabilities, with the possible exception of limited user-specific application configuration settings.

Platform as a Service (PaaS).

The capability provided to the consumer is to deploy onto the cloud infrastructure consumer-created or acquired applications created using programming languages, libraries, services, and tools supported by the provider. The consumer does not manage or control the underlying cloud infrastructure including network, servers, operating systems, or storage, but has control over the deployed applications and possibly configuration settings for the application-hosting environment.

Infrastructure as a Service (IaaS).

The capability provided to the consumer is to provision processing, storage, networks, and other fundamental computing resources in which the consumer is able to deploy and run arbitrary software, which can include operating systems and applications. The consumer does not manage or control the underlying cloud infrastructure but has control over operating systems, storage, and deployed applications, and possibly limited control of select networking components (e.g., host firewalls).

Mell and Grance (2011)

Two newer concepts not covered in NIST's original definitions are Data as a Service (DaaS) and what is being called Spatial Analysis as a Service or Cloud Analytics. DaaS is cloud-hosted storage of large datasets. These can be supplied by the user or, more commonly, they are curated datasets that are hosted by the cloud service. These include datasets such as Google Earth Ocean which provide worldwide bathymetry data collated from a variety of sources, statistical data from the UN via UNdata (data.un.org), and the National Aeronautics and Space Administration (NASA) Earth Exchange (OpenNEX) project to make 20TB of NASA climate data widely available (https://nex.nasa.gov/nex/).

Cloud Analytics include the ability to perform analyses utilizing a cloud resource. A prime example is Esri's ArcGIS Online (AGOL), which

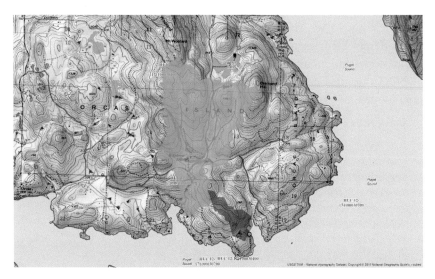

Figure 1.4 Two watersheds delineated using Esri's Watershed Delineation Tool. The user clicks on a point (green dot) and a cloud-based tool calculates the watershed for the point and displays the watershed as a polygon on the map (blue polygons).

provides GIS capabilities via web services. AGOL allows users in an organization or research project to share data, locate external data, create maps, and perform analyses online. The results can be shared with either the organization or the public. The spatial analysis service supports a variety of analyses such as aggregating, summarizing, buffering, calculating view or watersheds (Fig. 1.4), finding hot spots, and spatial interpolation. Open-source tools such as Hadoop provide analytics for big data, including scientific or business data, and less-structured data such as location of taxi pickups and drop-offs in New York City or tweets about severe weather. A chapter in this book by Sinha introduces Hadoop and describes ways in which it can be used for spatiotemporal data.

TYPES OF CLOUDS

Cloud infrastructure can be provided in a number of ways. A private cloud is a cloud resource that is for the exclusive use of a client or research group. Examples include banks, health care providers, or other institutions with high-security requirements, the military, or others with secrecy requirements, and scientific entities such as national weather services requiring high reliability and having timing constraints.

A community cloud is a shared cloud infrastructure. The cloud provider provides a pool of resources and a community uses and pays for the pool.

Within the pool, the resources are shared and guided by community rules. Examples include the GeoCloud Sandbox project through the FGDC (Nebert and Huang, 2013) which provided a community test-bed for researchers to experiment with deploying applications in the cloud using standard operating systems and libraries.

A public cloud is a fully public resource in which a customer just uses a small portion of a large resource. Examples include purchasing time on the Amazon Elastic Compute Cloud (EC2) or storing data on Amazon's Simple Storage Service (S3). Another example is Microsoft's Azure for Research (http://research.microsoft.com/en-US/projects/azure/default.aspx). The Open Science Data Cloud (https://www.opensciencedatacloud.org/) is a scientific community cloud that hosts projects in genomics, processing of satellite imagery, and knowledge complexity.

A hybrid cloud is a mixture of cloud infrastructures with the ability to easily move data and applications between the infrastructures as demand requires. These could also be local computing resources that use the cloud at times of peak demand.

NIST provides a definition of the models for deploying cloud resources.

Private cloud.

The cloud infrastructure is provisioned for exclusive use by a single organization comprising multiple consumers (e.g., business units). It may be owned, managed, and operated by the organization, a third party, or some combination of them, and it may exist on or off premises.

Community cloud.

The cloud infrastructure is provisioned for exclusive use by a specific community of consumers from organizations that have shared concerns (e.g., mission, security requirements, policy, and compliance considerations). It may be owned, managed, and operated by one or more of the organizations in the community, a third party, or some combination of them, and it may exist on or off premises.

Public cloud.

The cloud infrastructure is provisioned for open use by the general public. It may be owned, managed, and operated by a business, academic, or government organization, or some combination of them. It exists on the premises of the cloud provider.

Hybrid cloud.

The cloud infrastructure is a composition of two or more distinct cloud infrastructures (private, community, or public) that remain unique entities, but are bound together by standardized or proprietary technology that enables data and application portability (e.g., cloud bursting for load balancing between clouds).

Mell and Grance (2011)

A private cloud is very similar to a traditional centralized owned computing resource. An entity such as an agency or research university has complete control over the cloud and there is strong physical and logical security for the cloud. The actual hardware can either be at the client's location (in the same way that traditional computing resources are on site) or the hardware can be remotely located but logically and physically separated for the exclusive use of the client (akin to a centralized computing resource for an agency or scientific project). A private cloud provides the greatest degree of security, but also means that a single organization or entity is bearing all of the costs of the infrastructure. One aspect of a private cloud that differs from owning hardware is the ability to cloud burst. This is a tool to transfer less-sensitive functions to a public cloud during times of peak demand and create a hybrid cloud implementation. For example, collecting payment of payment information would remain on the private cloud, whereas browsing a catalog or airline timetable and ordering gifts or booking flights could be burst to a less-secure public cloud during the holidays or during travel disruptions in bad weather. This model is probably best only for government agencies or large corporations in which security is paramount.

A community cloud shares the costs among a group of organizations and is a logical mode for a consortium of universities or a large multinational research program that has long-term funding.

The public (or commodity) cloud is the archetypal cloud in that a commercial vendor creates the cloud resource and charges for its use. It is open to anyone who wishes to pay for resources. This would be a type of cloud well suited to smaller projects, short-term needs, or projects involving a number of widely scattered researchers. It is frequently used by startups as they start development before their computing needs are defined and they have the cash flow to support infrastructure.

SCIENCE IN THE CLOUD

The rest of this book is intended to be an introduction to the use of cloud computing in the atmospheric and oceanographic sciences with an emphasis on scientific applications and examples rather than on the infrastructure of cloud computing. In includes chapters on work done both within the United States and worldwide and includes both theoretical and applied discussions in the marine and atmospheric sciences. Theoretical discussions consider topics such as why cloud computing should be used in the atmospheric and oceanographic sciences and what the special needs and challenges from the atmospheric and oceanographic sciences are in using cloud computing. Applied examples cover the gamut from using large datasets, to practical examples, to examples of using the cloud to support interdisciplinary work.

ACKNOWLEDGMENTS

This chapter was greatly improved by reviews from Kevin Butler, Nazila Merati, Ann Matarese, Ivonne Ortiz, and Janet Duffy-Anderson. This research was supported, in part, with funds from the National Marine Fisheries Service's Ecosystems and Fisheries Oceanography Coordinated Investigations (EcoFOCI). This is contribution EcoFOCI-0854 to NOAA's Ecosystems and Fisheries-Oceanography Coordinated Investigations. The findings and conclusions in the paper are those of the author and do not necessarily represent the views of the National Marine Fisheries Service. Reference to trade names does not imply endorsement by the National Marine Fisheries Service, NOAA.

REFERENCES

FGDC, 2014. GeoCloud Sandbox Initiative Project Reports. https://www.fgdc.gov/initiatives/geoplatform/geocloud (viewed 27.12.14.).

Fujioka, E., Vanden Berghe, E., Donnelly, B., Castillo, J., Cleary, J., Holmes, C., Halpin, P.N., 2012. Advancing global marine biogeography research with open-source GIS software and cloud computing. Transactions in GIS 16 (2), 143–160.

Helix Nebula, 2014. Helix Nebula at Work. http://www.helix-nebula.eu/ (viewed 27.12.14.).

Humphrey, M., Beekwilder, N., Goodall, J.L., Ercan, M.B., 2012. Calibration of watershed models using cloud computing. In: 8th International Conference on E-Science (e-Science2012), pp. 1–8, 2012.

Krishnappa, D.K., Irwin, D., Lyons, E., Zink, M., 2013. CloudCast: cloud computing for short-term weather forecasts. Computing in Science & Engineering 15, 30. http://dx.doi.org/10.1109/MCSE.2013.43.

Mell, P.M., Grance, T., 2011. SP 800-145. The NIST Definition of Cloud Computing Technical Report. NIST, Gaithersburg, MD, United States.

Molthan, A.L., Case, J.L., Venner, J., Schroeder, R., Checchi, M.R., Zavodsky, B.T., Limaye, A., O'Brien, R.G., 2015. Clouds in the cloud: weather forecasts and applications within cloud computing environments. Bulletin of the American Meteorological Society, 96, 1369–1379. http://dx.doi.org/10.1175/BAMS-D-14-00013.1.

Nebert, D., Huang, Q., 2013. GeoCloud initiative. In: Yang, C., Huang, Q. (Eds.), Spatial Cloud Computing: A Practical Approach. CRC Press, Boca Raton, FL, pp. 261–272.

NOAA, 2014. Environmental Response Management Application. Web Application. National Oceanic and Atmospheric Administration, Arctic. http://response.restoration.noaa.gov/erma/.

Vance, T.C., Sontag, S., Wilcox, K., 2015. Cloudy with a chance of fish: ArcServer and cloud based fisheries oceanography applications. In: Wright, D., et al. (Eds.), Ocean Solutions, Earth Solutions. ESRI Press, Redlands, California.

CHAPTER 2

Analysis Patterns for Cloud-Centric Atmospheric and Ocean Research

K. Butler
Esri, Redlands, CA, USA

N. Merati
Merati and Associates, Seattle, WA, USA

INTRODUCTION

When Ferdinand Hassler began the coastal survey of the United States (US) in the 1800s, the necessary trigonometric calculations and maps were painstakingly calculated and drawn by hand. Today, ocean scientists have access to complex digital overlay software, which automatically detects shoreline change from Light Detection and Ranging (Lidar) and satellite imagery, and the results are stored in spatially aware databases. When Joseph Henry established the first network of volunteer weather observers in 1848, there were only 600 observers for the entire US, Latin America, and the Caribbean, and observations were painstakingly recorded by hand (Smithsonian, 2015). In contrast, atmospheric scientists today have access to petabytes of archived, near- and real-time collections of globally observed and modeled atmospheric data. Modern atmospheric and ocean sciences operate in a very different framework than these early pioneers. Modern science is computationally intensive, under increased pressure to be more interdisciplinary, challenged to analyze increasing volumes of data, and, in some instances, regulated to share data, research, and results with the public. Management of all of these ancillary pressures can potentially divert focus from primary research activities. This chapter uses a transformation of the scientific method, called e-Science, and a practice from software engineering, analysis patterns, to illustrate how cloud computing can help mitigate these challenges to modern scientific investigation. The benefits and liabilities of a cloud-based analytical framework for research in the atmospheric and ocean sciences are discussed.

Cloud Computing in Ocean and Atmospheric Sciences
ISBN 978-0-12-803192-6
http://dx.doi.org/10.1016/B978-0-12-803192-6.00002-5

Copyright © 2016 Elsevier Inc.
All rights reserved.

WHAT IS e-SCIENCE?

At the most basic level, e-Science is computing power applied to research. It is the application of modern computing technologies and infrastructure to the process of scientific investigation. However, a variety of other terms are also used to describe this process such as cyberscience and cyberinfrastructure (Jankowski, 2007). The term "e-Science" was introduced in 1999 in the United Kingdom (UK) by Dr. John Taylor, then Director General of Research Councils in the Office of Science and Technology (Hey and Trefethen, 2002). Initially the term had a very broad meaning: "e-Science is about global collaboration in key areas of science, and the next generation of infrastructure that will enable it" and "e-Science will change the dynamic of the way science is undertaken" (Taylor, n.d.). Nentwich (2003, p. 22) provided a similarly broad definition: "all scholarly and scientific research activities in the virtual space generated by the networked computers and by advanced information and communication technologies in general." Significant investments were made in both the US and the UK to provide infrastructure to support the forward-looking goals of e-Science. In the UK, £120M UK established the e-Science "Core Programme," computing infrastructure, and large-scale pilot programs (Hey and Trefethen, 2002). In 2003 in the US, the National Science Foundation (NSF) summarized its goals and a one billion US dollar annual budget request for cyberinfrastructure in a report entitled "Revolutionizing Science and Engineering through Cyberinfrastructure." The NSF report had far-reaching goals such as establishing "grids of computational centers, some with computing power second to none; comprehensive libraries of digital objects including programs and literature; multidisciplinary, well-curated federated collections of scientific data; thousands of online instruments and vast sensor arrays; convenient software toolkits for resource discovery, modeling, and interactive visualization; and the ability to collaborate with physically distributed teams of people using all of these capabilities" (Atkins, 2003, p. 7). These early investments in cyberinfrastructure spawned an ecosystem of distributed scientific production environments—"a set of computational hardware and software, in multiple locations, intended for use by multiple people who are not the developers of the infrastructure" (Katz et al., 2011, p. 1). Although there was an initial focus on computation, e-Science today embraces all aspects of the process of scientific investigation; data acquisition, data management, analysis, visualization, and dissemination of results (Fig. 2.1).

This is captured in Bohle's (2013) comprehensive definition: "E-science is the application of computer technology to the undertaking of modern

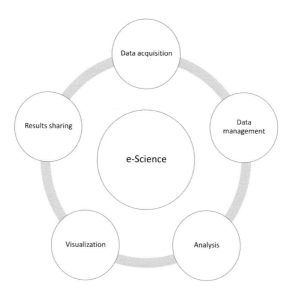

Figure 2.1 e-Science—the application of information technology to the process of scientific investigation.

scientific investigation, including the preparation, experimentation, data collection, results dissemination, and long-term storage and accessibility of all materials generated through the scientific process."

e-SCIENCE AND CLOUD COMPUTING

The commodification of computing resources, a deluge of data from increasingly inexpensive instruments, and solutions to previously intractable numeric simulations have changed the paradigm of scientific investigation. Since its empirical beginnings, science has expanded to include a theoretical branch with an emphasis on models and generalizations, a computational branch focusing on simulation, and, most recently, a data exploration branch (e-Science) which seeks to unify theory, experimentation, and simulation (Hey et al., 2009). Jim Gray provides a useful analogy to describe this transition to data–intensive, data–centric science: "people now do not actually look through telescopes. Instead, they are 'looking' through large-scale, complex instruments which relay data to data centers, and only then do they look at the information on their computers" (Gray, 2009, p. xix). Although this new paradigm provides tremendous opportunities for scientific investigation, it comes with a cost. The volume of data that needs to be stored and analyzed is intractable on a commodity scientific workstation. Further, management of the information technology keeps the scientist from focusing on the science. Utilizing Infrastructure as a Service (IaaS), Software as a Service (SaaS), and Data as a Service (DaaS) in the cloud changes the scale of scientific problems that can be addressed (Zhao et al., 2015). The benefits of the cloud as a scientific computing platform are summarized by Zhao et al. (2015, p. 3):

- Easy access to resources: Resources are offered as services and can be accessed over the Internet;
- Scalability on demand: Once an application is deployed onto the cloud, the application can be automatically made scalable by provisioning the resources in the cloud on demand, and the capability of scaling out and in, and load balancing;
- Better resource utilization: Cloud platforms can coordinate resource utilization according to resource demand of the applications hosted in the cloud; and
- Cost saving: Cloud users are charged based on their resource usage in the cloud, they only pay for what they use, and if their applications get optimized, that will immediately be reflected into a lowered cost.

Even with the resources of cloud computing, e-Science is not without its challenges. Jha and Katz (2010, p. 1) observe that "typically the time scales over which scientific applications are developed and used is qualitatively larger than the time scales over which the underlying infrastructure tends to evolve." In addition, good data curation, data provenance, and workflow documentation practices can be impeded when conducting analyses in a distributed environment (Gil et al., 2007). These challenges are not unique to e-Science. Software engineering—the application of engineering principles to the design, development, deployment, and maintenance of computer software—faces similar problems. One solution to the problem of documenting, describing, and reusing large complex software systems used successfully in software engineering is analysis patterns.

PATTERN LANGUAGE AND ANALYSIS PATTERNS

Christopher Alexander coined the term pattern language in the 1970s to describe a set of blueprints and directives for describing architectural elements. The language provides a formal way of describing problems and solutions and makes it possible for nonprofessionals to contribute to problem solving and improve conditions in communities. Pattern language was applied to computer programming and forms the core of object-oriented programming. Geyer-Schultz and Hahsler (2001) describe the use of pattern language in software engineering from the start of the project to the design of a solution. They also describe a "family of analysis patterns" for elements of the process from the use of a "pinboard" for team communication to virtual libraries of information for a project. Although some of these terms sound dated in the era of agile programming and Scrum, the basic functionalities are still relevant.

The term analysis pattern was introduced by Martin Fowler in 1997 to describe "groups of concepts that represent a common construction in business modeling" and that document "an idea that has been useful in one practical context and will probably be useful in others" (Fowler, 1997, p. 8). Analysis patterns can be used to abstract and document complex ideas and promote their reuse. They are abstractions of analytical processes rather than implementations in software. They describe "what" is to be accomplished, not "how" it will be done. Cloud computing provides a high-level abstraction of computing resources as services (Zhao et al., 2015) and analysis patterns provide a high-level abstraction of the process of data acquisition, data management, analysis workflow, visualization, and results dissemination within the cloud.

For example, an emerging pattern in the oceans and atmospheric sciences is collaborative data collection in which individual scientists, research institutes, and government agencies contribute scientific data into a large collection. Environmental Systems Research Institute (Esri) World Ocean Basemap is a real-world example of collaborative data collection. The map is compiled from a variety of best-available sources from several data providers, including General Bathymetric Chart of the Oceans (GEBCO), International Hydrographic Organization–Intergovernmental Oceanographic Commission (IHO-IOC) GEBCO Gazetteer of Undersea Feature Names, National Oceanic and Atmospheric Administration (NOAA), National Geographic, as well as several academic research institutes. When new contributions are received, a quality control process is initiated and the map is periodically republished. This type of collaborative data collection can be abstracted into a pattern called *Collector* (Fig. 2.2). In the collector pattern, two or more incoming sources converge without synchronization. If one or more sources are updated, the process following the merge is initiated.

For analysis patterns to be reusable, they must be described in a concise and consistent way. Geyer-Schulz and Hahsler (2001, p. 3) have devised a structured way to describe analysis patterns:

- **Pattern Name** (Gamma et al., 1995; Buschmann et al., 1996): A pattern's name precisely expresses the essence of a pattern. It becomes part of the vocabulary used in analysis.

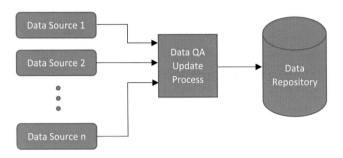

Figure 2.2 *Collector* analysis pattern. Two or more incoming sources converge without synchronization and trigger a data Quality Assurance and update process.

- **Intent** (Gamma et al., 1995): What does the analysis pattern do, and what problem does it address?
- **Motivation** (Gamma et al., 1995): A scenario that illustrates the problem, and how the analysis pattern contributes to the solution in the concrete scenario.
- **Forces and Context** (Alexander, 1979): Discussion of forces and tensions which should be resolved by the analysis pattern.
- **Solution** (Buschmann et al., 1996): Description of solution and the balance of forces achieved by the analysis pattern for the scenario in the motivation section. Includes all relevant structural and behavioral aspects of the analysis pattern.
- **Consequences** (Gamma et al., 1995; Buschmann et al., 1996): How does the pattern achieve its objectives and what trade-offs exist?
- **Design**: How can the analysis pattern be realized by design patterns? Sample design suggestions.
- **Known Uses** (Gamma et al., 1995; Buschmann et al., 1996): Examples of the pattern found in real systems.

The *Collector* pattern as applied to the oceans base map example introduced above can be described as follows:

Pattern Name: Collector

Intent: How can a group of geographically dispersed scientists collaborate to build an authoritative, high-quality collection of scientific data?

Motivation: You are working in large oceans research institute with the responsibility of producing high-quality maps for publication. Your research group conducts research across all of the oceans. You have high-quality bathymetric data for some of your research sites but not for all.

The process of creating high-quality, cartographically accurate base maps from your bathymetry data is very time-consuming.

Forces and Context:
- Efficient access to global data is required
- Base map must be authoritative and created with the best available data
- The base map must be cost free to its users

Solution: Use a cloud-based base map. This map is available at several cartographic scales and optionally includes marine water-body names, undersea feature names, and derived depth values in meters. It includes land features to provide context for coastal mapping applications. Land features can optionally include annotations for administrative boundaries, cities, and inland water names. The following actors use the system:

1. User: Retrieves maps from the cloud.
2. Information Provider: Adds new data to the cloud-based collection.
3. Administrator: Performs quality assurance tasks and republishes the map to the cloud.

Consequences: The benefits and liabilities of cloud-based collections of data.

1. *Data Curation*: The cloud-based map is curated and you can assume that it is based on the best-available data, relieving you of the responsibility of constantly searching for the most recent high-quality and authoritative bathymetry data for each of your research sites.
2. *Data provenance*: Relying on cloud-based data repositories brings up issues of data provenance—"where a piece of data came from and the process by which it arrived in the database" (Buneman et al., 2001, p. 1). For the cloud-based oceans base map, you do not have the ability to change the cartography applied to the data after it was collected.

Design: Follow any of the standard DaaS design patterns (see chapter: A Primer on Cloud Computing for a description of DaaS)

Known Uses: Esri Ocean Basemap (http://www.arcgis.com/home/item.html?id=5ae9e138a17842688b0b79283a4353f6)

e-SCIENCE ANALYSIS PATTERNS FOR THE CLOUD

Analysis patterns can be as numerous and diverse as the number of scientific methods (Fig. 2.3). This section describes five patterns which demonstrate the phases of the e-Science process. Each pattern is described using the structured approach of Geyer-Schulz and Hahsler (2001).

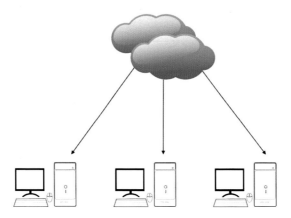

Figure 2.3 Analysis pattern *Cloud-based scientific data*: integrating cloud-based data into scientific workflow.

Pattern Name: Cloud–based scientific data—getting data from the cloud

Intent: Explores integrating the use of cloud-based data and how scientists can access large volumes of diverse, current, and authoritative data. Addresses the problem of locating and using large amounts of scientific data.

Motivation: You are working on a large, interdisciplinary science team and have been charged with finding data to support the analysis. There are many disparate data sources, and the datasets are too large to be stored locally. Some of the data change over time or are updated. You do not have the time or funding to update a local data repository. Using cloud-based datasets might alleviate the storage challenges while making it easier for all of the researchers to use the data.

Forces and Context:

- Efficient access to interdisciplinary data from a variety of sources of data is required
- Possible discomfort of scientists who are most familiar with using local data
- Concerns about reliable, fast data access
- Concerns about locating authoritative data
- Concerns about the persistence of the data

Solution: Use a cloud-based scientific data repository.

Consequences: The benefits and liabilities of using a cloud-based data repository are:

1. *Network latency*: For scientific workflows in which the analyses are performed outside the cloud, network latency can be a concern.

Although the I/O bandwidth of cloud-based IaaS systems has improved, it is still slower than local storage. Analysis workflows may be slower for data intensive applications.

2. *Data provenance*: Relying on cloud-based data repositories brings up issues of data provenance—"where a piece of data came from and the process by which it arrived in the database" (Buneman et al., 2001, p. 1).

3. *Data Curation*: With global scientific output doubling every nine years (Van Noorden, 2014), finding the highest quality data can be challenging. Cloud-based repositories are often curated. Data curation is the "on-going management of data through its lifecycle of interest and usefulness to scholarship, science, and education; curation activities enable data discovery and retrieval, maintain quality, add value, and provide for re-use over time" (Cragin et al., 2007).

Design: Plan for use of DaaS. See chapter "A Primer on Cloud Computing" for details on DaaS. See also other chapters in the book on integrating data from a number of services.

Known Uses: The Marine Geoscience Data System (MGDS) is "an interactive digital data repository and metadata catalog that offers a suite of services for data discovery and preservation. The Marine-Geo Digital Library currently provides open web-based access to 54.1TB of data, corresponding to over 635,000 digital data files from more than 2630 research programs dating back to the 1970s". The repository can be accessed through a geospatially enabled search interface and map-based data explorer, and is accessible through web services that follow the specifications of the Open Geospatial Consortium (OGC). The Marine-Geo Digital Library offers data preservation services including long-term archiving, the registration of data with digital object identifiers (DOIs)® for persistent identification and formal citation, and links to related scholarly publications. The Marine-Geo Digital Library provides reports summarizing data download activity to all contributing investigators twice annually; http://www.marine-geo.org/about/overview.php.

Jiang et al. in this book describe a Polar Cyberinfrastructure (CI) Portal that includes semantic searches and other tools to enable researchers to quickly locate the data they seek and ensure that they are getting the best data for their needs.

Pattern Name: Cloud-based management of scientific data—storing data in the cloud (Fig. 2.4).

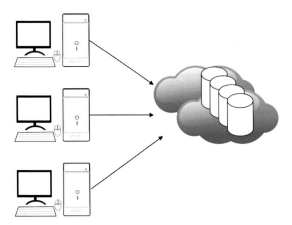

Figure 2.4 Analysis pattern *Cloud-based management of scientific data*: storing and managing data in the cloud.

Intent: Explores storing and managing data in the cloud. Addresses the problem of ever increasing data quantities with decreasing budgets for data management. Explores the secondary question of ways scientific projects can meet data access and dissemination requirements such as the US Public Access to Research Results (PARR) mandate (Holdren, 2013).

Motivation: You are the data manager for a fisheries oceanography project in a US federal agency. Your project includes ocean modelers who run circulation models and fisheries biologists who gather and analyze biological samples. The models generate hundreds of gigabytes per run in Network Common Data Form (netCDF) files. The biologists use a database to store their data and a Geographic Information System (GIS) to analyze their results. Due to the requirements of PARR, you must make your data available in a machine-readable format to the public within one year. How might storing and managing the data using a cloud resource have advantages over continuing to store the multiple formats of data locally?

Forces and Context:
- You are required to make your data publicly and easily accessible
- The data must be cost free
- Possible discomfort of scientists who wish to keep their data close at hand
- Concerns about reliable, fast data access
- Concerns about losing control of the data

- Security concerns if a government data provider. Required to meet data security requirements and to use an accredited cloud provider
- Concerns about the persistence of the data. Will the cloud provider stay in business?

Solution: Utilize a cloud-based scientific data platform that provides a storage system, metadata services, resource management, and security services, but retain local control of data loading, presentation, and archiving.

Consequences:

1. Storing the data in the cloud and making it available via a portal or server such as a Thematic Real-Time Environmental Distributed Data Services (THREDDS) server makes the data discoverable and usable by a variety of users.

2. The discovery tools inherent in creating a portal or server meet the requirements of PARR. Keeping responsibility for administration of the resource internal ameliorates some of the concerns about data ownership.

3. The architecture of a cloud-based portal or server enables the loading of disparate datasets—e.g., biological data and circulation model outputs in a single resource.

Design: Platform as a Service (PaaS) to host a THREDDS server or SaaS such as ArcGIS Server or ArcGIS portal.

Known Uses: ArcGIS Online (AGOL) is a cloud-based platform to make information available via discovery of data, quick visualization of the data, the creation of Story Maps to narrate datasets, and direct use of data in ArcGIS and other tools. NOAA is using AGOL to make a large number of datasets publically available (noaa.maps.arcgis.com). For a good summary of NOAA's implementation of AGOL see http://proceedings.esri.com/library/userconf/proc15/papers/1165_591.pdf. Mendelssohn and Simons provide a cautionary tale of the advantages and disadvantages of creating a data service in the cloud.

LiveOcean, described in the Fatland et al. chapter in this book, is an example of how to store and disseminate ocean modeling data and analyses based upon the data. Ramamurthy describes the plans for a THREDDS server at Unidata. Raoult and Correa describe dissemination of model outputs and multipetabyte datasets at European Centre for Medium-Range Weather Forecasts (ECMWF) and Sheets et al. present Environmental Resource Management Applications (ERMA) is a tool used by NOAA and its partners for gathering and disseminating: storing and managing data needed for response and restoration after oil spills and other disasters.

Pattern Name: Computing infrastructure for scientific research

Intent: Explores the ways in which cloud computing, in the form of PaaS or IaaS could be used as part of a research program and for teaching. It addresses the need for larger computational capabilities, especially under constrained budgets.

Motivation: Your university's computing infrastructure is insufficient for your analysis needs. When you look at the university plan for information technology for the next five years, you do not see any sign that the overall capabilities will improve and, in fact, you see a decreasing budget for infrastructure. You are in the midst of writing a grant proposal for a research project that will be linked with the development of new courses and techniques for your students to conduct their own research. What is the best computing platform for you to use for the work under your proposal?

Forces and Context:
- Your project needs better computing infrastructure than is currently available.
- There is no realistic local plan to provide improved infrastructure.
- Fears that scientists will need to become system administrators to create cloud infrastructure
- Concerns about integration with existing IT infrastructure
- Concerns about losing control of the infrastructure
- Need to meet data security requirements and to use an accredited cloud provider
- Concerns about the persistence of the platform. Will the cloud provider stay in business?

Solution: PaaS or IaaS depending on project needs. Might also include SaaS, DaaS, or even Analysis as a Service (AaaS) (Fig. 2.5).

Consequences:
1. A well-defined and provisioned infrastructure allow researchers to create resources as needed and gives them access to more powerful computing resources.
2. With time and effort, integration with existing local IT resources allows scientists "the best of both worlds".
3. The availability of a variety of operating systems allows researchers to use the platform that is best for their specific projects. Expansion to SaaS and AaaS allows researchers to share analyses.

Design: Various combinations of the services described in chapter "A Primer on Cloud Computing".

Known Uses: The Federal Geographic Data Committee (FGDC) Geo-Cloud Sandbox is an example of an infrastructure created to support a

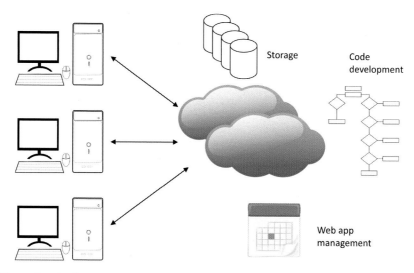

Figure 2.5 Analysis pattern *Computing infrastructure for scientific research*: which cloud computing, in the form of PaaS or IaaS, could be used as part of a research program and for teaching.

variety of projects. See http://cisc.gmu.edu/scc/presentation/Geo-Cloud-doug.pdf and https://www.fgdc.gov/initiatives/geoplatform/geocloud for details. The Global Earth Observation System of Systems (GEOSS) interoperability for Weather, Ocean and Water (GEOWOW) is a similar project based in Europe. See https://www.fgdc.gov/initia-tives/geoplatform/geocloud and the Raoult and Correa chapter in this book for details. Helix Nebula (http://www.helix-nebula.eu/) is "a new, pioneering partnership between big science and big business in Europe that is charting the course towards the sustainable provision of cloud computing - the Science Cloud." Unidata is planning to provide these kinds of resources, and their plans are described in Ramamurthy's chapter in this book. Additionally, James and Weber's chapter describes creating an infrastructure for teaching and student research. Li et al. describe Atmospheric Analysis Cyberinfrastructure (A2CI), a platform and services to support atmospheric research.

Pattern Name: Analysis in the Cloud (Fig. 2.6)
Intent: Explores conducting analyses in the cloud. Addresses the problem wanting to perform analyses on ever larger datasets and on datasets from multiple sources. Explores the secondary question of ways scientific projects can standardize analysis tools among geographically distributed researchers.

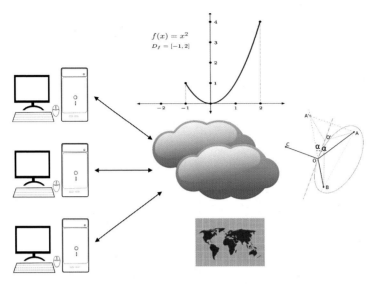

Figure 2.6 Analysis pattern: *Analysis in the cloud*: provides high-performance comput-ing resources for integrating, analyzing, and visualizing large scientific datasets.

Motivation: You are creating a regional climate assessment similar to the Intergovernmental Panel on Climate Change (IPCC) climate analy-sis. You will be gathering scientists from a number of countries and dis-ciplines to create regional analyses. You want the analysis methods to be repeatable, well described, and defendable. You have been asked to explore the possibilities of having a centralized cloud-based applications and analysis service available to all of the researchers.

Forces and Context:
- Efficient and standardized analysis tools are needed for your project.
- The tools should be the best available for the analyses being performed.
- You do not have the budget to install expensive software packages for all of your researchers.

Solution: Create and use a variety of cloud-based Data Analysis as a Service (DAaaS) tools. When needed, create your own tools and make them available as services.

Consequences: The benefits and liabilities of cloud-based analysis ser-vices include:

1. For scientific workflows in which the analyses are performed in the cloud, network latency can be a concern. Although the input/output (I/O) bandwidth of cloud-based DAaaS systems has improved, it is

still slower than local applications. Analysis workflows may be slower for data-intensive applications.

2. Scientists will have access to a standard suite of tools but there will need to be consensus on what tools are needed.

3. You may not be able to locate all the tools you need, and time and effort will be needed to write these tools and make them available as services. This will initially be more expensive and may require new skills from programmers and researchers.

Design: Explore the types of data analysis services needed and choose those needed for your project.

Known Uses: The Schnase chapter in this book describes a set of cloud-based analytical tools including the Modern-Era Retrospective Analysis for Research and Applications (MERRA) (http://gmao.gsfc.nasa.gov/research/merra/). McKenna describes using the cloud to support ocean forecast models and as a way to overcome challenges of widely distributed users and support staff when interactions are hindered by network latencies. Huang and Cervone describe the use of the cloud to analyze social media data, especially during natural and humanmade disasters. Sinha's chapter describes the use of Hadoop to analyze massive datasets, whereas Dhingra et al. address the use of cloud-based analytics for flood modeling and response. The Advanced Weather Interactive Processing System II (AWIPS II) as a tool for analyzing weather models and weather data is discussed in an educational setting by James and Weber and in a system provider context by Ramamurthy. Finally, Combal and Caumont describe CMIP5 and other analysis tools.

Pattern Name: Visualization.

Intent: Explores creating visualizations using cloud-based tools and making the visualizations available via the cloud. Addresses the need to visualize larger amounts of data and the opportunities provided by improved graphics processors and display devices such as Virtual Reality headsets.

Motivation: You are part of the analysis team for a multiyear oceanic and atmospheric modeling effort to look at the effects of aerosols. For your final product, you have been asked to create a series of visualizations for the public and policy makers. Because of the size of the datasets produced by the project, the visualizations will require large amounts of computing time and power and will need to be made available via the web for access by outside users.

Forces and Context:
- You need to produce high-level visualizations
- Visualizations must be comprehensive and created with the best available data
- Consumers expect to see the visualizations on their desktop or mobile device

Solution: Use visualization tools hosted on the cloud as SaaS and explore developments in DAaaS.

Consequences: The benefits and liabilities of using cloud-based visualization tools are:

1. A wider variety of high-end visualizations can be created. This may require learning new tools and will initially slow down the production of visualizations.
2. For scientific workflows in which the visualizations are performed in the cloud, network latency can be a concern. Analysis workflows may be slower for data-intensive applications.
3. Researchers will have access to larger computing resources for creating the visualizations and can expand their resources as needed without having to purchase expensive graphics hardware.

Design: Make use of various visualization tools available as cloud services. These would probably be available as SaaS but might also include DAaaS if the analysis tool includes visualization capabilities.

Known Uses: Robinson et al. describe the use of the cloud for visualization in their chapter and describe a cloud-based work environment to support the day-to-day aspects of their projects with mail, chat, code storage, and other capabilities. Ramamurthy describes a deployment of the Integrated Data Viewer (IDV) in the cloud, and Li et al. describe tools for visualizing sea surface temperature and tropical storm tracks.

Pattern Name: Results Dissemination in Real Time/Storytelling/ Outreach

Intent: Explores ways in which cloud-based platforms and tools can be used to reach new audiences. Addresses the need to make research results rapidly available and relevant to a wide variety of audiences—scientific and nonscientific.

Motivation: You have been asked to create public presence for a research project. The project will include field research, modeling, visualization of data, and the publication of scientific papers. Your specific area of responsibility is to convey the results of the project to the general public.

The ships and aircraft used for the project will have Internet connectivity. The modelers you are working with are adept at creating visualizations and animations. You also have a graphics department that is familiar with creating Esri Story Maps, short videos about scientific projects, and narrated videos explaining scientific results. You are also personally familiar with social media such as Twitter, Facebook, Snapchat, Instagram, and blogging networks.

Forces and Context:

* You need to engage users of social media to tell the story of your research
* You have a varied audience in both their scientific expertise and preferred medium for receiving information.
* You have considerable local talent but lack the computing resources and network bandwidth to host these visualizations locally.
* Your researchers may want to create and share visualizations while in the field.

Solution: Create a YouTube channel, Twitter, Facebook, Snapchat, Instagram, and blogging networks. Conduct cloud-based chat, video, and create a cloud-based repository for visualizations. Use tools that allow you to aggregate all content and create stories to talk about the project in a linear manner. These tools allow for two-way exchange of information—users and other scientists in the project can interact, add data, and create maps for all to see—for example, Google maps for forest fires, and the mPing project to collect crowd-sourced weather reports (http://mping.nssl.noaa. gov/). Take advantage of easy-to-set-up sites such as blogger.com that allow you to publish on the go and link to videos. The videos could be hosted on Vimeo and YouTube. These tools also allow you to publish to other social streams such as an official Twitter or Instagram page for the project, a Flickr site to share photos and a Facebook page for the project. Also explore tools that are used to let others discover your findings such as Tweetdeck and other social media trackers that are cloud-based.

Consequences: The benefits and liabilities of using cloud-based tools to disseminate your results are:

1. Increased access to and exposure of your research. Making your research known to a new, younger, audience.
2. Possibly being seen as "unscientific" for using nonstandard tools and platforms
3. Need stable network connectivity to ship or field station if hosting video chats.

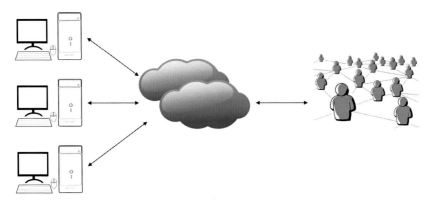

Figure 2.7 Analysis pattern: *Results dissemination*: making research results rapidly available and relevant to a wide variety of audiences—scientific and nonscientific.

4. Rapid dissemination of initial results that are shareable prior to formal publishing.

Design: Make use of a variety of cloud-based SaaS platforms to store and disseminate visualizations and other information about your research (Figure 2.7).

Known Uses: NOAA's Office of Ocean Exploration and Research (OER) hosts a Twitter page (https://twitter.com/oceanexplorer), a Facebook page (https://www.facebook.com/OceanExplorationResearch), and a YouTube channel (https://www.youtube.com/oceanexplorergov) to show the results of their expeditions.

Humanity Road is a crisis response nonprofit that activates a social media presence during and after disasters (http://humanityroad.org/, http://humanityroad.org/typhoon-etau-september-11/). The Weather Channel has a set of pages called Hurricane Central (http://www.weather.com/storms/hurricane-central) that provides summary information on active hurricanes and historical context. It also has Twitter and Facebook pages along with a Weather Channel app.

CONCLUSION

Cloud computing is the new paradigm in information technology driven primarily by the explosion in the scope and scale of scientific investigation. The cloud enables researchers to tackle larger and more complex problems without the time and financial burden of building and maintaining local computing infrastructure. Science has been relatively slow to adapt for several reasons but primarily out of fear and lack of a clear mechanism for adapting science to the cloud. Researchers are balancing their fears of missing out on the benefits of the cloud versus their fear of learning the next "new thing." Fortunately, transitioning to the cloud is becoming easier. Cloud providers are competing for consumer revenue and therefore try to drive cost down and value up. Part of this is making it easier to publish to the cloud; so that the cloud is becoming almost identical to traditional local server environments but with value added. The e-Science model coupled with analysis patterns enables researchers to move beyond issues of infrastructure and provide a mechanism for describing and reproducing successful research patterns in the cloud. Analysis patterns are a simplified way of communicating how science can be conducted in the cloud to fellow researchers, research administrators, and funding agencies. They can help streamline processes, guide development across disparate disciplines, and help managers understand the pros and cons of using the cloud.

REFERENCES

Alexander, C., 1979. The Timeless Way of Building, vol. 1. Oxford University Press, New York.

Atkins, D., 2003. Revolutionizing Science and Engineering through Cyberinfrastructure: Report of the National Science Foundation Blue-Ribbon Advisory Panel on Cyberinfrastructure.

Bohle, S., 2013. What is E-science and how should it be managed? Nature. com, Spektrum der Wissenschaft (Scientific American), http://www.scilogs.com/scientific_and_medicallibraries/what-is-e-science-and-howshould-it-be-managed.

Buneman, P., Khanna, S., Wang-Chiew, T., 2001. Why and where: a characterization of data provenance. In: Database Theory—ICDT 2001. Springer Berlin Heidelberg, pp. 316–330.

Buschmann, F., Meunier, R., Rohnert, H., Sommerlad, P., Stal, M., 1996. Pattern–Oriented Software Architecture, a System of Patterns. John Wiley & Sons Ltd, Chichester.

Cragin, M., Heidorn, P.B., Palmer, C.L., Smith, L.C., 2007. An educational program on data curation. In: ALA Science & Technology Section Conference. http://hdl.handle.net/2142/3493. Retrieved September 7, 2013.

Fowler, M., 1997. Analysis Patterns: Reusable Object Models. Object Technology Series. Addison–Wesley Publishing Company, Reading, PA.

Gamma, E., Helm, R., Johnson, R., Vlissides, J., 1995. Design Patterns: Elements of Reusable Object–Oriented Software. Addison–Wesley Professional Computing Series. Addison–Wesley Publishing Company, New York.

Geyer-Schulz, A., Hahsler, M., 2001. Software Engineering with Analysis Patterns. Working Papers on Information Systems, Information Business and Operations, 01/2001. Institut für Informationsverarbeitung und Informationswirtschaft, WU Vienna University of Economics and Business, Vienna.

Gil, Y., Deelman, E., Ellisman, M., Fahringer, T., Fox, G., Gannon, D., Goble, C., Livny, M., Moreau, L., Myers, J., 2007. Examining the challenges of scientific workflows. IEEE Computer 40 (12), 26–34.

Gray, J., 2009. Jim Gray on e-Science: a transformed scientific method. In: Hey, T., Tansley, S., Tolle, K.M. (Eds.), The Fourth Paradigm: Data-Intensive Scientific Discovery, vol. 1. Microsoft Research, Redmond, WA.

Hey, T., Trefethen, A.E., 2002. The UK e-science core programme and the grid. Future Generation Computer Systems 18 (8), 1017–1031.

Hey, T., Tansley, S., Tolle, K. (2009). Jim Gray on e-Science: a transformed scientific method. In: Hey, T., Tansley, S., Tolle, K., (Eds.), The Fourth Paradigm: Data-Intensive Scientific Discovery, Microsoft, Redmond, WA.

Holdren, J.P., 2013. Increasing Access to the Results of Federally Funded Scientific Research. Memorandum for the Heads of Executive Departments and Agencies. Office of Science and Technology Policy, Executive Office of the President, Washington, DC. Available: www.whitehouse.gov/sites/default/files/microsites/ostp/ostp_public_access_memo_2013.pdf (August 2013).

Jankowski, N.W., 2007. Exploring e-science: an introduction. Journal of Computer-Mediated Communication 12 (2).

Jha, S., Katz, D.S., 2010. Abstractions for Distributed Applications and Systems: A Computational Science Perspective. Wiley Publishing.

Katz, D.S., Jha, S., Parashar, M., Rana, O., Weissman, J., 2011. Survey and Analysis of Production Distributed Computing Infrastructures Technical Report CI-TR-7–0811. Computation Institute, University of Chicago & Argonne National Laboratory.

Nentwich, M., 2003. Cyberscience: Research in the Age of the Internet. Austria Academy of Sciences, Vienna.

Smithsonian, 2015. Meteorology. http://siarchives.si.edu/history/exhibits/henry/meteorology. Retrieved September 1, 2015.

Taylor, J.E., n.d. Defining e-Science. http://www.nesc.ac.uk/nesc/define.html. Retrieved September 1, 2015.

Van Noorden, R., 2014. Global Scientific Output Doubles Every Nine Years. http://blogs.nature.com/news/2014/05/global-scientific-output-doubles-every-nine-years.html. Retrieved September 1, 2015.

Zhao, Y., Li, Y., Raicu, I., Lu, S., Tian, W., Liu, H., 2015. Enabling scalable scientific workflow management in the Cloud. Future Generation Computer Systems 46, 3–16.

CHAPTER 3

Forces and Patterns in the Scientific Cloud: Recent History and Beyond

R.S. Wigton
Bin Software Co., Bellevue, WA, USA

The most valuable and pervasive technologies eventually recede into the background. For the most part, we do not labor over the complexity of the cellular network each time we make a call. Nevertheless, no such utility comes quickly. The cellular phone required several decades to move from invention to pervasive utility.

We are on a similar path with scientific cloud computing. The overarching promise of the scientific cloud is that scientists can intuitively reach for sophisticated, reliable parallel computing resources in a form that serves their work, and with effort that would not distract them. This paper surveys what we have learned about the forces and challenges at work as we have begun to move technical computing work flows to the fabric of the cloud. With that context, we then examine two challenges—collaboration and visualization—that represent opportunities for deeper exploitation of what the cloud has to offer.

2005 TO 2015: A PERIOD OF FIT AND RETROFIT

High-scale scientific computing has its roots in the grid, cluster, and high-performance computing (HPC) solutions of the 1990s, which were often deployed using time-sharing paradigms established in the mainframe period. In contrast, the most prominent commercial cloud systems of 2015 began as internal infrastructures. Before they were commercially accessible cloud services, these server racks and virtualization approaches were internally designed systems built largely for nonscientific computing.

In announcing Amazon's Elastic Compute Cloud (EC2) in 2006, for example, the company explained that "The reason we're doing the web services that we are doing is because they are things that we've gotten good

Cloud Computing in Ocean and Atmospheric Sciences
ISBN 978-0-12-803192-6
http://dx.doi.org/10.1016/B978-0-12-803192-6.00003-7

Copyright © 2016 Elsevier Inc.
All rights reserved.

35

at over the last 11 years in terms of building out this web-scale application called Amazon.com" (Amazon.com, 2006). The elasticity and economies of scale promised by such systems were immediately intriguing to everyone running a data center.

However, these early commercial cloud architectures had two dominant parents: transactionally distributed workloads (like online commerce) and data-parallelizable process flows (such as the embarrassingly parallel data-intensive computing workloads of the search and social web revolution). This legacy made the execution models and storage provisions of the early cloud more readily adoptable by customers with these workloads. Web sites, e-commerce properties, and data-parallel business intelligence operations were eager early adopters.

FORCES AND CHALLENGES IN SCIENTIFIC CLOUD ADOPTION

To their credit (and benefit), the major cloud service vendors like Amazon, Google, and Microsoft realized that the technical computing community not only required more, but that the demands of scientific computing would challenge and expand their infrastructure-as-a-service (IaaS) and platform-as-a-service (PaaS) clouds. Driven by varying strategies and economic realities, each has made significant inroads on the forces essential to scientific workloads. These forces include: (1) richer, more performant expressions of *parallelism*, including those utilizing a graphics processing unit (GPU); (2) relevant support for internode *communication* patterns (in both message- and agent-based execution), including advanced interconnect infrastructure; (3) *affinity* provisions reflecting the pivotal role of scientific data intermediates and interset dependencies; (4) fault tolerance beyond the forgiving bar often set for web data parallelism; (5) *elasticity* to serve the sometimes 'bursty' nature of computation; and (6) *load-balance* controls required by cost- and reliability-conscious global organizations.

By 2011, an increasingly credible set of services had emerged from these and other vendors, thanks to collaboration with scientific teams committed to engagement, research, and trial. This accelerated the quantitative and qualitative progress toward realizing large-scale scientific computation in the cloud, in both research and practice. The virtualization, scheduling, and performance challenges of cloud HPC workloads were tackled (Vecchiola et al., 2009; Wong and Goscinski, 2013; Hassani et al., 2014). Partitioning and locality strategies were vetted for both simply parallel workloads and more complex execution paradigms (Yuan et al., 2010; Zhao et al., 2011). Popular scientific workflow management systems of the period—e.g., Taverna, Kepler, Triana, PEGASUS, etc.—spawned numerous evaluations of their transferability to cloud deployments. And a growing set of ecosystem innovations—from the Amazon-compatible open-source software (OSS) Eucalyptus software stack, to unique models like the UberCloud Experiment—have accelerated experimentation and adoption across scientific domains.

Some notable achievements in cloud computing have occurred in this 10-year period, to be sure. Solid-state storage architectures, some key improvements in interconnect, and the acclimation of the scientific and business communities to cloud adoption, are relevant examples. However, for many scientific computing professionals, the period between 2005 and 2015 may best be characterized by two overarching questions: fit ("Can the specific requirements of my domain, data, and computational workloads be met in a commercial cloud deployment?") and retrofit ("What percentage of my existing intellectual capital—code, data, governance and security protocols, human understanding of workflow—can I reasonably shift to the cloud?").

LOOKING BEYOND FIT AND RETROFIT

This latter impulse to retrofit systems has yielded mixed results. Certainly, there are classes of scientific workloads that lend themselves seamlessly to the contemporary provisions of both IaaS and PaaS platforms. To the extent that they serve some portion of a composite workflow, patterns like MapReduce, scatter/gather, and the like are often straightforward retrofit candidates. They are also often a crucial part of data preprocessing or management.

Nevertheless, the demands of technical computing stretch well beyond this comparative simplicity. It is often best, in the words of Gannon and Fox, to "think differently and rewrite the application to support the new computational and programming models" afforded by the cloud (Fox and Gannon, 2012). This is often particularly true for geoscientific and oceanographic computation for a variety of reasons. These reasons include:

1. *Composite and ensemble workflow patterns* are common. These patterns, and the hierarchical nesting they require, can complicate load balancing as well as demand the ability to gang together worker processes to support the inelastic subpatterns with tightly coupled communication patterns.

2. *Conventional HPC* architectures are often required, and do not always port directly without strict consideration of affinity and performance.

3. *Demanding data processing* arrangements (such as real-time capture of global instrument output, spatial index optimization, or specialized reprojection) often dominate aspects of a workflow's storage and parallelization design.

4. The count of required *special-purpose tools* is often high. The geospatial toolset is a rich one, and has evolved over a long time. Essential subcomponents are not always cloud ready, even if the overall workflow lends itself to cloud deployment.

5. Geospatial extent can pose significant *scale* challenges. A watershed model for a single US county within the Mississippi watershed must scientifically scale to the full length of the Mississippi; so must the manageability of its data and compute given the primitives of the target cloud system (Larus, 2011).

6. Because of the sources of geospatial imagery/telemetry, or the purpose of the work at hand, data *provenance and security* are often paramount concerns.

Thankfully, through deep interdisciplinary engagement, good solutions are emerging in the technical-computing cloud ecosystem. Regarding the HPC challenge, for example, both Amazon's EC2 and Microsoft's Azure offer serviceable (though different) approaches for hosting HPC patterns. Particularly favorable are approaches in which the cloud is exploited as a source of additional on-demand compute-node capacity during 'bursts'. Nevertheless, even abstract cloud deployments of fully virtualized clusters have increased in availability and performance over the past 5 to 10 years (Zhao et al., 2011; Fox and Gannon, 2012).

Similarly, progress in addressing the complexity of composite and ensemble work flows has been promising—and has again underscored the value of rethinking architectures for the cloud. Dynamic heterogeneous work flows like the Error Subspace Statistical Estimation (ESSE) have been reshaped (particularly at the storage and interconnect levels) for deployment on cloud IaaS (Evangelinos et al., 2011). Moreover, the high-throughput demands of ensemble forecasts, as well as the metadata-harvesting requirements for provenance, have been favorably addressed in commercial PaaS implementations on Microsoft Azure (Chandrasekar et al., 2012).

These and similar examples indicate that successful deployment to the scientific cloud is within reach, and is best addressed through thoughtful adaptation of workflow architecture and assumptions.

COLLABORATION AND VISUALIZATION AS UNDERSERVED CHALLENGES

Our focus thus far has been to examine the ways in which the fundamental forces essential to scalable technical computing have been addressed in the 10-year evolution of abstract cloud platforms. But are there additional forces essential to computational insight for which cloud offers promise? And if so, are there any additional challenges relevant to cloud adoption in the geoscientific realm?

One force not emphasized earlier relates to how effectively a cloud-based system can facilitate the capture, retention, and exchange of data, state, and processing. For scientists and organizations to depend on computational analysis, researchers must be able to make adequate assurances of provenance and procedure. This general force, which can roll up under reproducibility, has a direct relationship to those already mentioned. For example, the fault-tolerance provisions within the scientific cloud may need to go beyond basal toleration of component failure. Its provisions for affinity may need to address not only the proximity of data to computation, but also enable very particular types of intermediate collection.

Because any specific computational run occurs in the larger overall scientific work-flow context of peer review and public communication, it is appropriate to demand that the scientific cloud enable collaboration at every level. It can be argued that this has taken a back seat to the crucial work of bringing technical computation itself online. Perhaps due to economic incentives, the most promising cloud-focused work in collaboration has been driven not by the commercial platform vendors, but by the academic and open-source communities. Efforts like the iPlant Collaborative not only serve as hubs, but (as in iPlant's Atmosphere virtual machine (VM) persistence and sharing capability) they have driven those principles into the low-level function of the cloud substrate (Skidmore et al., 2011). This illustrates a key point for the continued evolution of the cloud: collaboration, and the underlying reproducibility force it serves, can and should be facilitated at every level in the scientific cloud stack, from VM through shared storage, from improved provisions for metadata harvesting to PaaS run-time abstractions that make acquisition, integration, and exchange even more natural.

A second force and associated challenge worthy of broader attention in the coming decade relates to work-flow composability and the challenge of visualization. Of the various unit operations essential to scientific computing work flows, visualization may be the task component most tightly coupled to execution context, stage of process, and run-time constraint. This has traditionally relegated visualization to specific, latter-stage positions in scientific work flows. Yet the use of iterative, even interactive, visualization as an integrated discovery mechanism is increasingly essential to work flows in medicine, oceanography, atmospheric science, proteomics, and modeling/simulation as a whole.

The improved accessibility and fitness of GPU computing has been one enabling backdrop for the increased viability of more composable, interactive visualization. The notable advancements in both video streaming and affordable display technology—including the coming wave of Augmented Reality (AR), Virtual Reality (VR), and mixed-mode displays—has also contributed. In addition, the increased access to expressive, well-thoughtout visualization libraries within the technical computing language family is a third promising development. The result is that scientists and domain specialists have new computational tools with which to manage multidimensional data (Ahrens et al., 2010), visually inspect workload provenance (Santos et al., 2012), and deeply explore and inspect geoscientific models (Hermann and Moore, 2009).

The challenge of making visualization a more loosely coupled, composable operation within scientific work flows is decidedly nontrivial. But like the other evolutionary force/challenge stages surveyed here, it warrants continued research and innovation.

CONCLUSION

The platform provisions and underlying technologies of the commercial cloud services have evolved rapidly over the past decade. That evolution has been driven by much positive joint work between cloud vendors and the global research community. As the viability of scientific cloud work-flow deployments continues to solidify, it is worthwhile to turn some increased attention to challenges like collaboration and visualization, which show promise given the foundational advancements behind us.

REFERENCES

Ahrens, J., Heitmann, K., Petersen, M., Woodring, J., Williams, S., Fasel, P., Ahrens, C., Hsu, C.-H., Geveci, B., 2010. Verifying scientific simulations via comparative and quantitative visualization. IEEE Computer Graphics and Applications 6, 16–28.
Amazon.com, I., October 24, 2006. Amazon.com, Inc (AMZ) Q3 2006 Earnings Call Transcript. From: http://seekingalpha.com/article/19142-amazon-com-q3-2006-earnings-call-transcript?part=qanda.

Chandrasekar, K., Pathirage, M., Wijeratne, S., Mattocks, C., Plale, B., 2012. Middleware alternatives for storm surge predictions in windows azure. In: Proceedings of the 3rd Workshop on Scientific Cloud Computing Date, ACM.

Evangelinos, C., Lermusiaux, P.F., Xu, J., Haley Jr., P.J., Hill, C.N., 2011. Many task computing for real-time uncertainty prediction and data assimilation in the ocean. Parallel and Distributed Systems, IEEE Transactions 22 (6), 1012–1024.

Fox, G., Gannon, D., 2012. Cloud programming paradigms for technical computing applications. In: Proc. Cloud Futures Workshop.

Hassani, R., Aiatullah, M., Luksch, P., 2014. Improving HPC application performance in public cloud. IERI Procedia 10, 169–176.

Hermann, A.J., Moore, C.W., 2009. Visualization in Fisheries Oceanography: New Approaches for the Rapid Exploration of Coastal Ecosystems. Computers in Fisheries Research. Springer, pp. 317–336.

Larus, J.R., 2011. The cloud will change everything. In: Proceedings of the Sixteenth International Conference on Architectural Support for Programming Languages and Operating Systems. ACM, Newport Beach, California, USA, pp. 1–2.

Santos, E., Koop, D., Maxwell, T., Doutriaux, C., Ellqvist, T., Potter, G., Freire, J., Williams, D., Silva, C.T., 2012. Designing a Provenance-based Climate Data Analysis Application. Provenance and Annotation of Data and Processes. Springer, pp. 214–219.

Skidmore, E., Kim, S.-J., Kuchimanchi, S., Singaram, S., Merchant, N., Stanzione, D., 2011. iPlant atmosphere: a gateway to cloud infrastructure for the plant sciences. In: Proceedings of the 2011 ACM Workshop on Gateway Computing Environments, ACM.

Vecchiola, C., Pandey, S., Buyya, R., 2009. High-performance cloud computing: a view of scientific applications. In: Pervasive Systems, Algorithms, and Networks (ISPAN), 2009 10th International Symposium on, IEEE.

Wong, A.K., Goscinski, A.M., 2013. A unified framework for the deployment, exposure and access of HPC applications as services in clouds. Future Generation Computer Systems 29 (6), 1333–1344.

Yuan, D., Yang, Y., Liu, X., Chen, J., 2010. A data placement strategy in scientific cloud workflows. Future Generation Computer Systems 26 (8), 1200–1214.

Zhao, Y., Fei, X., Raicu, I., Lu, S., 2011. Opportunities and challenges in running scientific workflows on the cloud. In: Cyber-enabled Distributed Computing and Knowledge Discovery (CyberC), 2011 International Conference on, IEEE.

CHAPTER 4

Data-Driven Atmospheric Sciences Using Cloud-Based Cyberinfrastructure: Plans, Opportunities, and Challenges for a Real-Time Weather Data Facility

M. Ramamurthy
University Corporation for Atmospheric Research, Boulder, CO, USA

Data are the lifeblood of the geosciences. Rapid advances in computing, communications, and observational technologies are occurring alongside parallel advances in high-resolution modeling, ensemble, and coupled-systems predictions of the Earth system. These advances are revolutionizing nearly every aspect of our atmospheric science field. The result is a dramatic proliferation of data from diverse sources, data that are consumed by an evolving and ever-broadening community of users and that are becoming the principal engine for driving scientific advances. Data-enabled research has emerged as a Fourth Paradigm of science, alongside experiments, theoretical studies, and computer simulations (Hey et al., 2009).

For more nearly 30 years, Unidata, a data facility funded by the US National Science Foundation (NSF) and managed by the University Corporation for Atmospheric Research (UCAR) in Boulder, Colorado (CO), has worked in concert with the atmospheric science education and research community to support data-enabled science to understand the Earth system.

Unidata's mission is to transform the geosciences community, research, and education by providing innovative, well-integrated, and end-to-end data services and tools that address many aspects of the scientific data life cycle, from locating and retrieving useful data, through the process of analyzing and visualizing data either locally or remotely, to curating and sharing the results. To that end, the Unidata program:

- Acquires, distributes, and provides access to a wide array of real-time meteorological data;

Cloud Computing in Ocean and Atmospheric Sciences
ISBN 978-0-12-803192-6
http://dx.doi.org/10.1016/B978-0-12-803192-6.00004-9

Copyright © 2016 Elsevier Inc.
All rights reserved.

- Develops and provides software for accessing, managing, analyzing, visualizing, and effectively using geoscience data;
- Provides comprehensive training and support to users of its products and services;
- Facilitates, in partnership with others, the advancement of tools, standards, and conventions;
- Provides leadership in cyberinfrastructure and fosters adoption of new tools and techniques;
- Fosters community interaction and engagement to promote sharing of data, tools, and ideas;
- Advocates on behalf of the academic community on data matters, negotiating data and software agreements;
- Grants equipment awards to universities to enable and enhance participation in Unidata;

Unidata's familiarity in helping geoscientists incorporate data-centric techniques into their scientific work flows positions it well to assist its community to take advantage of cloud-computing capabilities in a rapidly changing scientific landscape. Unidata's plans call for the program to adapt to and capitalize on the rapid advances in information technology, enabling the community to incorporate the dramatic explosion of data volumes into their research and education programs. This will ensure that the next generation of students and young researchers in universities and colleges become leaders using state-of-the-art tools, technologies, and techniques.

The overarching goal embodied in Unidata's strategic plan is the creation of a scientific ecosystem in which "data friction" (Edwards, 2010) is reduced and data transparency and ease of use are increased. In such an environment, scientists will expend less effort locating, acquiring, and processing data and more time interpreting their data and sharing knowledge.

To accomplish the goals set forth in our strategic plan, Unidata has been working to build and provide infrastructure that makes it easy to discover, access, integrate, and use data from disparate geoscience disciplines. This allows investigators to perceive connections that today are obscured by incompatible data formats or the mistaken impression that the data they need for their investigations do not exist.

In the following section, we provide both the context and the drivers for Unidata's cloud vision.

SCIENCE

Environmental challenges such as climate change, extreme weather, and the water cycle transcend disciplinary as well as geographic boundaries, and require multidisciplinary approaches to ameliorate their effects. These approaches increasingly underscore the importance of data-driven environmental research and education. In response to these "grand challenge" problems, the geoscience community is shifting its emphasis from pure disciplinary research to a more balanced mix that advances disciplinary knowledge while looking to apply research results to interdisciplinary questions touching both science and society.

Given the changing scientific landscape and emphasis, it has become clear that Unidata develops and supports cyberinfrastructure that not only enables researchers to advance the frontiers of science, but transcends traditional disciplinary boundaries. The main challenge for any data service is to provide easy access to the right data, in the right format, to the end-user application.

EDUCATION

For decades, the research community has harnessed the power of data and computers to better understand Earth System problems through complex models, visualizations, and analysis techniques. Educators are increasingly integrating data-driven exploration into the learning environment, and easy access to data for student exploration is crucial to making this transformation. Professors realize that authentic learning and effective pedagogy focus on solving real-world problems that engage student interest and increase understanding of phenomena and processes. Investigation, research, analysis, and discovery of natural phenomena make the geosciences an ideal platform for integrating authentic learning activities into the curriculum, and many atmospheric science programs have successfully done so.

Data-enabled learning encourages student projects to enhance the collection and analysis of data through smart tools and sensors that automate the capture, recording, processing, and sharing of results. As an example, mobile phone operating systems provide support for sensors that can monitor a variety of environmental properties. With these sensors and the geolocation facilities of their cell phones, students can engage in micro field studies and data–collection activities. Intercomparison of the crowd-sourced data gathered with such sensors is an ingenious way to advance data literacy and active learning.

It is critical to the health of the geosciences that future science and engineering leaders understand modern cyberinfrastructure, including cloud computing, and be trained using state-of-the-art tools, technologies, and techniques. Fluency is required not only in the geosciences and mathematical/statistical disciplines, but also in computational and information technology areas. Twenty-first century scientists must be data literate through many aspects of the data life cycle, including collection, management, analysis, and sharing of scientific data.

DATA

As mentioned earlier, data-intensive science has emerged as the Fourth Paradigm of scientific discovery after empirical, theoretical, and computational methods. This is particularly true in the geosciences, in which data have become increasingly important in scientific research. Modern data volumes from high-resolution ensemble prediction/projection/simulation systems and next-generation remote-sensing systems like hyperspectral satellite sensors and phased-array radars are staggering. For example, Coupled Model Intercomparison Project Phase 5 (CMIP5) alone will generate more than 10 petabytes of climate projection data for use in assessments of climate change. The National Oceanic and Atmospheric Administration (NOAA)'s National Center for Environmental Information (formerly National Climatic Data Center) projects that it will archive over 350 petabytes by 2030 (Harper Pryor, 2012).

For researchers and educators, this deluge, and increasing complexity, of data brings challenges along with the opportunities for discovery and scientific breakthroughs. Retrieving relevant data in a usable format from an archive should not be more time-consuming and arduous than the scientific analysis and investigation the fetched data make possible. At the other end of the spectrum, the majority of agency-funded research is conducted by scientists in relatively small projects with one lead researcher, typically a faculty member with a part-time commitment to the project, and one or two graduate students or part-time postdoctorals. Although great care is frequently devoted to the collection, preservation, and reuse of data on large, multiinvestigator projects, relatively little attention is paid to curating and sharing data that are being generated by these smaller projects (Heidorn, 2008), resulting in large amounts of "long tail" or "dark" data. As a result, it is difficult to discover unpublished "dark data," which might remain unshared, underutilized, or in some cases even be lost. By some estimates, only 5% of the data generated by individual project investigators in the geosciences is shared with the broader community (Killeen, 2011).

Publication of research data presents unique challenges, sociological and technical, for the science community. Society has begun to demand increased scientific transparency, but researchers face a lack of incentives, inadequate resources, intellectual property issues, a culture of protectiveness, and the absence of supporting cyberinfrastructure, tools, and data repositories for sharing data. In 2011, NSF issued a new guideline requiring that every proposal submitted include a Data Management Plan describing how the project will conform to NSF policy on the dissemination and sharing of research results. Recently, the Office of Science and Technology Policy directed federal agencies to develop a plan to support increased public access to the results of research funded by that agency (read online at http://1.usa.gov/VBkFJv). Among other elements, such a plan must provide a strategy for improving the public's ability to locate and access data resulting from federally funded research.

CAMPUS INFORMATION TECHNOLOGY INFRASTRUCTURE

Educational enterprises are in a state of rapid transformation. Since the introduction of personal computers in the 1980s and high-speed networking on college campuses in the 1990s, there has been a sea change in the Information Technology (IT) environment in which students and faculty operate. The Internet and the availability of ubiquitous wireless access, smartphones, tablets, and cloud-based services have accelerated the shift. The expectations, modalities, and skills of students have dramatically changed. As Conford (2008, http://bit.ly/18PahAH) argues, "in the past, universities and colleges were often the institutions that provided students with their first experiences of networked information technology services such as email and easy access to the web. Today, however, students arrive at universities and colleges with years of experience of these technologies. As a result, the ways in which individuals use technologies, and their expectations about how they are going to use those technologies, are already well established. Institutions no longer introduce users to information technology; instead, information technology is often the main context in which users are introduced to the institution."

Increasingly, students are not only allowed but encouraged to bring their own devices and connect them to campus networks. Students expect content and services to be delivered through their devices; they will not accept a step back in technology when they step onto campus. They expect to use interactive, intuitive, and collaborative tools to learn and communicate just as they do in their off-campus lives.

The computing environment in departments and faculty research labs is also changing. Anecdotal evidence suggests that reduced university budgets mean diminished system administration and maintenance support for many departments' scientific computing infrastructure. University administrators are realizing that the traditional model, in which discrete, dedicated computer systems perform specialized tasks, cannot be sustained and that new ways of delivering information technology and data services are needed. Choices driven by budget pressures include consolidation or centralization of IT systems, increased virtualization, and the adoption of cloud-computing technologies to deliver services.

VISION FOR THE FUTURE: MOVING UNIDATA'S SERVICES AND SOFTWARE TO "THE CLOUD"

We have identified some of the challenges universities are facing: shrinking budgets, rapidly evolving information technologies, growing data volumes, multidisciplinary science requirements, and high student expectations. Most faculty and researchers would prefer to focus on teaching and doing science rather than setting up computer systems. These changes are upending traditional approaches to accessing and using data and software; Unidata's products and services must also evolve to support modern approaches to research and education. In this section, we present a vision for Unidata's future that will provide a transformative community platform for collaborative development and an array of innovative data services to our users.

After years of hype and uncertainty, cloud-computing technologies have matured. Their promise is now being realized in many areas of commerce, science, and education, bringing the benefits of virtualized and elastic remote services to infrastructure, software, computation, and data. Cloud environments can reduce the amount of time and money spent to procure, install, and maintain new hardware and software, reduce costs through resource pooling and shared infrastructure, and provide greater security. Cloud services aimed at providing any resource, at any time, from any place, using any device are increasingly being embraced by all types of organizations. NOAA, the National Aeronautics and Space Administration (NASA), and other federal science agencies are establishing cloud-computing services. Universities are no exception; the University of Washington, University of Illinois, Cornell University, and George Washington University are some of the universities that have already set up cloud services for scientific and academic computing.

Given this trend and the enormous potential of cloud-based services, Unidata is working to gradually augment Unidata's products and services to align with the cloud-computing paradigm. Specifically, we are working to establish a community-based development environment that supports the creation and use of software services to build end-to-end data work flows. The design encourages the creation of services that can be broken into small, independent chunks that provide simple capabilities. Chunks could be used individually to perform a task, or chained into simple or elaborate work flows. The services we envision will be loosely coupled to meet user needs rather than tightly coupled into a monolithic system. The services will be portable, allowing their use in researchers' own cloud-based computing environments.

Unidata recognizes that its community is not monolithic. Our users have diverse needs and access to a range of cloud-computing resources. Users will be able to implement these services in conjunction with their own work flows in ways they want, for example, by leveraging the capabilities of Python, R, or other work-flow systems they use. This approach permits greater flexibility and interoperability. We envision users being able to invoke and use the services from an array of computing devices, including laptops, high-powered workstations, tablets, and even smart phones. The proposed vision is not about building a system, but an environment with a collection of capabilities, using a standardized approach.

CATEGORIES OF SERVICES

Here we present a list of candidate cloud-based data services that can be enabled as extensions of Unidata's current activities. In no particular order, these include:

- Remote access to real-time data streamed via Internet Data Distribution (IDD), as a cloud service (e.g., Unidata's Thematic Real-Time Environmental Distributed Data Services (THREDDS) server in a cloud)
- Data discovery, access, and extraction services (e.g., subsetting services, catalog search services)
- Catalog and metadata generation services
- Data manipulation and transformation services (e.g., decoders, unit conversion)
- Brokering services for data and metadata (e.g., GI-Cat from Earth and Space Science Informatics Laboratory)
- Server-side data analysis and operations (e.g., Network Common Data Form (netCDF) + operators, time-series generation, other mathematical calculations)

- Data-proximate display and visualization services that provide products to thin clients
- Data-publishing services allowing users to publish results, analyses, and visualizations
- Subset subscription services, providing delivery of specified custom products

COMMUNITY COLLABORATION

Unidata is not planning to build all of the needed "chunks" itself, but provide a platform for creating a range of cloud-enabled geoscience data services. Our objective is to harness the community's vast scientific and technical expertise in making this transformation. In partnership with the community and collaborators, Unidata will define the broad directions and design the general architecture/framework for how services can be contributed, provide a platform for developing services, establish a governance process for the envisioned environment, and develop and add component services to the suite incrementally and systematically.

As always, Unidata has been actively engaging its community and encouraging them to contribute applications and services to the collection by facilitating the "wrapping" of scientists' applications into web-enabled, pluggable services. Simultaneously, Unidata is also developing and demonstrating end-to-end prototype data services and making them available to the community.

Unidata recognizes that the governance of this process is extremely important and will pose some challenges. We have already put in place key elements to enable this transformation: by developing several web services, moving to Open Source environments like Github and Redmine for collaborative development of software, and providing remote data access for more than 15 years.

MANAGING CHANGE FOR OUR COMMUNITY

Unidata and its community are at a crucial juncture in history. The IT landscape is changing dramatically even as users face rapidly shrinking budgets; our supporting community is expecting Unidata to lead by innovating and enabling new capabilities that let them do more with the resources they have. It is imperative that Unidata present a bold vision in response to these needs. In doing so, Unidata assumes the responsibility to help community members make the transition as their work flows and *modus operandi* change. To maintain the trust of our community, we firmly believe we must follow the spirit of the medical profession's Hippocratic Oath: *First, do no harm.*

As part of its five-year proposal to the NSF, Unidata 2018: Transforming Geoscience through Innovative Data Services (http://www.unidata.ucar.edu/publications/Unidata_2018.pdf), Unidata has proposed to align its future activities around this vision, but take a measured and disciplined approach that eases the community's evolutionary transition to newer, more powerful ways of working. Toward the goals stated in the NSF proposal, Unidata has been investigating how its technologies can best leverage cloud computing for providing its data services. In the following section, we describe some of the ongoing cloud-computing efforts to deploy a suite of Unidata services and tools in the Amazon Web Services and Microsoft Azure cloud environments, including the progress of those efforts as well as the initial lessons we have learned thus far.

CURRENT UNIDATA CLOUD-RELATED ACTIVITIES

The Use of Docker Container Technology at Unidata

In the last few years, there has been considerable momentum behind "containerization" of applications. A container like Docker (https://www.docker.com) provides an environment for the application to run within the containers, which is then portable from platform to platform, for example, to a cloud environment. In essence, a container, rather than the application contained within it, handles the underlying platform, and, as such, that approach is driving considerable interest in cloud computing, especially for porting legacy applications to cloud-computing environments.

From a development and deployment point of view, Docker can be characterized as lightweight virtualization, as opposed to a virtual machine (VM); plus, they can run on multiple VMs. Docker containers, therefore, can be spun up much faster than VMs and are more cost-effective in a cloud environment.

With the goal of better serving our core community and in fulfillment of objectives articulated in our five year proposal, Unidata has been investigating how its applications and services can best take advantage of container technologies and cloud computing. To this end, Unidata has been employing Docker container technology to streamline building, deploying, and running Unidata technology offerings in cloud-based resources. Specifically, they have created Docker containers for the Integrated Data Viewer (IDV), Repository for Archiving, Managing and Accessing Diverse Data (RAMADDA), THREDDS, and the Local Data Manager (LDM), and they have been deploying Docker containers in the Microsoft Azure and Amazon Elastic Compute Cloud (EC2) cloud-computing environments.

In addition, with the objective of gathering Unidata tools and services related to Python under one environment, Unidata has created a Docker container that contains Python libraries for netCDF, MetPy, and Siphon as well as many Python libraries that work well in conjunction with these libraries such as Jupyter Notebook technology (https://jupyter.org/).

Finally, in addition to using Docker for the deployment of tools and services, Unidata has also been employing the Docker container technology for the Regression Testing of its software such as netCDF. When a corresponding Docker image runs, it performs regression testing against the netCDF-C library, as well as the netCDF-Fortran and netCDF-C++ (CXX) interfaces which depend upon it. The image is fairly sophisticated and can be used for a wide range of tests, using either local code or pulling down source code from the GitHub repositories. It is an efficient way to make sure a change in one library does not break something in a different library. Docker's strength, when used for regression testing, is that it lets Unidata test multiple environments simultaneously and in an efficient manner.

Product Generation for Ingesting into Internet Data Distribution and Additional Experimentation

Unidata continues to operate midsized virtual machine instances in both the Amazon EC2 and Microsoft Azure cloud environments for generating image products for the IDD Next Generation Weather Radar (NEXRAD) Level III national composites and Geostationary Operational Environmental Satellite (GOES)-East/West image sectors data streams. The generated products, which are currently the primary source of the data streams to Internet Data Distribution network participants, are streamed first to Unidata-operated data facilities in Boulder, CO, and from there they are distributed to users that have subscribed to those data streams. A midsized VM instance on Microsoft Azure is being used to investigate running the IDV in the cloud. RAMADDA has been installed and can generate noninteractive IDV displays using the X Virtual Frame Buffer (Xvfb) for the needed X Window environment.

Deployment of AWIPS II in the Cloud Software

The Advanced Weather Interactive Processing System, version II (AWIPS II) is a weather forecasting, display, and analysis tool that is used by the

National Oceanic and Atmospheric Administration/National Weather Service (NOAA/NWS) and the National Centers for Environmental Prediction (NCEP) to ingest analyze and disseminate operational weather data. The AWIPS II software is built on a Service Oriented Architecture, takes advantage of open source software, and its design affords expandability, flexibility, and portability.

Because many university meteorology programs are eager to use the same tools as those used by NWS forecasters, Unidata community interest in AWIPS II is high. The Unidata staff has worked closely with NCEP staff during AWIPS II development to devise a way to make it available to the university.

The Unidata AWIPS II software was released in beta form in 2014, and it incorporates a number of key changes to the baseline NWS release to process and display additional data formats and run all components in a single-server stand-alone configuration. In addition to making available open-source instances of the software libraries that can be downloaded and run at any university, Unidata has been testing *small-footprint data-server side* of AWIPS II, known as the Environmental Data Exchange system (EDEX), on both Microsoft Azure and Amazon EC2 environments. EDEX is the AWIPS II backend data server, which ingests live data via the LDM and writes it to Hierarchical Data Format, version 5 (HDF5) for processed data files and Postgres for metadata. In the past, Unidata has deployed EDEX as a stand-alone server on an Amazon EC2 instance, with encouraging results, with the most limiting factor for users being bandwidth/time to display products queried from the cloud. In this setup, universities receive all of the data from remote cloud instances, whereas they only have to run the AWIPS II client, known as the Common AWIPS Visualization Environment (CAVE), to analyze and visualize the data. During the prototyping phase in Spring 2015, 22 universities were running CAVE visualization clients and connecting to Unidata-operated EDEX instances on the Microsoft Azure cloud environment. Additionally, the emphasis on the development of a reduced footprint version of AWIPS II by Unidata is deliberate, because most universities do not have the resources to install the kind of hardware that is typically deployed for AWIPS II at a typical National Weather Service Forecast Office.

Separately, another AWIPS II instance was created cooperatively by Unidata and Embry Riddle Aeronautical University (ERAU) in Prescott, Arizona (AZ), as part of their equipment award from Unidata, on Amazon EC2. The ERAU AWIPS II cloud instance was set up as a private instance for use in education and research at the university and for their collaborations with the nearby NWS office in Flagstaff, AZ. Further details on that project are described by James and Weber in another chapter of this book.

INTEGRATED DATA VIEWER APPLICATION-STREAMING CLOUD SERVERS

The Integrated Data Viewer (IDV) is a state-of-the-art, desktop-client visualization application that gives users the ability to view and analyze a rich set of geoscience data, including real-time data, in a seamless and integrated fashion. Rewriting existing packages for new platforms, such as the cloud, often requires significant investment in terms of development time and developer expertise. In many cases, porting older software to modern devices is neither practical nor possible.

As an example of bringing legacy applications to a cloud environment, Unidata has been working toward deploying the IDV in a cloud setting with two objectives in mind:

1. Creating a "zero deployment" experience for users so that they can run the IDV from any device they choose using application-streaming technology.

2. Running the IDV in batch mode for the purpose of testing the IDV. Such a batch-mode capability could expand to users who wish to use the IDV on a server, proximate to data, and generate images and products in a batch mode and accessible via web browsers.

With these goals in mind, Unidata has created virtual images of the IDV using the Docker container approach for deployment in the cloud and evaluating application streaming as a strategy for making the IDV available to a new generation of users and computing devices, such as tablets and smart phones. According to Technopedia, "application streaming refers to an on-demand software delivery model that works based on the fact that the majority of applications need just a small portion of their total programming code for operation." This indicates that there is no need to fully install a program on a client machine; however, portions of it could be offered across the network whenever required. Application streaming is transparent to the end user and requires only a small fraction of the information to be transmitted to trigger and run the application.

To achieve the above goals, Unidata is using the Microsoft Azure cloud platform to explore cloud-based IDV-as-a-service instances to the user community on an on-demand basis. That experimentation has help Unidata gain a better understanding of how the IDV works in cloud environments and changes that are needed to improve the performance of IDV as a service.

Among the lessons learned, Unidata experienced several challenges in building IDV Docker containers, including handling the visualization capability of the IDV in a "headless" environment and using Xvfb to handle the remote visualization.

After overcoming those challenges, Unidata has successfully created a proof of concept of the cloud-based IDV by instantiating IDV instances dynamically in the cloud and streaming the resulting displays to tablet devices. Current efforts are focused on creating a web dashboard which will allow users to register and manage IDV-streaming requests.

COMMUNITY ENGAGEMENT, EDUCATION, AND LEADERSHIP

Internet technologies have amply demonstrated the compounding benefits of a global cyberinfrastructure and the power of networked communities as people and institutions share knowledge and resources. Effective cyberinfrastructure is a powerful tool for knowledge creation—one that is shaped with foresight and strategy, while considering human, organizational, and social factors. Unidata recognizes that providing robust, comprehensive, and persistent cyberinfrastructure is essential to transforming geosciences education and research, as well as the community's culture. By developing and providing state-of-the-art data services, tools, and middleware, Unidata has become a leader in the establishment of a geosciences cyberinfrastructure. That leadership role goes beyond the day-to-day operations and management of the program, and includes advocating both within and outside of the Unidata community for adoption and diffusion of new innovations and approaches to data access, sharing, and use.

Unidata has always taken an active role in building a virtual, distributed knowledge community and bringing stakeholders together to address important cyberinfrastructure issues by organizing national and international meetings. Over the years, Unidata, through its commitment to developing and sharing new knowledge, has served as an intellectual commons, providing stimulation of ideas, and appropriate models for community interaction, while developing a strong culture for technological change in academia and government. Our plans and ongoing activities for delivering cloud-based solutions for data delivery and analysis are no different. Timely and effective communication and provision of comprehensive support for users on Unidata's current and emerging tools and services remain integral to its mission. Strategic communication is not only crucial in informing and educating the community about the program's plans and activities, but is required to continually engage community members and other stakeholders as active participants.

As a part of its mission, Unidata, working closely with its User's Committee, holds summer workshops every 3 years on topics of interest to a broad community of users in the atmospheric and related sciences. The 2015 Unidata Users Workshop, which was held June 22–25 in Boulder, CO, had the theme "Data-Driven Geoscience: Applications, Opportunities, Trends, and Challenges" to educate as well as communicate contemporary ideas and approaches to scientific computing. The workshop topics were specifically chosen to further Unidata's goals for transitioning its data services to a cloud-computing environment. The workshop drew nearly 75 participants from across the atmospheric and other geosciences communities, and they explored how modern approaches in cloud computing and Python-based work flows can be leveraged by scientists to interact with and manage ever-growing data volumes.

CLOSING REMARKS

Data services, software, and committed support are critical components of geosciences cyberinfrastructure that can help scientists address problems of unprecedented complexity, scale, and scope. In this chapter, we have presented innovative ideas, new paradigms, and novel techniques to complement and extend Unidata's offerings. Our goal is to empower users so that they can tackle major, heretofore difficult problems.

Our long-term goal is to create rich, self-sustaining scientific research ecosystems in which data, services, and tools are nurtured by an engaged geosciences community. The full realization of this vision will likely take a decade or longer, but we expect to make demonstrable progress by the end of the period of performance on 5-year award from the National Science Foundation. We believe that the cyberinfrastructure environment that will arise from the steps we are proposing has the potential to profoundly transform the conduct of research and education in the geosciences and beyond.

The goals and the activities outlined in Unidata's future plans are in furtherance of and congruent with the NSF's strategic plans "Empowering the Nation Through Discovery and Innovation: The National Science Foundation Strategic Plan for Fiscal Years 2011–2016" (2011), "GEO vision"

(2009), "Strategic Frameworks for Education & Diversity, Facilities, International Activities, and Data & Informatics in the Geosciences" (2012), and "A Vision and Strategy for Data in Science, Engineering, and Education—Cyberinfrastructure Framework for the twenty-first Century" (2012). We are building on the foundation Unidata has laid over the years, providing cyberinfrastructure for researchers to address frontier science questions, enabling new discoveries, and building the geoscience community's capability to educate the next generation of scientists. Our commitment goes beyond technology. By creating an environment in which community members can work together, we hope not only to lower technological barriers to solving multidisciplinary grand challenge problems, but to develop the profession's human capacity and transform the conduct of science.

We acknowledge that this is an ambitious plan with many interrelated goals, but we believe bold thinking is required to address the emergent scientific, educational, and cyberinfrastructure challenges facing the community. As a cornerstone geoscience cyberinfrastructure facility with an established record in creating ground-breaking software and services that are in use far beyond the intended audience in academia, the geoscience community expects Unidata to both lead users by providing innovative solutions as well as follow the community's direction in setting priorities and being responsive to its current and emerging needs. Yet even as we strive to engage these broad challenges, the Unidata program remains firmly committed to meeting its responsibilities to and addressing the evolving needs of its core atmospheric sciences community. Sustained and strong engagement by our community, close partnerships, and collaboration with geoscience data providers, tool developers, and other stakeholders, and informed guidance from our governing committees will all be important catalysts for Unidata's success.

Our community's desire for revolutionary ways of wringing knowledge from an ever-expanding pool of Earth System science data presents Unidata with multiple, quickly moving targets. At the same time, the reality of constrained resources means we must choose the problems we will tackle with care and prudence. To succeed in dramatically improving the way data-centric geoscience is conducted will require an approach that is flexible in the face of evolving technologies and shifting priorities. As a result, the underlying theme of our long-term planning is to remain nimble. We must parlay creative, out-of-the box thinking and ongoing collaboration with the community we serve into pragmatic projects that solve today's scientific problems while setting the stage for future advancements.

Achieving these goals will help our community realize the vision of geoscience at the speed of thought. This simple statement asks Unidata, in partnership with the community, to work toward a transformation in the practice of data-intensive research and education in the geosciences, enabling researchers and educators to carry out their work in more innovative, efficient, and productive ways, pushing beyond the boundaries of their current knowledge and approaches. In the process, we envision a future that dramatically reverses today's situation in which a researcher may spend 80% of his or her time dealing with data discovery, access, and processing, and only 20% "doing science" by way of interpretation, synthesis, and knowledge creation (Michener, 2012). Unidata is firmly committed to work toward realizing this transformation.

ACKNOWLEDGMENT

The Unidata Program at UCAR is funded by the National Science Foundation grant AGS-1344155. The author is grateful to the Unidata team, who largely contributed to the work that is described in this chapter.

REFERENCES

Conford, J., 2008. "Emergent" or "Emerging" Technologies: What's the Difference and Does it Matter? (Blog Post). Institutional Responses to Emergent Technologies. http://bit.ly/18PahAH.

Edwards, P.N., 2010. A Vast Machine, Computer Models, Climate Data, and the Politics of Global Warming. The MIT Press.

Heidorn, P.B., 2008. Shedding light on the dark data in the long tail of science. Library Trends 57 (2), 280–299 (The Johns Hopkins University Press).

Harper Pryor, 2012. Big Data in NASA and beyond. 2012 Summer Short Course for Earth System. Modeling and Supercomputing. https://nex.nasa.gov/nex/static/media/other/Summer_School_Big_Data_Talk_071812.pdf.

Hey, T., Tansley, S., Tolle, K. (Eds.), 2009. The Fourth Paradigm: Data-intensive Scientific Discovery. Microsoft Research, Redmond, WA.

Killeen, Ti., 2011. November: Speech at First EarthCube Charrette. Washington, DC.

Michener, B., 2012. Grand Challenges and Big Data: Implications for Public Participation in Scientific Research [PowerPoint Slides]. Retrieved from: http://www.slideshare.net/CitizenScienceCentral/michener-plenary-ppsr2012.

CHAPTER 5

Supporting Marine Sciences With Cloud Services: Technical Feasibility and Challenges

B. Combal
IOC-UNESCO, Paris, France

H. Caumont
Terradue Srl, Rome, Italy

INTRODUCTION

Under the growing pressure of human activities on natural ecosystems, interest in the analysis and reanalysis of Earth monitoring data and models goes beyond the studies of physical and ecological systems. Decision makers increasingly acknowledge that human activities apply stresses of different natures and magnitudes on biophysical systems, changing the ecosystem's state, which can in turn affect the delivery and value of services to human communities (Fig. 5.1). In this cycle, human activities apply pressures on natural systems which may result in a change of their natural states and functions. The causal chain of interactions, going from human applied stressors, and leading to a potential change in ecosystem services involves physical, chemical, and biological factors, then loops back to the human society potentially affecting its activities. In addition, the interactions between human related activities, stressors resulting from those activities, governance activities, and the exploitation of ecosystem services are strongly mediated by socioeconomic factors.

Assessing the risk to which a human society is exposed is an increasing demand from policy makers. The risk can be defined as the combination of hazard, exposure, and vulnerability (Cardona et al., 2012), which relates respectively to biophysical, socioeconomic, and governance variables. A better understanding of the risk exposition is required to improve the governance of human activities from local to global scales (Fanning et al., 2007). The time projection of the natural ecosystem stresses and the corresponding states is crucial for supporting political decisions. For the ocean, prominent

Cloud Computing in Ocean and Atmospheric Sciences
ISBN 978-0-12-803192-6
http://dx.doi.org/10.1016/B978-0-12-803192-6.00005-0

Copyright © 2016 Elsevier Inc.
All rights reserved.

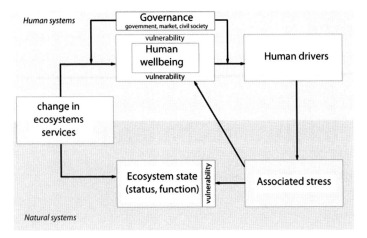

Figure 5.1 Relationship between human and natural systems from the point of view of ecosystem services and its consequences for people expressed as human well-being. The top half of the diagram is the human system, the bottom half the natural system. *Adapted from IOC-UNESCO, 2011. Methodology for the GEF Transboundary Waters Assessment Programme. In: Methodology for the Assessment of Large Marine Ecosystems, vol. 5. UNEP, viii + 115 pp.*

stresses are related to industrial fisheries, pollution, loss of biodiversity, and climate change.

Risk assessment in such a context requires merging data and scientific expertise from different disciplines. In particular, the projections of ocean key variables have ultimately to be combined with socioeconomic and governance indicators.

Integrating stressors such as climate–change data into assessments has raised some practical problems: estimates or outputs from ocean climate models in particular, and Earth observation data in general, usually require technical and thematic knowledge to be properly processed and interpreted. Teams of collaborators from various expertise fields, as is required in a risk assessment scenario, do not always have the needed knowledge to properly process and integrate datasets.

Such datasets generally correspond to quite large volumes, requiring specific software or Application Programming Interfaces (APIs) for their processing, and dedicated technical preparations (such as reprojections, interpolations, recalibrations, etc.). This is particularly true for data products delivered by satellite-based Earth observation systems that are continuously acquiring large collections of datasets, quantifying a significant number of essential oceanic variables. Some substantial data-processing capacity and skills in computer sciences are generally required. As a result, the integration of information from Earth monitoring or climate-model outputs to a risk assessment activity is typically hampered by domain-specific technical issues.

In the last decades, an important effort was made by the Group on Earth Observation (GEO) to establish the Global Earth Observation System of Systems (GEOSS) Common Infrastructure (GCI) allowing users to retrieve and access data published from different Earth observation systems (Percivall, 2010). Such an approach relies on the interoperability of the metainformation describing datasets, actually hosted on different platforms, through a unique portal. Though crucial, interoperability and retrieval of datasets is not enough to allow use of Earth observation and climate models data sets beyond their community with only minimal efforts and under the best processing practices.

The "GEOSS interoperability for Weather, Ocean and Water" project (GEOWOW, 2011–14), co-funded by the Seventh Framework Programme, or "FP7," of the European Commission (Community Research and Development Information Service (CORDIS), 2009), had the goal to enrich the GCI with related essential variables, and new data access and processing functionalities. For the ocean component of the project, we explored the possibility of providing a data-processing service discoverable from the GCI. Our objective was to support research and development of marine assessments, through the improved availability of information derived from ocean data, and an easier combination with model data to feed decision support systems.

To test the idea against real-world constraints, we developed a climate-model processing service based on well-established cloud-computing technology assets. The processing algorithm itself is conceptually simple and consists of computing the ensemble mean, as done in Intergovernmental Panel on Climate Change (IPCC) Assessment Reports, of models outputs for a given variable and climate-change scenario (the so-called Representative Common Pathway (RCP), defined further down in this chapter). The demonstration does not target high-complexity computations, but rather how such a service can be created in a scientific community, and then shared and reused beyond this community. We address here the following questions: how to share technical and scientific know-how about data processing, how to share the processing tools, including the algorithm and the processing framework, and how to make the processing feasible for a larger audience of users. In our experiment, we found that cloud-computing technologies allowed us to address those issues; first, because the virtualization decouples software (including its installation and library dependencies) from hardware, but also lets users trigger large data-processing tasks from their premises, where the Internet bandwidth is for them insufficient to download efficiently the datasets (adding to this the constraint of local storage of the repatriated data). This opens up the sharing scenario to a much broader user base. In addition to those properties, applications, together with their datasets and their required data-processing resources, can be indexed altogether, as a "Service," through systems such as the GEOSS-GCI or even mainstream Internet search engines. Another clear advantage for us of cloud computing-based technology was that the processing system is paid per process, removing the cost of buying and maintaining processing facilities, which is of utter importance in developing countries where Earth observation information is of critical importance and processing infrastructure not available.

BRIDGING TECHNICAL GAPS BETWEEN SCIENTIFIC COMMUNITIES

In the last decade, substantial efforts were made to allow geospatial data discovery and interoperability, in particular spurred by the normalizing effort led by the Open Geospatial Consortium (OGC) (Percivall, 2010). Nowadays, Geographical Information Systems (GIS) offer functionalities compliant with OGC Web services standards, allowing users to display maps, share geographic features, or access data catalogs from remote servers, making data services interoperability across the Internet a reality.

Data interoperability and Web services allow users to explore the exponentially growing amount of data available, with a reduced cost in local storage. However, this is not an optimal situation for allowing experts from different fields to access data and integrate them into their research.

An illustration can be given through the Transboundary Water Assessment Programme (TWAP), funded by the Global Environment Facility (GEF) and led by the United Nations Environment Program (UNEP) (UNEP, 2015). The United Nations Educational Scientific and Cultural Organization-Intergovernmental Oceanographic Commission (UNESCO-IOC), coordinating the Open Ocean component of TWAP, worked on the development of risk indicators integrating biophysical indicators, under the form of a Cumulative Human Impact (Halpern et al., 2008) indicator, socioeconomic, and governance factors. Some of the biophysical parameters are projected using ensemble means of climate model projections of ocean variables, allowing an estimate of future risk. The evaluation of the risk in itself requires merging information from different fields, such as biology, the environment, socioeconomics, and government policies, and furthermore, integrating this information, compressing it, and turning the data into indicators. The highly technical tasks often involved in the processing of Earth observation or climate models adds unnecessary burden to the risk assessment, generally solved by requesting specific support to technical experts. In many cases, that is a technical-expert exercise done solely in the community which is managing the actual observations or modeling. However, extracting information from the different sources of data, in our case calculating a risk indicator, requires the data to be exploited by a community of scientific experts who do not share the same culture and practices in the fields of observations and models management, making these phases of data processing sometimes too complex for them. As a result, socioeconomists and environmental scientists often rely on project funding or on the production of ad hoc datasets, for example, to integrate Earth observation data or climate models projections in their assessments, making the overall assessment a slow process highly dependent on external expertise.

Data and Processing Sharing

A basic idea tested in the GEOWOW project was to understand how a group of subject matter experts can share their data and their research experiments (including data-processing algorithms) in a fashion allowing them to reach the largest number of the scientists and experts beyond their own community (see Table 5.1). The tested approach involved the following concepts:

- Open Source software-based solutions as the norm;
- Compute-intensive data processing to be done on low-cost, community-independent computing resources (leveraging commercial cloud-computing providers);
- Both data-processing technology resources (software frameworks and components) and data-analysis software codes (algorithms) can be shared and reviewed within the community of users.

Moreover, the GEOWOW project offered us the means to experiment with new types of collaboration, delivering multiple gains for several types of stakeholders:

- Authors of marine assessments gain access to development environments that are more flexible;

Table 5.1 A vision for GEOSS as part of the GEOWOW project approach

Present situation	Future GEOSS infrastructure?
Expert scientific advice identifies the required datasets	Expert scientific advice identifies the required datasets
Discovery: we can look in the GEO portal, but most data required is already known	Assembly: needed datasets are assembled on demand into the GEOSS infrastructure (into a cloud?)
Data-usage rights are dealt with each individual provider if there are restrictions	Data-usage rights are globally registered for scientific use of cloud-computing infrastructures
Downloads of individual datasets to local computing (repeatedly in case of evolving indicators)	Data as a service can live and remain "in the cloud," for the application of algorithms and calculation of indicators
Calculations: on individual datasets, mostly for each needed spatial reference system transforms	Calculations: modular infrastructure components allow the exchange and reuse of scientist's workspaces
Heavy algorithm management and chaining for each host workflow and data input	Repeatable environments allow for easy updates (in time or with new concepts) and scientific sharing
Web platforms designed to specific audiences, but live only as long as the project	Interactive "live" Web platform output maps and graphics readily part of the GCI: a Legacy for other users

- Decision makers in the "Ecosystems" Societal Benefit Areas benefit from more frequently updated and more specific indicators;
- Value-adding companies can get new opportunities to provide their customers with additional services by implementing extension points (such as data gateways, cloud-brokering services), built on existing services, and in turn adding new capabilities for GEOSS users.

If we take the example of an application that is processing the outputs of climate change models, the life cycle of such an application can be described as follows: the community of modelers decide to create an application for processing its own dataset, say, the processing of the ensemble means of variables output by all models. An Open Source processing chain is written, possibly collaboratively reviewed and elaborated on a social coding platform (e.g., GitHub), and bundled, on demand, into a virtual machine, together with the required software libraries to make it readily operational. For this, the task of solving the library dependencies must be done by the expert writing the processing chain. In our case, this code has been integrated as an Hadoop MapReduce processing chain, a well-adopted parallel computing technology (especially in the Web industry) that we selected with the perspective to natively offer a highly scalable data-processing workflow, able to run on low-cost commodity hardware, easily rented, on demand, from commercial cloud-computing providers. One example of our approach, from another domain, is the integration of the Caltech/ Jet Propulsion Laboratory (JPL) Repeat Orbit Interferometry Package (ROI_PAC) processor, https://github.com/Terradue/InSAR-ROI_PAC, developed in the area of Interferometric Synthetic Aperture Radar (InSAR) data processing, that we ported to Hadoop to enable users with massive parallel-processing capabilities. The result is deployable on cloud infrastructures. This type of approach, generalized across multiple domains, is allowing users to tackle challenges of large experiments involving climate-model outputs and/or collections of satellite-based Earth observation data. The processing system knows how to retrieve large collections of data and how to process it efficiently accordingly to the provisioned compute power. Finally, the input data (i.e., the climate-model products), the processing-system endpoint, the processing-chain resources, and the set of published results are exposed to the GCI, so they can be discovered by the broader GEO community (http://bit.ly/1Q2P1yp). Discoverability beyond GEOSS can go through other channels, like the submitter setting up a Web page with resource access links on his own portal. Provided for proper Web page publishing (selection of header tags, cross-linking with social media services like blog posts or Twitter posts), this page will be indexed and well ranked by Web search engines in a matter of hours or few days.

The GEOSS users can therefore retrieve the published datasets, and also reuse the data-processing resources in mainly two complementary ways: either by invoking a deployed instance, delivered "as-a-service," interacting in this case with a client portal to select sources and parameters and trigger the processing chain, or by retrieving not only the climate model's outputs, but also the associated data-processing chain itself. This allows the user to modify, configure, and redeploy the full experiment on their own, in principle with the sustained goal of community sharing and feedback gathering.

Such an approach offers many benefits. Although most scientific publications would only provide a conceptual description of their algorithms, this approach offers not only the source code, but also the overall experiment context, including the APIs, and the processing system (a virtual machine template with the software framework and the required libraries for scalable data processing), all tested and described.

Conversely to an approach based on publishing a Web service endpoint only, in which the service provider has to maintain a processing infrastructure and to scale it out depending on the demand, it is up to each user (or each user's sponsor organization) to provision an account to a cloud-computing provider (e.g., based on a pay-as-you-go service subscription model) and deploy the solution for the required duration of his own experiment.

Overall, anyone can review the processing code and reference it in scientific publications: both data and processing can be allocated a Digital Object Identifier (DOI), attached to a publication, and made available as additional material, allowing a full reproduction of the experiment. In addition, experienced users can bring improvements and corrections, so the service can evolve and grow thanks to support from inside and outside its community of origin.

Massive Processing on the Cloud

We aimed our work at supporting scientists who are addressing the needs from decision makers and scientific communities interacting worldwide through use of ocean forecasting products, developed under the umbrella of global initiatives, such as GEOSS or the Global Ocean Data Assimilation Experiment (GODAE).

One key aspect in serving the community is supporting the transition to cloud-computing environments, and especially commercial cloud providers. Whether it comes to the staging of data collections or the burst-loading onto a cloud cluster of dedicated Virtual Machines embedding a processor, it remains fundamentally about doing data retrieval, ingestion, processing, and dissemination operations, but with the goal that everyone can do it at scale. To realize that objective, we wanted to allow users to run the calculations of environmental indices or indicators globally, at very low cost, and with a flexible deployment process. As a corollary, we were also looking at releasing users from issues with software library management and how to compile them, and ultimately to remove the burden for them to buy dedicated hardware to conduct their research. Data storage was important as Coupled Model Intercomparison Project Phase 5 (CMIP5) model output ensemble can be larger than several Terabytes (TB) for a single three-dimensional (3D) variable.

To implement this approach, we exploited the Cloud Platform operated by Terradue. The company offers cloud services for the scientific communities exploiting satellite-based acquisitions of Earth observation data (Caumont et al., 2014), in particular the communities served by the European Space

Agency (ESA). Terradue develops, maintains, and operates several cloud services for the benefit of scientific communities in the field of Earth observation (e.g., supporting scientists in the Maritime Research and Experimentation domain and promoting results through the GEOSS Architecture Implementation Pilot), including for the ESA (e.g., for massive InSAR processing over Greenland for the ESA Climate Change Initiative's Ice Velocity products study).

The Terradue Cloud Platform basically allows cloud orchestration, storage virtualization, and Virtual Machine provisioning, as well as application burst-loading and scaling on third-party cloud infrastructures (See for example presentations on http://www.slideshare.com/terradue Open-NebulaConf'2013: http://www.slideshare.net/terradue/opennebulaconf2013 and EGI Conference 2015 http://www.slideshare.net/terradue/terradue-egi-conference-2015).

Typical cloud services are marketed and delivered under the following operational models:

- Infrastructure as a Service (IaaS): computing resources provided as commodity services, accessed by users on demand via pay-as-you-go pricing models, and with optimizations of cloud resources usage (due to continuous, rolling, user consumption) that are driving prices continually downward.
- Platform as a Service (PaaS): interoperable platforms tailored for very specific applications (e.g., Hadoop MapReduce, Web Development Frameworks). Such services develop high interoperability and manageability of cloud resources from a developer user point of view, as well as facilitate engineering support activities for solutions providers (e.g., Microsoft Azure for software developers, SalesForce PaaS to build Customer Relationship Management Applications, etc.).
- Software as a Service (SaaS): Web-accessible applications based on a user subscription model (e.g., monthly fee), with the main challenge being related to the standardization of APIs, so that users can interact with other environments (cloud-storage services, email services, etc.) from their SaaS application. More specifically, APIs are critical to handle data stage-in and stage-out of the cloud, to enable interoperability for user communities, and prevent vendor lock-in.

Within the Terradue Cloud Platform, the Developer Cloud Sandbox service provides a Platform-as-a-Service (PaaS) environment for scientists to prepare data and processors (including development and testing of new algorithms), designed to automate the deployment of the resulting environment to a cloud-computing facility that will allow consuming cloud-compute resources. This technology is integrated with content management tools that accompany the development and testing process. A user dashboard, embedded in the user's Virtual Machine, provides a graphical interface for the management functions of the Cloud Sandbox.

From the Platform, registered users can access dedicated virtual resources for application development and integration, and then template these resources for scalable hosted processing on the cloud. With this mechanism, an application developer or a service integrator is equipped with a Sandbox mode to build a self-contained application, and has a direct access to a Cluster mode, automatically scalable on large number of computing elements, as a result of his work. From the Sandbox mode developments, users can request the Cloud Platform to configure a cluster mode deployment, benefitting from all the packaging work done previously to scale the solution on demand.

This capability relies in particular on a technical process (generally referred to as "contextualization"), which can be optimized through the definition of software baselines that are deployable from validated software packages. In other words, a single software package (specifically using the Linux RedHat Package Manager (RPM) format) containing all the application resources and the specification of all software dependencies. Following the proposed approach, the application packaging includes all the software dependencies, improving in that way the scientific reproducibility of the application, one of the key drivers of our approach (Hey et al., 2009), and providing a more robust identification and management of such dependencies. It also allows a better maintainability on the side of the Cloud Operations support team, especially avoiding the cumbersome loading on the Platform's Cloud Controller of several versions of the application's virtual machine images, and enhances the portability of the developed applications to different cloud providers.

Another key component of the approach is the Hadoop MapReduce Framework, used to automate distributed computing over large datasets on clusters of virtual machines. Leading Web companies like Google Inc. operationalized the now open-source (Apache Hadoop) and well-adopted framework for data-intensive computations. The MapReduce (MR) programming model is a simple yet powerful way to execute on clusters of commodity computers (Dean and Ghemawat, 2004). The framework targets applications that process large amounts of textual data (as required, for example, when generating and updating the index of a global Web search engine), which are parallelized following a master/worker principle.

Scalability and robustness are supported through features like distributed and redundant storage, automated load balancing, and data locality awareness.

In a Developer Cloud sandbox, the implementation of MR is used in a simple way, to map URLs of input data to be processed over single shell command execution on virtual machine, and reduce the output data URLs to results list. This Framework is then deployed according to the virtual machine mode (sandbox or cluster). Terradue supported IOC in the ramp-up and learning curve required when learning how to develop with the MR programming model. User support and hands-on exercises were provided for:

- Importing data on the Hadoop Distributed File System (HDFS) from existing sources;
- Processing data with Hadoop MR Streaming jobs;
- Oozie workflow as orchestrator of MR Streaming, to optimize the processing time and ensure scalability.

Oozie is a workflow scheduler system to manage Hadoop jobs. It is a server-based Workflow Engine specialized in running workflow jobs with actions that run Hadoop MR. Oozie is implemented as a Java Web Application that runs in a Java Servlet-Container. A workflow is a collection of MR jobs arranged in a control dependency DAG (Directed Acyclic Graph). A "control dependency" from one action to another means that the second action cannot run until the first action has completed.

The process of data distribution with the embedded tools is simple: the Map part is accomplished by the Hadoop JobTracker, dividing computing jobs into defined pieces, and shifting those jobs to the TaskTrackers on the cluster machines in which the needed data is stored. Once the job is completed, the correct subset of processed data is Reduced back to the central node of the Hadoop cluster, combined with all the other datasets produced on each of the cluster's machines. To share a file with the next (or another) parallel job in the processing chain, it has to be published on HDFS. The ciop-publish tool returns an HDFS Universal Resource Locator (URL), accessible by all the jobs in a processing chain because it is on a distributed file system.

A typical workflow in using Hadoop is as follows:

- Load data slices into the Hadoop Distributed File System (HDFS) of the cluster;
- Send the first job to the cluster workers to analyze these input data slices (MR tasks on each worker machine of the cluster);

- Store results into the cluster's HDFS;
- Send the second job to the cluster workers to analyze these intermediary data.

In our specific context, users could interact from a Developer Cloud Sandbox with the ESGF Gateway using the Earth System Grid Federation (ESGF) client, and were able to automate the parallel download of products from the "ESGF Data Search," getting as their global experiment input a list of a single product's URLs to be distributed over a cloud-computing cluster.

The "esgf-tools" are provided as a Redhat linux package. After accessing his or her Cloud Sandbox, a developer can install the esgf tools using the standard "yum" command line ("yum install esgf-tools"), follow installation instructions, and test the service with sample queries.

The ESGF client being designed to perform data search and access from the ESGF, the first task consists of querying the ESGF Gateway on its Resource Description Framework (RDF) endpoint. This query returns RDF metadata about the subset of data filtered by using the given parameters: time_frequency, experiment, ensemble, and institute. The ESGF client then parses the responses and retrieves the list of OPeNDAP online resources.

Starting from this list, the ESGF client builds a new list of query strings by adding the rest of the options. To build valid OPeNDAP queries, the time search and level parameters are converted to a set of indexes needed to make a spatial and temporal query.

A template bash script, corresponding to a generic version of the ESGF download scripts, is used to generate the specific download script needed to download the list of OPeNDAP online resources obtained from the former step.

The access to the remote servers are protected by OpenID, so the user has to give his credentials by passing "--openid" option and "--password" options to the wget command (wget is a standard unix command to download data from a server) the generated download script. The security certificates are automatically downloaded by the script.

OPeNDAP servers impose a size limit to data download. For the ESGF, the limit is set to the default value (500 MB, as described in http://www.unidata.ucar.edu/software/thredds/current/tds/reference/Thredds-ConfigXMLFile.html#opendap). Should the query reach this limit, the server would return a 403 error (the server refuses to take any further action for this request) and the download fails. The corrective strategy consists of downsizing the requested data, by defining a small temporal dataset (to avoid a post-processing reconciliation of tiles, we chose to consider only slices along the time axis, though the script offers to process only a window of the data grid).

Once the OPeNDAP files (including .dods, .das, .lev, .lat, etc. components) are downloaded, an internal script converts the .dods binary format file to a Network Common Data Form (NetCDF) file, matching the .das Dataset Attribute Structure description file. Eventually, the download script deletes the downloaded temporary files.

The next step in the MR streaming consist in averaging each model, regridding the average to a common grid and then averaging the model averages for each time step. The processing is described in section Climate Model Output Processing.

To allow operational models of type SaaS, an OGC-compliant Web Processing Service (WPS) interface to the developed processing service was set up (Fig. 5.6). It was meant to access ESGF data and process two-dimensional (2D) and 3D datasets. A version of the WPS for processing a 2D dataset was assembled and successfully deployed on a cloud-computing cluster.

The following software components were involved in the development:

- ESGF Gateway: an OGC OpenSearch Catalog server used as a Gateway to the ESGF OPeNDAP-based federated Web services endpoints: caching ESGF metadata and proxying OPeNDAP URLs to be accessed directly by the processing software;
- ESGF Client: an OGC OpenSearch client developed by Terradue, used for discovery, access, and download of data from ESGF;
- Ultrascale Visualization Climate Data Analysis Tools (UV-CDAT)-cdms2 "Climate Data Analysis Tools": a python library for processing CMIP5 datasets;
- Developer Cloud Sandboxes based on Oozie and the Hadoop MR framework;
- GEOSS Common Infrastructure (potentially): the processing system is providing an OGC Web Processing Service (WPS) endpoint, and can thus be exposed to the GEOSS-GCI. Note that for doing so, a permanent infrastructure must be funded, which was beyond the scope of the GEOWOW project;
- GitHub: for having a public version of the processing code, that can be reviewed and updated by any contributor. Open access is a cornerstone of TWAP assessment: all methods and datasets public.

CLIMATE MODEL OUTPUT PROCESSING

A simple use case was chosen for the processing of the ensemble mean of ocean variables projections under climate change. Though the outcome was of immediate interest for the TWAP project, the intention of this discussion is not to concentrate on the sole production ensemble means from climate models. Actually, it could be envisaged that the modeling groups themselves could directly provide such ensemble means products. However, that is not the type of output product which was the driver of our work, but rather the overall approach, which can be adapted to different types of processing.

CMIP5: A Heterogeneous Dataset

The Coupled Model Intercomparison Project Phase 5 (CMIP5, Taylor et al., 2012) is a project of the World Climate Research Programme (WCRP) for providing IPCC AR5 (Fifth Assessment Report, IPCC 2013) with time-projected environmental variables. Virtually all climate modeling teams in the world contributed to this work (Taylor et al., 2012). A large repository of output physical variables and their time series were generated by this initiative, made available to the scientific community, and made accessible from the ESGF portal (Program for Climate Model Diagnosis and Intercomparison (PCMDI), 2014).

The CMIP5 database allows access to the variables output from each individual model. The outputs are generally available as time series, with various time steps, including daily, monthly, and yearly time steps; from 2006 to 2100 (some models even propose outputs projections up to 2300).

The models' variable time projections are computed for different RCPs, like RCP 2.6, RCP 4.5, RCP 6.0, and RCP 8.5, named after the target energy forcing by 2100, in W/m^2. Each RCP contains the same categories of data, but the values can vary a great deal, reflecting different emission trajectories over time as determined by the underlying socioeconomic assumptions (which are unique to each RCP).

The sea surface temperature variable, short named TOS (Temperature Of Surface) in CMIP5, is a 2D variable (gridded against latitude and longitude, no depth axis). This variable is available from the output of 35 models, listed in the Annexes. This variable is also projected for different time steps, including a monthly time interval, from 2006 to 2100.

Most climate models are run with different sets of input ensemble parameters, numbered r<N>i<M>p<L> corresponding to different settings of the models, to get the domain of variability of each model output. The triad of integers (N, M, L), for example r3i1p21, distinguishes among closely related simulations by a single model (Fig. 5.2). The so-called "realization" number (a positive integer value of "N") is used to distinguish among members of an ensemble typically generated by initializing a set of runs with different, but equally realistic, initial conditions. Models used for forecasts that depend on the initial conditions might be initialized from observations using different methods or different observational datasets. These should be distinguished by assigning different positive integer values of "M" in the "initialization method indicator" (i<M>). If there are many closely related model versions, which, as a group, are generally referred to as a perturbed-physics ensemble, then these should be distinguishable by a "perturbed physics" number, p<L>, in which the positive integer value of L is uniquely associated with a particular set of model parameters (e.g., r3i1p78 is a third realization of the seventy-eighth version of the perturbed-physics model).

The common method to compute the ensemble mean consists of averaging the output of a model corresponding to the various input ensemble parameters to get a model output average (Fig. 5.2), before averaging all the

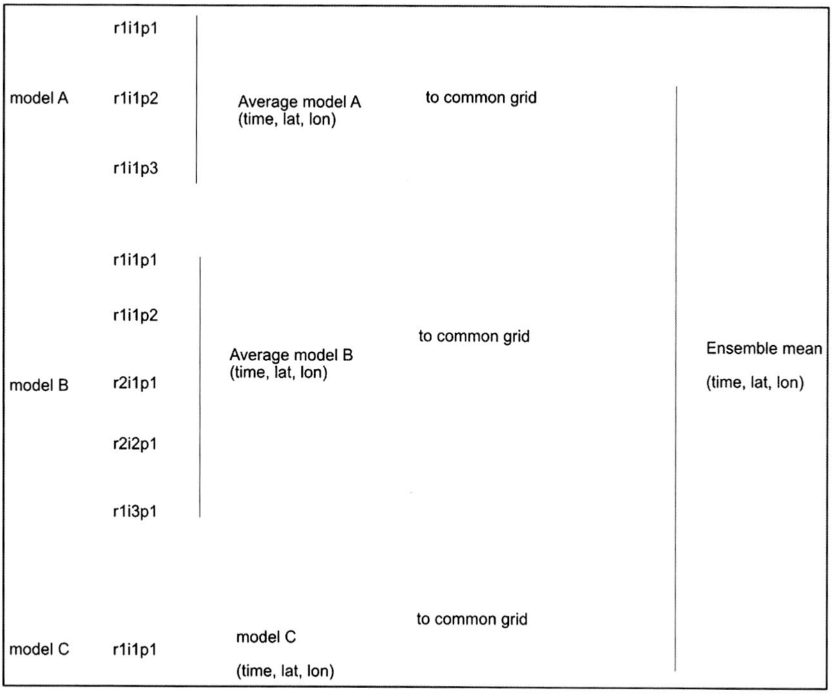

Figure 5.2 Each model generates outputs for a series of input ensembles r<N>i<M>p<L>. The ensemble mean requires averaging individual model outputs, reprojecting to a common grid, and averaging the results.

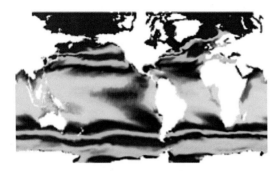

Figure 5.3 Geophysical Fluid Dynamics Laboratory (GFDL) (left) and Community Climate System Model 4 (CCSM4) (right) output grids.

model averages (Oldenborgh et al., 2013). This procedure avoids giving more weight to a model for the sole reason that it would have been run against many input ensemble parameters.

Model outputs are made available with a model grid, specific to each model. Fig. 5.3 shows an example of TOS variables from two different models in which the landmasses (in white) clearly outline the different grids: not only is the spatial resolution different, but the grid mesh is not evenly spaced and is significantly different between model outputs. The polar regions are not equally represented either. The grids are different in terms of their spatial sampling, but also in term of their description in the NetCDF format. Different kinds of grid systems are used, including rectangular and curvilinear, all matching the NetCDF Climate Model Output Rewriter (CMOR) norm adopted for CMIP5.

For the sake of the tests, a common grid was defined as a rectilinear grid (grid points evenly spaced along the latitude and the longitude axes), with a spatial resolution of ½° × ½°, latitude ranging from −85° to 85°, and longitude from 0° to 360°. This horizontal grid was designed with a higher resolution than most global average products found in the literature, which often have a spatial resolution of 1° × 1° (which has the advantage of smoothing any discrepancies between the datasets), to keep finer local resolution found in some models. However, the spatial resolution is controlled by a parameter and can easily be changed to 1°.

As the NetCDF grids do not start and end at the same longitude and do not have the same spatial resolution, some areas are under sampled. Two meridians (around 40° West and 80° East), corresponding to projection limits in several models are underrepresented. Continental shores show also large discrepancies due to the significant difference in land mass gridding in between the models. Grid cells with a smaller number of samples for computing the ensemble mean may slightly differ from the neighboring open ocean. A code (filter_verticalLines.py) was written to replace visible boundaries seams due to under-sampling with linear interpolated values (https://github.com/IOC-CODE/cmip5_projections/releases/tag/1.0). However, data are provided under their raw format (Combal, 2014a,b,c), to let users apply their favorite filtering operators or analyze the properties of the ensemble mean itself.

Data Slicing Along the Time Dimension

CMIP5 datasets are accessible through OPeNDAP servers of the ESGF PCMDI node data portal (http://pcmdi9.llnl.gov/).

OPeNDAP is an Open Source project for a Network implementation of the Data Access Protocol, or DAP (Gallagher et al., 2007), used in many scientific communities. The DAP was originally developed by the oceanographic community. Since then, it has been extensively used by a number of oceanographic, meteorological, and climate groups. There is, for example, a growing convergence over the Internet protocols of NetCDF-Climate and Forecast (CF) practices (a NetCDF providing machine-independent data formats for array-oriented scientific data, and applying a specific set of CF conventions) and OPeNDAP practices in ocean observing networks, such as Argo, OceanSITES, the OceanDataPortal, or the National Virtual Ocean Data System (NVODS).

The ESGF portal stores CMIP5 datasets as NetCDF files; the Web interface allows searching for a model, an output variable, an RCP, an input ensemble r<N>i<M>p<L>, a time frame, etc. As a data search result, the ESGF Web interface lets the user download a shell script, containing the wget commands required to download a copy of the stored files matching the search. Though extremely simple to use, this approach has the disadvantage of downloading entire files, making the download process a heavy task. The data portal also allows connecting through OPeNDAP and to any subset (a slice along a dimension of the datacube) from the stored datasets.

The CMIP5 datasets have two (longitude and latitude) or three (depth, longitude, and latitude) geographical dimensions, and a time dimension. The ensemble mean process is independent on those dimensions; it only requires averaging voxels between models, for the same set of coordinates (time, depth, longitude, and latitude); there are thus many possibilities for handling the dataset volume by slicing it up.

An interesting way to reduce the data volume, and, this way, ease the data-processing work, is to slice the N-dimensions dataset into independent slices (Fig. 5.5).

Slicing the dataset via OPeNDAP queries offers a handy way to manage multiple small data chunks that can all be processed the same way. This provided us the baseline approach to develop our distributed data-processing chain, with the time dimension of the climate model projections being our axis of performance optimization. A data connector, acting as a gateway to the ESGF Portal, based on the ESGF download scripts and using OPeNDAP libraries, was developed to allow a selective slicing and access to data from our processing chain. Technical details are found in the next section.

In summary, slicing along the time dimension offered several advantages. The ensemble mean is time independent; slicing along time does not affect the processing. Slicing along the time axis avoids cutting the latitude–longitude–depth grid, and thus avoids the postprocessing reconstruction effort. There is also no need for seaming together small chunks as a postprocessing of all the datasets, each parallel processor writing in output the ensemble mean for each time step, for its own chunk of time series. In addition, the approach works for bidimensional (longitude, latitude) and 3D (depth, longitude, and latitude) datasets. Framing the dataset (cropping data to a geographical window) is still feasible to reduce the final output.

To test the ESGF Gateway connector developed for the needs of GEO-WOW, the ensemble mean of the time series of thetao (sea potential temperature) was considered. Thetao is a 3D (latitude, longitude, and depth) time series, with a monthly time step. The dataset was sliced against the time dimension only. The computation of the ensemble mean being time independent, each worker node of the Hadoop-enabled processing chain only processes a succession of data slices extracted from the larger time series. The approach is completely scalable, supporting scientific workflows from integration and testing (use only a few data slices with minimal parallel processing capacity), and up to massive processing campaigns over several decades of data, without modifying the software logic at all.

The ensemble mean of the ocean surface temperature is of direct interest in many integrative assessments, such as TWAP (http://geftwap.org). A complete ensemble mean, involving all models outputs for each of their input ensemble r<N>i<M>p<L> was created with the same processing code. Three datasets were created. The ensemble mean of the CMIP5 time-projected TOS (Temperature Of Surface, equivalent to the Sea Surface Temperature), one-month time step, was processed for RCP 4.5 and RCP 8.5, for the time period 2010 to 2059, covering five decades from 2010 to 2050 (Combal, 2014a). Fig. 5.4 shows the last value in the time series for RCP 8.5, from file ensemble_tos_rcp85_205,912.nc.

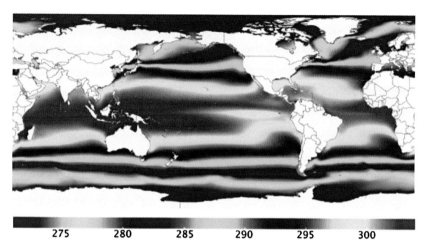

Figure 5.4 Ensemble mean of variable TOS for December 2059.

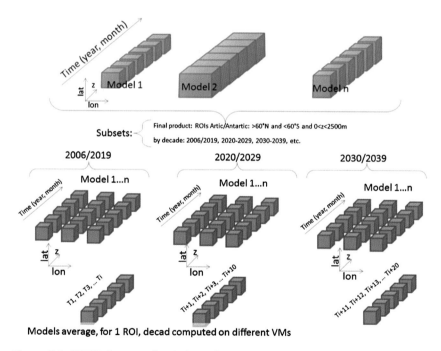

Figure 5.5 CMIP5 data are sliced along the time axis, average per model, regridded, then averaged in between model averages.

The outputs are available as NetCDF CF-1.0 files. Each file corresponds to 1-month ensemble mean, one-month being the input data time step. Four variables, gridded with the common $1/2° \times 1/2°$ grid described above are found in each file: the TOS ensemble mean (variable "mean_mean_tos"), the count of models used to compute the ensemble mean per grid cell (variable "count"), the minimum of the ensemble mean for this date (variable "minimum") and the maximum (variable "maximum"). Missing values are encoded with 1×10^{20} (missing count is encoded on 999999).

The ensemble mean of historical runs does not depend on any RCP, and is available for the period 1971 to 2000, at a monthly time step (Combal, 2014b). The archive starts on 197,101 and ends on 200012.

The TOS climatology (equivalent to the Sea Surface Temperature climatology) was computed by computing the monthly average of the former dataset (Combal, 2014c). The files are CF-1.0 formatted and contain only one variable, "TOS", in Kelvin, with 1×10^{20} for the missing values. This latter dataset allows computing deviations of the time projections to the climatology, under scenarios RCP 4.5 and RCP 8.5, which is base information for computing other environmental indicators.

SCALABLE DATA PROCESSING: NUTS AND BOLTS

Data Staging: Some Optimization Challenges

Developing a system for processing the CMIP5 database clearly required appropriate tooling and strategies, both for accessing the portal using OPeNDAP and for processing the datasets. In particular, due to the complexity of the NetCDF files, showing rectangular or curvilinear grid systems, and with different ways of storing the axis information, there was a need to develop a processing code based on a general purpose library. Most of such libraries provided with scripting languages limit the interaction with NetCDF datasets to the extraction of a matrix of values from the file format, making the generalization quite difficult to encode, so we preferred the library "cdms" (for Python), which allows accessing the NetCDF format and its components as objects, so the same methods apply to a variety of data types.

Processing CMIP5 datasets, for those who do not model climate, is rather challenging: we could access data; however, it was difficult to know how the datasets were generated, if all of them should be considered, or how to handle the little differences in the data formats. Also, at the time of this work, the ESGF Portal support was strictly related to the infrastructure (accessing the portal, downloading data), not the science. The only solution was to contact individuals, found in metadata records—but this did not always translate to a useful response, nor did it solve generic issues about processing. Also problems occurred with the APIs: generally, in this scenario, the user is left alone. The access and data formatting is complex and suffers from the paradigm described previously: a belief that always using more generic data and metadata formatting will address all issues. Instead, responsibilities are shifted from the expert communities to the users. To process CMIP5, the user must understand many details about the experiments (such as those described in this section), have enough scientific background to understand and process the datasets, be an excellent developer with an ability to install the APIs (while recompiling and working with them), write a code generic enough to cope with all the cases, and find his/her way in the documentation, scattered in different places. Such lengthy requirements typically prevent the use of the data. We did not originally intend to allocate so much effort, but rather to search for existing datasets matching our needs. As the needed datasets did not exist, we had to create them. It was through this process that we discovered the complexity of CMIP5 datasets and the ESGF portal. The lesson learned here was the realization that making data available is not enough for exploiting it, and thus it was a limitation to reach the goals set in GEO.

These difficulties encouraged the partners in GEOWOW to build a solution based on a shared cloud service, delivered as a data-access gateway, able to enhance the user experience when accessing ESGF resources through the OPeNDAP protocol, and then to further explore the potential of a service layer that would be cross-cutting among communities, and discoverable through the GEOSS Common Infrastructure. Providing such cloud service, discoverable through the GCI, made sense because it would offer a completely packaged data-access point, following interoperability protocols as promoted in GEOSS.

This data-staging step was a significant achievement, upon which we could start working on performance gains for data processing. To sum it up, we started with the idea of being able to slice the potentially huge volumes of climate data into smaller chunks to distribute them over many worker machines, and have each of all these workers automatically coordinated to process in parallel its share of the input data, which is in line with Hadoop's functionality. This enables fast heavy processing, at a relatively low cost, and with automated handling of failure, including the automated rescheduling of individual failed tasks without hampering the overall process (see next section).

Generally speaking, data-staging operations consist of storing data resources to be processed by worker machines, like on a cloud cluster. It makes use of a temporary storage area (e.g., a cloud storage) used to host multiple sources of information (e.g., from different Data e-Infrastructures) that are to be processed, generating results to be stored into a corporate storage area dedicated to support the equivalent of an in-house data market, or a "Datamart". Data staging is considered an intermediate step in a business process workflow, as the hosted content can be deleted after being successfully processed, and the resulting information products loaded back in the Datamart. A data-staging service can be useful for:

- Collecting data from different curated sources (operational data centers), and making these copies available for processing at different times;
- Loading information from a curated database, then freeing the temporary storage when there is no more planned usage and new data needs to come in for other tasks. All the data processing can then occur without interfering with the data-center operations;
- Precalculation of data aggregates or running data-conflation operations.

In our case, working with distributed data repositories over the Web, but also working with distributed partners across global initiatives, we were engaging in an approach in which data staging would happen on the cloud, becoming the enabler of as many virtual, reallocatable corporate Datamarts as needed.

For IOC/UNESCO, the ESGF Gateway was deployed as a front end for users, offering a standard OGC Opensearch interface to be queried by a client software (Goncalves et al., 2014) sitting on top of the distributed ESGF OPeNDAP servers, and enhanced to manage transactions for the OpenID user credentials required by the ESGF servers. In addition, this gateway enabled a GEOSS-compatible, standard way (via the OpenSearch interface) to automatically connect the ESGF resources to the GCI via the GEO Discovery and Access Broker (DAB) service, so they can be discovered and more easily further exploited by the GEO community partners. It is enhancing the user experience when accessing ESGF resources through the OPeNDAP access protocol.

On the client side, the ESGF client software is interacting with the developed ESGF Gateway. It was also provided to GEOWOW partners (IOC/UNESCO) for the data access to the ESGF resources from within an algorithm integration environment. Basically, this environment resides in a Developer Cloud Sandbox service, also contributed to the GEOWOW partners. The developed ESGF client is able to make OPeNDAP queries using time and level parameters, to leverage the data-staging capabilities offered by the Gateway service.

Moreover, as the GEOWOW partners focused on having the data discovery and access to rely only on the OPeNDAP ESGF resources, the developed workflow soon offered a fine-grained level of data access, and could run different instances of the ESGF Client with different spatial subsets of data (by one region, one time frame, or decade, etc.). The ESGF Client also allowed users to convert a bounding box in OPeNDAP indexes, and then convert the downloaded binary files, resulting in NetCDF files containing the requested subset grid as expected by our processing chain.

In conclusion, as we needed to test new processing workflows, a number of data-staging issues toward the ESGF emerged, were analyzed, and were solved during the project's lifetime: problems with the ESGF certificate server, changes in the ESGF portal structure, difficulties to find ESGF APIs and documentation, and requirements for improved metadata exploitation of the retrieved NetCDF files.

The development of client software for this ESGF Gateway was addressed to slice each downloadable resource according to Time and Level parameters. This resulted in splitting the data in tiny grids. In addition, the z variable was added to the ESGF Gateway, and made accessible to the Catalog service and Client software.

Our GEOWOW experience with building and exploiting the ESGF Gateway suggests that this is a precious resource for the communities of GEOSS users, and other types of users that do not need the full ESGF models. With the Gateway, they can focus on performing the download of any user-defined data slice, without fighting with certificates and group management.

Overall, the ESGF Gateway offers a simple tool to deal with the combination of ESGF OPeNDAP queries and OpenID authentication.

It also supports a direct registration of these "data slices" resources in the GEOSS Common Infrastructure, as they are made available through the OpenSearch Geo & Time extension protocol supported by the GEO DAB.

BUILDING A SHARABLE DATA-PROCESSING CHAIN

Our experiment helped us to test and analyze the concept of a processing system in line with open science principles: open data, open source, open methodology, open peer review, open access, and open educational resources (OpenScienceASAP, 2014). A large number of popular applications for science were built on those principles (Zenodo, ResearchGate, Figshare, etc.), but most of them concentrate on scientific data sharing and publication.

The concept of open data is now well understood, accepted, and supported by the scientific community. In Europe, the European Union's Research and Innovation program "Horizon 2020" (http://ec.europa.eu/programmes/horizon2020) is strongly encouraging the generalization of open data for publicly funded research; the current work is in line with this working direction. Beyond the scientific public policy, we also consider the data retrievability through GEO/GEOSS, with the DAB (openSearch).

The data cataloging leveraged the standard OGC OpenSearch Geo & Time extension interface, which is a significant improvement for supporting end-users in the registration of results in the GCI, via the DAB.

One major outcome of our work for the benefit of researchers was the prototyping of tools and workflows for scientific experiment reproducibility, testing a new concept of operation for the elaboration of environmental indicators. By integrating access to and interaction with GitHub repositories from within their Cloud Sandbox environment (Fig. 5.6), researchers had the tools and protocols to simply describe, share, and reproduce an experiment, e.g., opening the door for more in-depth collaboration with peer reviewers or with other research laboratories.

In the context of GEO/GEOSS, which was central in the GEOWOW contribution to the community, the support provided to IOC data-production tasks, by proof-testing the deployment of the processing chain on a cloud cluster, consisted of systematic cataloging of results, and custom repatriation of processing results to a storage workspace for the scientist.

Ultimately, it shaped a process in which the GEOSS researcher would benefit from a good level of automation in linking and cross-referencing, through the use of Digital Object Identifiers (DOIs), their work on Cloud Sandboxes with their work on scientific publications.

CONCLUSION

The exchange of scientific data and processing between scientific disciplines are emerging needs, especially in the field of the assessment of the impact of human activities on ocean services and the impact of the changes in these services on human wellness. Such assessment requires that experts in ecosystems can access and process ocean monitoring datasets and models outputs.

We tested the idea of a data-processing service based on cloud technology. For the sake of a real-size case, we chose to develop a service processing

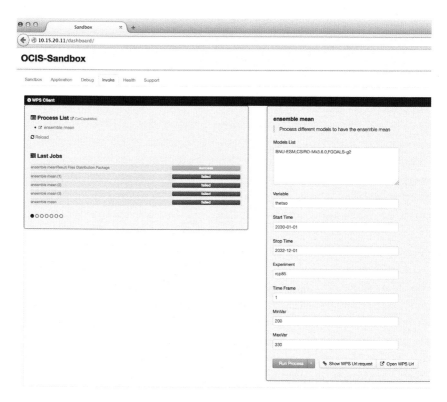

Figure 5.6 Web interface to the WPS.

climate model outputs. We applied the processing to compute the ensemble means of oceans temperature projections (from 2006 to 2050): the ensemble means and derived indicators are used in a UNEP assessment.

The experiment has shown us that the technologies for building such a system are already available. The software components can be obtained at no charge under open-source licenses. Hadoop allows implementing a MR workflow, allowing us to scale up the data processing. Most of the development effort was dedicated to the data staging and to the connection (through OPeNDAP) to the data provider (ESGF). Practically speaking, it is likely that a similar effort in developing an ad hoc data connector will be required for each specific application. However, operations internal to the cloud compute (i.e., once the data are obtained from the data provider) may use an approach similar to the one described in this experiment.

We found that developing processing services based on cloud technologies offers many advantages. In the first place, it allowed sharing a fully operational processing system. We suggest sharing the source code on a revision control platform (e.g., GitHub), which allows reviewing the code and improving it. Such an idea is central for the scientific world: such a processing system can be attached to a publication, as additional material. Reviewers and readers can reexecute the processing, check the algorithm, which increases the reproducibility of the experiments. However, we could not test in practice the impact of a massive adoption: scientific reviewers are not necessarily expert in Information Technology or algorithmics, which make the review of the processing service difficult if not impossible. How should an editor consider a published processing service? If it is a part of a demonstration, who can review it? Though we think that sharing the processing system is of interest, the best practice for its integration in scientific publications review is still to be defined.

Developing a processing service on a cloud undoubtedly allows reduced costs of exploitation, especially for sporadic use. In addition, technologies such as MR allow upscaling processing in terms of data storage and processing power. This can put heavy processing usually reserved for technicians specialized in modeling or Earth observation in the hands of nontech-savvy users. The technology we described easily wraps in a WPS, which allows the service provider to develop a Web interface, making trivial the interaction with the service for the less tech savvy, whereas more advanced users can directly interact through command lines or even submit new versions of the code.

The service is meant to be offered in a marketplace, ready for deployment in a cloud. The advantage for the processing service provider is that there is no need to maintain a scalable processing hardware (conversely, to a classical OGC WPS, which may have to face a peak of activities following the demand). With this approach, the cost of the cloud service to deploy the application is left to the user, who can either deploy on a private or public cloud.

Our experiment was limited by the fact we could not test the distribution of the application and its deployment in a cloud architecture. The concern is not about the technology, which exists, but rather about the ergonomics for potential users.

We could test neither the business model, especially the cost for accessing cloud infrastructure, nor the possible funding for scientific applications. At this stage, we could identify technologies and principles, that we believe, make possible to share processing beyond a scientific community. However, technologies and principles do not define a product as long as the complete business model is not identified and endorsed by all stakeholders, the industry, users, and funders (which are mostly governments in the case of public science).

In addition, the question of where the processing is actually done and data are stored can be sensitive for some institutions: though not technically problematic, the possibility to choose a specific processing cloud and to know clearly under which laws it is managed is a requirement for most public entities.

ACKNOWLEDGMENT

This project has received funding from the European Union's Seventh Programme for research, technological development and demonstration under grant agreement No 282915.

REFERENCES

Cardona, O.D., van Aalst, M.K., Birkmann, J., Fordham, M., McGregor, G., Perez, R., Pulwarty, R.S., Schipper, E.L.F., Sinh, B.T., 2012. Determinants of risk: exposure and vulnerability. In: Field, C.B., Barros, V., Stocker, T.F., Qin, D., Dokken, D.J., Ebi, K.L., Mastrandrea, M.D., Mach, K.J., Plattner, G.-K., Allen, S.K., Tignor, M., Midgley, P.M. (Eds.), Managing the Risks of Extreme Events and Disasters to Advance Climate Change Adaptation. A Special Report of Working Groups I and II of the Intergovernmental Panel on Climate Change (IPCC), Cambridge University Press, Cambridge, UK, and New York, NY, USA, pp. 65–108.
Caumont, H., Brito, F., Boissier, E., 2014. Big earth sciences and the new 'platform economy'. In: Proceedings of the 2014 Conference on Big Data from Space (BiDS'14). http://dx.doi.org/10.5281/zenodo.12728; https://zenodo.org/record/12728-.VV5JSVWqqko.
Combal, B., 2014a. Time Projections of Sea Surface Temperature, for RCP 4.5 and RCP 8.5, for Decade 2020, 2030, 2040 and 2040. IOC-UNESCO. http://dx.doi.org/10.5281/zenodo.12781; https://zenodo.org/collection/user-ioc-code-cmip5.
Combal, B., 2014b. Ensemble Mean of CMIP5 TOS, for the Period 1971 to 2000. IOC-UNESCO. http://dx.doi.org/10.5281/zenodo.12843; https://zenodo.org/collection/user-ioc-code-cmip5.
Combal, B., 2014c. Monthly Climatology of CMIP5 Models Historical Run, for 1971-2000. IOC-UNESCO. http://dx.doi.org/10.5281/zenodo.12943; https://zenodo.org/collection/user-ioc-code-cmip5.
CORDIS, 2009. GEOWOW Project Page. http://cordis.europa.eu/project/rcn/100182_en.html (last accessed 19.05.15.).
Dean, J., Ghemawat, S., 2004. MapReduce: Simplified Data Processing on Large Clusters. Google Inc. https://www.usenix.org/legacy/publications/library/proceedings/osdi04/tech/full_papers/dean/dean_html/index.html.
Fanning, L., Mahon, R., McConney, P., Angulo, J., Burrows, F., Chakalall, B., Gil, D., Haughton, M., Heileman, S., Martinez, S., Ostine, L., Oviedo, A., Parsons, S., Phillips, T., Arroya, C.S., Simmons, B., Toro, C., 2007. A large marine ecosystem governance framework. Marine Policy 31, 434–443.

Gallagher, J., Potter, N., Sgouros, T., Hankin, S., Flierl, G., 2007. The Data Access Protocol –
DAP 2.0, ESE-RFC-004.1.2, NASA Earth Science Data Systems Recommended Standard, 2007-10-10.

Goncalves, P., Walsh, J., Turner, A., Vretanos, P.A., Nebert, D.D. Open Geospatial Consortium
(OGC), Catalogue Service for the Web (CS-w) – OpenSearch Geo & Time Extensions,
an OGC Standard since April 2014. http://www.opengeospatial.org/standards/
opensearchgeo.

Halpern, B.S., Walbridge, S., Selkoe, K.A., Kappel, C.V., Micheli, F., D'Agrosa, C., Bruno, J.F.,
Casey, K.S., Ebert, C., Fox, H.E., Fujita, R., Heinemann, D., Lenihan, H.S., Madin,
E.M.P., Perry, M.T., Selig, E.R., Spalding, M., Steneck, R., Watson, R., 2008. A global
map of human impact on marine ecosystems. Science 319 (5865), 948–952. http://
dx.doi.org/10.1126/science.1149345.

Hey, T., Tansley, S., Tolle, K. (Eds.), 2009. The Fourth Paradigm: Data-intensive Scientific
Discovery. Microsoft Research, Redmond, WA. ISBN: 9780982544204. http://research.
microsoft.com/en-us/collaboration/fourthparadigm/.

IOC-UNESCO, 2011. Methodology for the GEF Transboundary Waters Assessment Programme. Methodology for the Assessment of Large Marine Ecosystems, vol. 5. UNEP.
viii + 115 pp.

Oldenborgh, G.J., Doblas Reyes, F.J., Drijfhout, S.S., Hawkins, E., 2013. Reliability of regional
climate model trends. Environmental Research Letters. 8 (1). http://dx.doi.org/
10.1088/1748-9326/8/1/014055; http://iopscience.iop.org/1748-9326/8/1/014055/.

OpenScience ASAP, 2014. Was Ist Open Science? http://openscienceasap.org/open-sci-
ence/ (accessed 11.06.15.).

Programme for Climate Model Diagnosis and Intercomparison. http://www-pcmdi.llnl.
gov/about/index.php (accessed 19.11.14.).

Percivall, G., 2010. The application of open standards to enhance the interoperability of geoscience information. International Journal of Digital Earth 3 (S1), 14–30. http://dx.doi.
org/10.1080/17538941003792751.

Taylor, K.E., Stouffer, R.J., Meehl, G.A., 2012. An overview of CMIP5 and the experimental
design. Bulletin of the American Meteorological Society 93, 485–498. http://dx.doi.
org/10.1175/BAMS-D-11-00094.1.

UNEP, 2015. TWAP, Open Ocean Component. http://www.geftwap.org/water-systems/
open-ocean (accessed 13.05.15.).

ANNEXES

List of Models Providing "TOS" Variable, for Ocean Sea Surface Temperature Projections Analysis

```
ACCESS1-0      r1i1p1.
ACCESS1-3      r1i1p1.
bcc-csm1-1     r1i1p1.
bcc-csm1-1-m   r1i1p1.
BNU-ESM        r1i1p1.
CanESM2        r1i1p1 r2i1p1 r3i1p1 r4i1p1 r5i1p1.
CCSM4          r1i1p1 r2i1p1 r3i1p1 r4i1p1 r5i1p1 r6i1p1.
CESM1-BGC      r1i1p1.
CESM1-CAM5     r1i1p1 r2i1p1 r3i1p1.
```

```
CESM1-WACCM    r2i1p1 r3i1p1 r4i1p1.
CMCC-CESM      r1i1p1.
CMCC-CM        r1i1p1.
CMCC-CMS       r1i1p1.
CNRM-CM5       r10i1p1 r1i1p1 r2i1p1 r4i1p1 r6i1p1.
CSIRO-Mk3-6-0  r10i1p1  r1i1p1  r2i1p1  r3i1p1  r4i1p1  r5i1p1  r6i1p1
    r7i1p1 r8i1p1 r9i1p1.
EC-EARTH       r10i1p1 r11i1p1 r12i1p1 r14i1p1 r1i1p1 r2i1p1 r3i1p1
    r6i1p1 r7i1p1 r8i1p1 r9i1p1.
FIO-ESM        r1i1p1 r2i1p1 r3i1p1.
GFDL-CM3       r1i1p1.
GFDL-ESM2G     r1i1p1.
GFDL-ESM2M     r1i1p1.
GISS-E2-H      r1i1p1 r1i1p2 r1i1p3.
GISS-E2-R      r1i1p1 r1i1p2 r1i1p3.
HadGEM2-AO     r1i1p1.
HadGEM2-CC     r1i1p1 r2i1p1 r3i1p1.
HadGEM2-ES     r1i1p1 r2i1p1 r3i1p1 r4i1p1.
inmcm4         r1i1p1.
IPSL-CM5A-LR   r1i1p1 r2i1p1 r3i1p1 r4i1p1.
IPSL-CM5A-MR   r1i1p1.
IPSL-CM5B-LR   r1i1p1.
MIROC5         r1i1p1 r2i1p1 r3i1p1.
MPI-ESM-LR     r1i1p1 r2i1p1 r3i1p1.
MPI-ESM-MR     r1i1p1.
MRI-CGCM3      r1i1p1.
NorESM1-M      r1i1p1.
NorESM1-ME     r1i1p1.
```

List of Cited Projects and Technologies

GEOSS Interoperability for Weather Ocean and Water (GEO-WOW) project, co-funded under the European Community's Seventh Framework Programme FP7/2007–2013, grant agreement no. 282915 in response to call ENV.2011.4.1.3-1 "Interoperable integration of Shared Earth Observations in the Global Context." The project kicked off September 1, 2011, for a three-year duration with the consortium partners: ESA, Terradue, European Commission Joint Research Centre (JRC), Centro Nazionale delle Ricerche (CNR), European Centre for

Medium-Range Weather Forecasts (ECMWF), German Federal Institute of Hydrology (BfG), UNESCO, University of Bonn, 52° North, KISTERS AG, MET OFFICE, METEO-FRANCE, Karlsruher Institut für Technologie (KIT), Brazilian Space Agency (INPE), and the University of Tokyo. http://www.geowow.eu/.

CHAPTER 6

How We Used Cloud Services to Develop a 4D Browser Visualization of Environmental Data at the Met Office Informatics Lab

N. Robinson, R. Hogben, R. Prudden, T. Powell,
J. Tomlinson, R. Middleham, M. Saunby, S. Stanley, A. Arribas
Met Office Informatics Lab, Exeter, UK

INTRODUCTION

The Met Office is a world-leading center for weather forecasting and climate science. Its activities encompass weather observations, 24/7 operational forecasting, and leading climate research. The Met Office was created over 150 years ago and employs around 2000 staff, most of them based at its headquarters in Exeter, United Kingdom.

Like many other large organizations, the Met Office has different departments (e.g., Science, Technology, Business, HR, etc.), each with its own culture and structures—structures that, in most cases, have developed over many years to ensure product delivery and business continuity. Resilience is obviously important to an organization that is responsible for providing real-time environmental forecasts, critical for ensuring safety of life and property 24 h a day, 365 days a year (e.g., National Severe Weather Warning Service). To protect this resilience, any changes to these systems are subject to strictly controlled processes.

However, this has negative consequences for innovation: the speed of change is limited by these processes and structures, and a risk–averse culture is quickly developed around the avoidance of failure. Furthermore, although senior management wants innovation, middle management must meet today's targets. This raises conflicts of priorities and often means that staff

Cloud Computing in Ocean and Atmospheric Sciences
ISBN 978-0-12-803192-6
http://dx.doi.org/10.1016/B978-0-12-803192-6.00006-2

Crown Copyright
© 2016 and Elsevier Inc.
All rights reserved.

89

with the relevant skills are fully occupied with business-as-usual activities, leaving little time and space for innovation. As a result, the Met Office often struggles to transform new science into services or respond quickly enough to requests from users to make our data relevant to them.

Finally, rapid improvements in computing power are making things worse: the Met Office is increasing the already vast amounts of data that it produces. For example, in 2014 the Met Office's operational forecasting systems generated around 30 terabytes of data every day, not to mention research data. In 2016 we will generate more than 10 times this amount, a truly staggering quantity.

It is with this background that the Informatics Lab was created in April 2015 with two well-defined purposes:
- our external purpose is to help make environmental data and science useful
- our internal purpose is to help improve working practices at the Met Office

The Lab tries to achieve these aims through the rapid delivery of prototypes that demonstrate what is possible. The Lab also facilitates, supports, and helps others inside and outside the Met Office. By doing this we can create communities that can build on what we do to maximize positive impact.

The Lab comprises nine members with expertise in technology, science, and design. Four of the members of the team are permanently in the Lab to ensure continuity. The rest of the members of the core team are seconded from Science, Technology, and Design (and possibly from other areas in the future) for periods between 12 and 24 months. On top of this, and to further facilitate the connection with other teams across the office, an additional group of "Associates" to the Lab was created. This is an entirely voluntary group that contributes to the Lab in their spare time, and who are given access to all the Lab's tools and facilities. Associates can contribute to Lab prototypes, or work on projects of their own.

The Informatics Lab has the freedom—and the obligation—to work differently from the business-as-usual approach: to improve something you need to try something different. One of the most important implications of this is that the Lab works entirely outside of the Met Office IT networks, meaning we have complete flexibility to create and deploy solutions as required.

This freedom is overseen by a board comprising members of Met Office Executive and relevant external partners, who govern which projects are tackled by the Lab, and how the work and best practices of the Lab can be linked to the rest of the Met Office.

THE GENERIC LAB APPROACH

As mentioned in the introduction, the two aims of the Lab are to improve existing ways of working in the Met Office, and to make environmental data and science useful. Of course, these two aims are related: by improving the ways of working in the Met Office, we will make more environmental data and science useful to people outside the Met Office. However, one thing is clear, we cannot achieve this by simply repeating what we are already doing. Therefore, we need a different approach.

Our Approach to People

We are engaged in work that requires significant collaboration across disciplines within our team and with others outside the Lab. The Lab combines expertise from technology, science, and design, allowing us to quickly complete an iterative cycle of developing, testing, and deploying prototypes. This multidisciplinary approach contrasts with the rest of the Met Office in which many teams are made up of experts of a single discipline. We foster a working environment in which experiments can be completed quickly and with little risk. Indeed, the "removal of fear" has been cited previously as fundamental to engendering creativity (Sheridan, 2013). Failing fast means we can learn lessons fast, and prototypes can progress rapidly.

To get the right blend of skills we decided that the ideal size of the core team would be no more than 10 people (there is evidence for this number, for example, Wittenberg (2006) asserts that research on optimal team numbers is "not conclusive but it does tend to fall into the 5 to 12 range").

There are plenty of talented people in the Met Office trying to achieve the same end goal—we recognize this and actively seek to work with these people. The structure of the Lab team could be likened to an onion: we have an inner core, in the next layer we have a group of Met Office Associates, and in the outer layer we have a group of non-Met Office Associates. Essentially, we are open to collaborate with anybody who wants to make environmental data and science useful.

Our Approach to Data and Source Code

We aim to be open with as much of the work we do in the Lab as possible to encourage fruitful collaborations. The majority of our code is publicly available on Github (https://github.com/met-office-lab/) on which it can be reused and augmented by anyone who is interested. We also work to make Met Office data more widely available to the public, who in many cases have an explicit right to it as the ultimate financial backers. If we use any Met Office data in our prototypes, we do it in such a way that it is publicly available to allow others to easily build on top of our initial concepts.

Our Approach to (Physical and Virtual) Workspace

The team is fully mobile, with all members working on laptops which are not linked to, nor dependent on, the Met Office's corporate network. This was a deliberate decision so we could avoid the security restrictions imposed by the corporate network, and conversely, to avoid being a potential security or continuity threat to the rest of the organization.

Therefore, we fully rely on cloud technology to store and process data (more specific details on this are provided in section: Our Approach to IT Infrastructure), and to communicate and work with each other. After some testing and trials, we have converged on the following tools:
- GitHub for joint code development and code sharing
- Slack for instant messaging
- Gmail and Google Docs for email and group documents
- Appear.in for video conferencing and e-working
- Amazon Web Services for infrastructure and deployment

Mobile working has many advantages: it allows us to work at partner sites, making it easier to collaborate with others; and it also gives team members the flexibility to work from home when required, facilitating life–work balance. However, there are disadvantages to a fully mobile team as nothing can really substitute human contact. So, to facilitate contact time we agreed on a weekly schedule:
- Monday–Wednesday we work in our main project as a team, normally at Met Office HQ
- "Open Thursdays" are dedicated to work with others in a variety of projects
- Fridays are reserved for individual work or work with external collaborators, normally outside Met Office HQ

Therefore, although we make extensive use of technology to work remotely when needed, we are mindful of the advice of Wittenberg (2006): "teams that rely solely on electronic communication are less successful than those that understand why communication in person is important. Email is a terrible medium…It does not relate sarcasm or emotion very well, and misunderstandings can arise. There is something very important and very different about talking to someone face-to-face[sic]." In addition, and, although it may sound counter-intuitive, having a structure determining what we work on and where we work encourages creative and innovative work; it protects precious time and facilitates prioritization of tasks.

We have adopted several working practices that have been found effective in other creative organizations. We put a large emphasis on understanding the problem we are trying to solve before solidifying the concept using paper prototyping. We have also found that paired programming (two coders sharing the same computer), while not suitable for every task, has been the origin of some of our most efficient, effective, and innovative work (as in Williams et al., 2000; see section: Mobile Controls).

In terms of physical space, the Lab is a converted meeting room at the center of the Met Office HQ building in Exeter (Fig. 6.1). The choice of this room is intentional to facilitate the contact of the Lab with the rest of the organization and avoid the creation of an us–and–them culture. We do not have any arcade machines, bean bags, or ping–pong tables, but the space is ideal for collaborative work: essentially an empty room with a big table that we can move around to seat ourselves in whatever configuration best suits what we are doing; a few storage cupboards to hide kit and a huge TV to do demos, training, and presentations. We have found that having a space over which the team members can feel some sense of ownership fosters a safe environment which encourages creative ideas.

Figure 6.1 The Lab workspace: mixing discussion, paper prototyping, paired programming, and solo working.

In 2016, the Lab will also use new facilities currently under construction at the Exeter Science Park. The Exeter Science Park will be home to a number of knowledge-based businesses, and we expect that this will open the door to new opportunities for collaboration.

Our Approach to IT Infrastructure

As mentioned, we fully rely on cloud services to do our work. From an infrastructure point of view, we build all our servers, databases, and data analysis facilities in the cloud. For the project described in this article, we used Amazon Web Services (AWS).

We use servers from Amazon's Elastic Cloud Compute service (EC2) in a few different ways. The simplest use is for testing and development. EC2 allows us to create a new server in a matter of minutes. We can then use to test deployment of prototypes under active development. It also gives us a space to play with libraries, modules, and technical approaches before we decide if they are useful for the problem at hand. This has been instrumental in our technology choices as it allows us to quickly explore something and then destroy the server once we are finished, saving us money.

One such example of this is our use of the Thematic Real-Time Environmental Distributed Data Services (THREDDS) data service. THREDDS allows us to host large data sets on a web server and, importantly, it allows subsets of these data to be extracted without having to download the whole dataset. Using EC2, we have been able to very quickly experiment with different versions and ways of installing THREDDS, which has allowed us to deploy it efficiently with minimal support to keep it running.

Our second and more robust use of EC2 is to pair it with containerization and orchestration tools such as Docker and Kubernetes. Docker allows us to automate the deployment of our different cloud microservices, and define the amount of resource which they should be allocated. Kubernetes then creates and manages the EC2 servers to host a set of these containers. Kubernetes can scale the deployment of these contained microservices dynamically to meet demand using the concept of a "replication controller." For instance, the might mean increasing the number of containers if demand goes up, or moving containers from one server to another if it becomes over subscribed or slow. Replication controllers also increase resilience by automatically redeploying a service if it crashes.

Another interesting use of Kubernetes is to allow the flexible parallelization of nontime sensitive processing. An example of this is the way we convert raw weather data into files which can be displayed in our browser application (see section: The Project: Interactive 4D Browser Visualization of High-Resolution Numerical Weather Prediction Data). We receive a huge volume of raw data every 3 h. If we were to process this using a full-time dedicated EC2 instance, it would sit idle for a large fraction of time, waiting for the next influx of data. Kubernetes monitors the amount of data to be processed and scales the infrastructure accordingly. In this instance, it creates multiple servers every 3 h to handle the massive data load but then destroy them all once they have finished, reducing costs.

We have chosen to expose the original forecast data in the form of Network Common Data Format (NetCDF) files, a format which stores gridded data and associated metadata. It was initially developed for geospatial data by NASA, and is recognized by the Open Geospatial Consortium, meaning it is widely supported.

We use Iris (www.scitools.org.uk/iris/), the open-source Met Office data analysis tool kit written in Python, to analyze gridded meteorological and climatological data. Although primarily used by scientists for postprocessing model output, it has proved essential for preparing our atmospheric data on the cloud. As Iris is a Python module, it can easily be incorporated into server-side scripts, in which it can sit alongside other useful Python modules, for instance, Boto (which allows us to interface with AWS) or PyPng (which writes images from data arrays). In addition, Iris is conversant in many data formats, meaning it can convert any atmospheric data to standard NetCDF files.

The reader needs to bear in mind that we did not know what technologies would be most appropriate (and how to best apply them) when we started this project. Our approach is to create and run experiments that allow us to test different technologies and designs, before choosing the most effective.

THE PROJECT: INTERACTIVE 4D BROWSER VISUALIZATION OF HIGH-RESOLUTION NUMERICAL WEATHER PREDICTION DATA

The Concept

The first project tackled by the Lab is the development of a four-dimensional (4D) (i.e., animated three-dimensional (3D) space) visualization of high-resolution weather forecasts (Fig. 6.2). To make this technology widely accessible, we have set ourselves the challenge of delivering this via a web browser on a standard specification computer. There are many motivations for

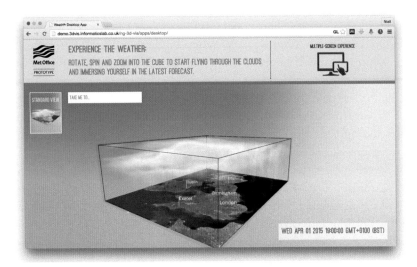

Figure 6.2 Forecast cloud fields being rendered and animated in a web browser.

developing this concept: public engagement; improved data access; scientific analysis; etc. In addition, the Met Office celebrates the 50th anniversary of our first Numerical Weather Forecast in 2015, and our 4D visualization prototype will be unveiled as part of the celebrations.

Forecasting the weather is a complex process. First, up-to-date observations of meteorological parameters such as temperature and wind speed need to be taken at millions of locations around the world and collated at the Exeter HQ. These observations are then fed into our forecasting model on our supercomputer. This model relies on fundamental physical equations to dynamically calculate the future weather.

From these relatively simple equations naturally emerge the complex features of our dynamic atmosphere: storms appear and disappear; a jet stream forms; and winds flow. However, the supercomputer abstracts these tangible phenomena away to a stream of numbers. Our fundamental goal is to turn these meaningless numbers into information that is intuitively understandable.

We chose this as our first project for two reasons. The most obvious reason is that we hope it is something that will inspire people to think about data visualization—bread and butter to and organization such as the Met Office—in a new, more ambitious light.

The second and perhaps more important reason is that this project encapsulates a core challenge facing Earth science at the current moment.

The volumes of data generated are staggering—indeed, we expect to have ~1 exabyte (EB) of data archived at Met Office HQ in the near future. Nevertheless, more than this, Earth science data are highly multidimensional. Science faces the significant and increasing challenge of allowing users of these data (be that customers or scientists themselves) to explore this data in a way that is effective, efficient, and intuitive. If we hold to the status quo of exploring this vast and rich data with scatter plots and static 2D color maps, we will flounder and, ultimately, miss rich and useful information hidden in the data. We believe this will be a core theme in the Lab's work.

There are many possible audiences for our 4D visualization, but we are directly tackling two:

1. Operational Met Office Meteorologists use a range of visualizations to enable them to transform raw forecasts data into meaningful advice. They can do this more effectively if we can give them a faster, more in depth understanding of the forecast model output.
2. The general public largely interact with weather data through 2D forecast maps. We believe that, by exposing the complexity and beauty of our data, we can inspire the public, and hopefully communicate more in depth messages about the coming weather, or weather science in general.

To reach as many people as possible, we wanted our visualization to work without relying on expensive hardware or software: ideally, all you should need is an Internet connection and a web browser. Our aim is to make our science and data interesting and easy to understand, and to make use of the new experiences available through devices such as smart phones.

Implementation

For the 4D visualization project, we decided to use data from our highest-resolution forecast model over the United Kingdom (UK). Our first task was to expose appropriate data from inside the secure Met Office corporate network to our cloud platform (Fig. 6.3). We sourced data for our highest-resolution UK forecast, which is issued every 3 h and provides a forecast extending 36 h into the future. Each forecast field comprises $36 \times 70 \times 810 \times 621$ (time × altitude × latitude × longitude) data. Such a field amounts to ~5 GB of data, and these fields are stored in the internal Met Office archive for hundreds of physical phenomena, from wind and precipitation to ice-crystal fall rate and wet-bulb potential temperature.

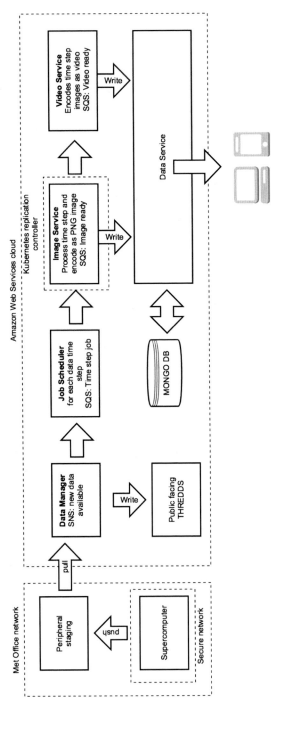

Figure 6.3 Process schematic.

Obviously, we needed to choose a subset of these phenomena to export to our cloud for visualization, so, due to its tangibility, we have initially chosen "cloud fraction" (i.e., the fraction of any spatial grid box classified as being cloud). As the forecasts are created, the data are pushed from the supercomputer inside the secure corporate network to a less secure perimeter network. They can then be pulled to the cloud. The transfers to and from the perimeter network are performed via file transfer protocol (FTP). Although we would rather use more modern protocols such as secure shell (SSH) or hypertext transfer protocol (HTTP), we found FTP to be the most practical solution to implement within corporate constraints.

We then make extensive use of AWS Simple Queue Service (SQS) and Simple Notification Service (SNS) to orchestrate a series of microservices. The original data are stored and made publicly available via a THREDDS data server.

These data are processed further so they can be used in our 4D visualization. We use Iris to do various data standardization: we regrid the data onto an (orthogonal) latitude × longitude grid, and restratify the altitude from "model levels" (an esoteric model-specific definition) to height above sea level.

We then make extensive use of image and video coder–decoders (codecs) to encode the data, as the compression is very good and there is widespread support for streaming data to web browsers. To encode our data as a two-dimensional (2D) image, we restructure the 3D spatial data into an image comprising a series of latitude × longitude tiles. We tile these altitude slices across horizontal (i, j) dimensions, but we also store separate sets of tiles in the red, green, and blue channels of the image (which, of course, no longer bear any relationship to color; Fig. 6.4).

The atmospheric data, now encoded as pixels in a video, are then streamed to the client in which they are rendered using a technique known as volume rendering. Note that we have consciously chosen to render the field on the client browser, rather than a render farm, so that any application would be highly scalable.

The videos are rendered in a hidden HTML5 canvas, in which they are periodically sampled as a static image, which is sent to the graphics processing unit (GPU) for rendering as a 3D volume (Fig. 6.5). There is no standard implementation of volume rendering, so we wrote our own routines using:

- WebGL, a Javascript 3D visualization library which facilitates the use of much of the 3D capability of Open Graphics Library (OpenGL) but in the browser

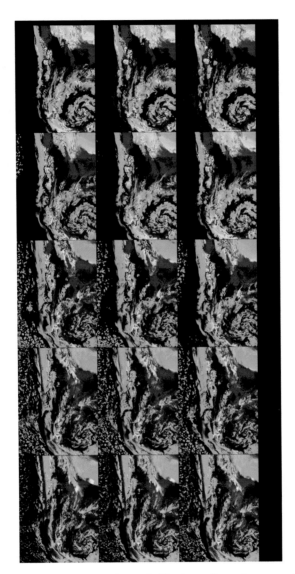

Figure 6.4 Horizontal data-field slices, tiled horizontally, and across color channel.

- three.js which wraps WebGL, making it easier to setup 3D environments
- a series of bespoke GPU routines written in OpenGL Shader Language (GLSL)

These routines sample the data along straight lines starting at the camera, moving out in all directions. These lines are analogous to rays of light, so this approach is known as ray tracing.

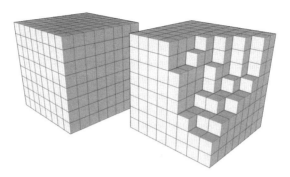

Figure 6.5 Our data are gridded in three spatial dimensions. These 3D grid boxes have a 2D analog in pixels of an image. As such, these 3D grid boxes can be thought of as volumetric pixels, or "voxels."

Our GLSL routines calculate what values of data each ray passes through before it reaches a pixel on the screen, and then color the pixel accordingly. The rays enter the volume of data at the points marked a. The rays exit the volume of data at the points marked b. n represents the number of times we sample the data field on the journey through the data.

Applying vector mathematics, we calculate that the ray moves d to get from one sample point to the next sample point, in which

$$\vec{d} = \frac{\vec{i} - \vec{f}}{n}$$ [6.1]

A ray starts with an intensity of one and a color of nothing (Fig. 6.6). We begin at a and take steps of d through the data. At each step, we accumulate some color and lose some intensity based on the data value until the ray reaches the other side of the data, or the ray is extinguished. The power of the graphics processor allows us to do this for all rays at the same time which is known as ray marching.

The value of d is provided by the GPU environment by the three.js library. To obtain the back position it is necessary to render the scene twice: the first render (Fig. 6.7) is not displayed in the application, but the result is passed in and used for the second rendering. We start by rendering the x, y, z coordinate of the back face of our volume as an r, g, b color.

We then take this rendered rainbow image, and pass it into the second rendering. For a given start point, the corresponding end-point x, y, z values are the r, g, b values of this rainbow image indexed at that start point. Combining the two rendering passes means we are using the 3D environment to

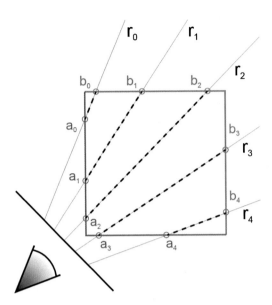

Figure 6.6 A 2D schematic of ray tracing (which we perform in 3D). The observer (bottom left) views pixels on a screen which are derived from sampling data along a series of straight lines which are analogous to light rays.

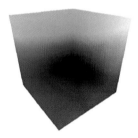

Figure 6.7 The initial volume render pass. Each x, y, z coordinate of the back face of the cube is encoded as r, g, b color values, respectively. This is not displayed, but passed to the second stage of volume rendering.

calculate what the corresponding start and end points are by taking into account the field orientation.

WebGL only supports transmitting data to the GPU as a 2D array. As the data are encoded as a video, they are already tiled into an image. As such, the GPU routines need to detile the data so that it can be indexed as if it were a 3D array.

We can make a big improvement to this volume rendering by using light sources to illuminate the field. This means that at every step of the ray

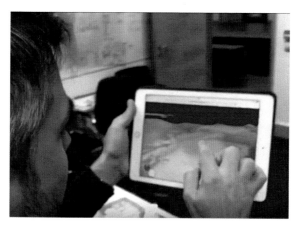

Figure 6.8 A user testing an early iPad version, which utilizes the device accelerometer to give an intuitive "fly-through" experience.

march, we need to march another ray from the light source to the current position to calculate what the contribution is. This creates shading on our object, making the 3D rendering much more effective. The shadow rendering proved too expensive to calculate in detail. Instead, we settled upon a more cursory shadow calculation that only samples a few points, which preserves performance.

Mobile Controls

One of the most popular and innovative features we have developed emerged over the course of a few hours of paired programming. We created a version of the application which can run on an iPad, in which we utilize the device accelerometer to allow the user to intuitively move through the field. The user simply moves the device as if it were a window inside the model field; that is, to look above you, you hold the device above your head; if you want to turn around and look in the other direction through the model field, you literally turn around (Fig. 6.8). Forward movement is controlled by touching the screen.

This device can be synchronized with another by entering a common four-digit code. Rather than synchronizing the screen images, the devices simply share the position of the 3D visualization "camera" via a WebSocket server hosted on the cloud, which dramatically cuts the amount of data transfer. We can see this approach being useful for briefing operational forecasters in the field, who often have very limited bandwidth, for instance, in the Falklands or Afghanistan.

Finally, we improved the mobile application (app) by adding stereo-scopic 3D functionality. Once the 3D environment is established, this is simply a case of replacing the "camera," with a pair of "cameras," side by side. Delivering the perspective of each camera to each eye of the user gives the impression of 3D depth. We run this app on a mobile phone encased in a Google Cardboard headset, resulting in cheap but effective virtual reality.

The point we want to make here is that, by removing the fear of failure, working in very close collaboration, and creating a playful environment, we implemented these features in approximately 14 person hours.

COLLABORATION AND OUTREACH

Internal Impact

One of the great challenges for the Lab is how to ensure that the knowledge we acquire is disseminated within the organization so it can be of wider benefit. Although the Lab is young, and still learning the most effective approach, we have already started this process.

We have regular meetings with Executive members and relevant directors to talk about the working practices that we have found most useful. We hope that the effectiveness of these practices is apparent by our increased productivity and creativity. There are already plans to implement the Lab working model in some operational project teams, in the expectation that it is as effective there. Although culture change will take time in such a large organization with operational responsibili-ties, we are strongly advocating for changes to the physical workspace and wider adoption of practices such as paired programming and multi-disciplinary teams.

As with all research, we only expect a small proportion of the work we do to crystallize into fully operational products: this is the price to pay for truly innovative progress. However, one of the most useful services the Lab has provided so far is to scope emerging technologies and techniques that could improve future operational products. For instance, the use of d3.js (an interactive data visualization module for Javascript) in a side project has sparked a move to use this technology to create more engaging publicity material on the corporate website. We see this kind of scoping of emerging technology as being a core way of disseminating the knowledge encapsu-lated in the Lab prototypes.

External Collaboration and Public Engagement

Since 2012, the Met Office has been a significant contributor to the UK hackathon scene, through the hosting of events, support of events led by others, and encouraging staff participation. The intensive collaboration at hackathons is important to the Lab in several ways: we can find out about new technologies and working practices, and learn about them from people who are already using them; it helps raise the profile of the Lab; and finally it also is a source of inspiration for new ideas.

To effectively design useful prototypes it is important that we seek feedback from the eventual users as soon as possible. They inevitably look at our products from a completely different viewpoint, and with a perspective that can be lacking when you are deeply involved with development. We arranged with a local museum and art gallery, the Royal Albert Memorial Museum (RAMM) in Exeter, to attend on two occasions a month to talk to visitors and gather feedback. These visits started at an early stage in the development process which was important to incorporate comments, criticisms, and suggestions into what we do. For example, early on, visitors to the RAMM commented how much they liked having a realistic representation of the land surface in our visualization which encouraged us to continue working to add real-time data (e.g., snow) into the application. Later on, visitors, although extremely positive about the use of mobile devices to fly through the data, did not like the controls we had developed. Therefore, we changed them following their suggestions.

We also support and participate in other local initiatives such as the Exeter Initiative for Science and Technology (ExIST), a group that aims to share best practice and to raise Exeter's profile as an internationally renowned location and gateway for excellence in science, research and innovation. On the back of this, we are planning minihack sessions with interested businesses hosted at the Exeter Science Park.

CONCLUSIONS AND FINAL REMARKS

Cloud computing has freed us to rapidly and flexibly develop our prototype applications. Once we have extracted data from the corporate Met Office network, processing on the cloud allows us to simply use the technology that we think is best suited to the task, rather than that which is currently implemented and supported by the organization. By packaging our cloud microservices in Docker containers and replication controllers, we have created an implementation which can dynamically scale to meet demand, minimizing cost, and maximizing resilience.

We believe that being afforded this freedom has yielded gains in efficiency and quality of work (not to mention job satisfaction and motivation). We are keen that by becoming the change we want to see, these working practices and principles will be extended to more operational sections of the organization, bringing about cultural change. Ultimately, we believe that this is the only way to achieve our ultimate aim of making environmental science and data as useful to the world as it can be.

REFERENCES

Sheridan, R., 2013. Joy Inc.: How We Built a Workplace People Love. New York: Portfolio, 288 pages.
Wittenberg, E., 2006. Is Your Team Too Big? Too Small? What's the Right Number? (Online) Available at: http://knowledge.wharton.upenn.edu/article/is-your-team-too-big-too-small-whats-the-right-number-2/ (accessed June 2015).
Williams, L., Kessler, R.R., Cunningham, W., Jeffries, R., 2000. Strengthening the case for pair programming. IEEE Software 17 (4), 19–25 (accessed September 2015).

CHAPTER 7

Cloud Computing in Education

C.N. James
Embry-Riddle Aeronautical University, Prescott, AZ, USA

J. Weber
University Corporation for Atmospheric Research, Boulder, CO, USA

INTRODUCTION

Cloud computing is the new technological frontier for teaching, learning, and research in higher education. In the face of recent declines in external funding and the increasing demand for online courses and learning technologies, colleges and universities are turning to cloud computing as a flexible and affordable solution (e.g., Sultan, 2010). E-learning, or the use of electronic media to facilitate learning, is transforming the educational experience at colleges and universities (e.g., Bora and Ahmed, 2013; Kats, 2013). A plethora of e-learning materials and scientific tools are already available online for students in the atmospheric and oceanic sciences (e.g., Cooperative Program for Operational Meteorology, Education and Training [COMET]'s MetEd Website: http://www.meted.ucar.edu/; National Oceanic and Atmospheric Administration [NOAA]'s Ocean Explorer: http://oceanexplorer.noaa.gov/edu/). More and more, these online tools and learning modules are being integrated into e-learning ecosystems requiring tremendous computational and storage resources requiring the kind of flexibility and infrastructure provided by cloud-computing providers (Dong et al., 2009). E-learning ecosystems are slowly replacing traditional textbooks and assignments, and traditional courses are being replaced by hybrid or online courses. According to Allen and Seaman (2013), the number of college students in the US taking at least one online course in 2012 was 6.7 million (or 32%), and this number continues to increase.

Atmospheric and oceanic science certificates and degrees are presently being offered online. For example, Pennsylvania State University offers an online Weather Forecasting certificate (http://www.worldcampus.psu.edu/). The University of West Florida has an online Ocean Sciences degree. The University of Arizona claims to have the first online bachelor's degree

Cloud Computing in Ocean and Atmospheric Sciences
ISBN 978-0-12-803192-6
http://dx.doi.org/10.1016/B978-0-12-803192-6.00007-4
Copyright © 2016 Elsevier Inc.
All rights reserved.

in atmospheric science (http://bas.atmo.arizona.edu/), offered to transfer students with associate degrees from Air University.

To reach much larger student audiences on current science issues like climate change, some private companies (e.g., Coursera) and universities are offering Massive Online Open Courses (MOOCs). Massachusetts Institute of Technology (MIT)'s Global Warming Science course is one such MOOC offered free of charge to the public. These MOOCs certainly achieve high numbers of matriculates from around the world at relatively low cost, and provide a self-paced learning environment.

However, MOOCs often have large dropout rates and poorly achieved learning outcomes (http://facultyrow.com/profiles/blogs/mooc-s-a-failed-plan). Despite their flexibility and ease of accessibility, online courses require carefully designed content to be effective (http://www2.ucar.edu/atmos-news/in-brief/13103/flipping-classroom-paradigm). The overall trend in higher education in the geophysical sciences seems to be the flipped course, in which structured learning takes place before class through online lectures or other e-learning materials. Then, class time is designed to build on these concepts through faculty-guided active learning activities and discussions. A variety of pedagogical studies indicate that flipped courses can be very effective, especially advanced courses in which students are highly motivated and honing employable skills.

The proliferation of online learning and educational technology has motivated universities to adopt web-hosted software packages called Learning Management Systems (LMSs), which link student databases to online learning content. LMSs provide virtual spaces in which students can access course materials, interface or collaborate with faculty members and fellow students, and achieve learning objectives. Kats (2013) provides an exhaustive list of the LMSs that have been adopted by colleges and universities in the US. According to recent surveys of over 540 campuses, over 90% had already adopted an LMS (Green, 2011/2012). For reasons that will be enumerated in this chapter, many of these LMSs are being hosted in the cloud, either as a paid Software as a Service (SaaS) or as a free open source code on a Platform as a Service (PaaS; Bora and Ahmed, 2013).

As of 2014, nearly half (47%) of educational institutions reported having an LMS in the cloud. Other computing services for students were also available in the cloud, but only about 10% of university administrators believed that the cloud was secure enough to provide "high value" computing such as financial information and student records (campuscomputing.net). It seems that concerns about cybersecurity have, therefore, caused the migration to cloud computing in higher education to be rather gradual. In fact, only about 8% of universities have been utilizing cloud resources for research and high-end computing (HEC; Green, 2011/2012).

Nevertheless, cloud computing is slowly emerging as a powerful, economical, and flexible solution for *in silico* research and collaboration at institutions of higher learning. Academic institutions that are financially strapped by reduced government funding will especially find that cloud computing can be a powerful and viable computing alternative (Sultan, 2010). The atmospheric and oceanic sciences, which deal with large geophysical datasets at high resolution in four-dimensional space and time, require HEC resources. Faculty and students need the capability to run computer models, manipulate data, and render large datasets for visualization. Most of the existing scientific software in these fields is built for UNIX or Linux operating systems and require specialized system administrators to configure and maintain hardware, operating systems, and software. Skilled, reliable UNIX/Linux administrators are a rare find, and the median salary for a Linux administrator ($69,380) is comparable to that of a tenured associate professor ($71,500; www.payscale.com). Moreover, HEC hardware is costly to purchase and maintain and must be housed in expensive climate-controlled server facilities.

Cloud-computing providers also supply Infrastructure as a Service (IaaS), which gives scientists the flexibility to provide their own virtual machine configuration of operating system, processing, and storage resources which can be tailored to various atmospheric and oceanic applications. These virtual machine configurations can be saved and later restored in the cloud or shared publically with others in the science community, making research analyses more reproducible for peer collaborators (Dudley and Butte, 2010). Unlike grid computing, which is more hardware specific and has been available since the 1990s for scientists to share HEC resources and data storage, cloud computing comes with much more flexibility, less cost, more ease, and customizability (Ostermann et al., 2014).

In this chapter, we review the many benefits and strengths of cloud computing for geophysical teaching, learning, and research at educational institutions (the section entitled Cloud-Computing Benefits for Education). Then, in the section entitled Cloud-Computing Challenges for Education, we explore some of the cloud-computing challenges facing higher education, and in the section entitled Sample Cloud Instance, we cite sample instances of cloud computing that have been successfully implemented and tested.

CLOUD-COMPUTING BENEFITS FOR EDUCATION

There are many benefits to cloud computing—SaaS, PaaS, and IaaS—in higher education. The left-hand column of Table 7.1 contains a list these benefits. Commercial cloud providers continue to become more available, and the services and applications that are being offered continue to grow. There are many reasons for migrating an educational computing environment to the cloud, and financial benefits are only one consideration.

The cost of operating in a cloud-computing environment can be more feasible economically than obtaining and maintaining a physical and local computing environment. Pricing for cloud computing has been quite volatile over the 2014–2015 time frame and will likely continue to be so until a consistent supply and demand situation exists. There have been swings of over 50% in pricing over the past year, and there is a secondary market, within cloud providers, that offers an auction process to bid for computational resources at a fraction of the market price on an as available basis. Flexibility on demand will always be the most economical approach, but for research and education there are certain critical needs. The labs need to have resources consistently available for student access along with other baseline

Table 7.1 Benefits and challenges of using cloud computing in education

Benefits of cloud computing	Challenges of cloud computing
Cloud computing is reasonably priced compared to the costs of hardware, system support, and maintenance.	The dynamic cost structure of cloud computing requires consistent oversight and monitoring.
IaaS provides computing resources that can be expanded or adapted to meet the changing needs of the users.	Massively parallel processing can be slowed due to lack of physical proximity within cloud networks.
IaaS does not require hardware maintenance or upgrades.	It is still necessary to manage the OS and software in the cloud.
Software can be containerized in the cloud, allowing sets of applications to be executed on practically any operating system.	Cloud resources still require effort to configure the needed hardware and maintain the OS and software.
Software and data stored in the cloud are triplicated, and this data redundancy eliminates the need for costly and time-consuming data backups.	Users may have a sense of insecurity when storing data remotely in the cloud.
Access to the cloud is easy and convenient, from any location where Internet connectivity is available.	Computing in a distributed cloud computing environment is subject to latencies when/where Internet connectivity is unreliable.

needs for research. Given the multitude of use cases, and a dynamic cost environment in cloud-computing resources, it is not obvious that cloud computing is always the most economical choice to make.

However, some of the greatest cloud-computing benefits can be achieved by utilizing the flexible, elastic nature of the cloud, which is a main benefit behind the framework of IaaS. The ability to ramp up computing resources quickly, without the need to purchase new hardware, is a great benefit achieved through cloud computing. Thus, scientists can run high-resolution geophysical models and perform analysis and visualization that would be far more challenging to maintain on local hardware due to the highly elastic demand. This large-scale access to computational resources, on demand, is a huge benefit to science and should continue to grow in use and capacity.

Although the ease of ramping up hardware in the cloud is very attractive, the ability to install and use software easily may be an even greater benefit that the cloud can provide via proprietary cloud applications and *containerization*. Containerization is the practice of separating a portion of a computing environment for specific tasks performed within another operating system. Operating in the cloud enables the installation of any desired operating system, and within that operating system, a series of software applications may be configured to operate in unique environments via containerization. These benefits are significant and offer a greater flexibility and capacity to accomplish many computing needs in support of science and education. Containerization is a new concept, and many system support tasks will take on a different character than traditional computing when operating in the cloud environment.

To this end, the cloud user will need to become comfortable with the concept of not having the data on a local machine. The data residing in the cloud will need to be in a form that allows the end user to operate on the data on the server side, and return back low-volume products to a visualization client. This process is made easier with currently available software that facilitates rapid cluster building in a cloud environment. Software can configure a cluster, allow the job to complete, and then bring the cluster down to minimize financial costs. As software choices become more varied, ease of use, installation, and maintenance becomes increasingly important.

Educators and researchers are responsible for more systems administration regarding computing resources, as many universities migrate to centralized computing support services while eliminating those services at the department level. Methods now available for creating a virtual environment allow operating system deployment across a wide range of platforms, and this plays well in the SaaS and IaaS environments. Single or multiple configurations, or images, can be saved and reused or shared for the specific needs for which they were intended to run. This facilitates the reproducibility aspect of computational science. This concept is not solely for cloud environments, as virtualization is widely used currently on local hardware. However, the use of virtual machines in the cloud environment facilitates collaboration among multiple users across a variety of platforms. Moreover, the ability to move easily among operating system and software versions with the selection of an appropriate virtual machine continues to advance the utility of the cloud in education and research. Some software requires certain operating systems, even specific versions thereof, and the ability to run virtual machines takes away the need to keep physical machines around simply to support specific operating systems or capabilities. Although virtualization lets one keep an image snapshot of the operating system and hardware configuration for reproducing that environment, containerization allows individual applications to be cordoned off from dependencies on the operating system. For example, software that needs a specific kernel version of an operating system can run in a container with that version, while running on a physical or virtual machine that continues to update the kernel.

CLOUD-COMPUTING CHALLENGES FOR EDUCATION

The right-hand column of Table 7.1 lists the challenges of cloud computing associated with each of the benefits tabulated in the left-hand column. The cloud is not a panacea; it offers opportunity for access to large amounts of computing resources for indeterminate amounts of time. These computing resources may not be optimized for performance. For example, when creating a cluster environment, there is no guarantee all nodes of the cluster will be collocated. This will still work for most cases, but it will be slower than if the nodes of the cluster were collocated.

Some cloud providers are using proprietary data-storage protocols, and it is possible to get locked into one provider once you begin. Because cost structures are dynamic, it is also easy to lose sight of costs that are accruing in a given month. Care should be taken to monitor activity in the cloud, to prevent being blindsided by an unexpectedly large monthly billing statement from the provider. The "pay after use" paradigm of cloud computing makes it difficult to budget for computing resources and seek external grant funding without being able to estimate actual costs or invoices in advance. There is also the issue of data storage and transfer, both into and out of the cloud, and the associated costs and data formats. Significant cost savings can be achieved by limiting large data transfers out of the cloud and using collocated instances as much as possible, rather than transferring large datasets out of the cloud for visualization and analysis. A more optimal solution would be to analyze and visualize data using software installed on collocated instances. The graphical images rendered by these visualization tools and the results of the analysis are much less data intensive and less expensive to transfer out of the cloud than the full-resolution digitally stored datasets.

Security is always a concern, and some people feel their data and software are less safe and private in the cloud. One loses physical control over hardware in the cloud, leading to a sense of vulnerability to hackers. Moreover, scientists using cloud-computing resources should take care to ensure compliance with International Traffic in Arms Regulations (ITAR) Part 125 restrictions on both classified and unclassified technical data (http://www.pmddtc.state.gov/regulations_laws/itar.html). Licenses may be required when making unclassified technical data or software available to foreign users via the cloud. In addition, the transmission of classified data overseas must be authorized by the Directorate of Defense Trade Controls; however, it is the State Department's recent opinion that ITAR-controlled technical data may be stored on foreign cloud servers provided that the data are protected using tokenization and entirely received and used by US persons (http://www.pmddtc.state.gov/documents/Tokenization_clarification_statement_DDTC.pdf). To remain on the safe side, it is a good rule of thumb to avoid storing sensitive or restricted information in the cloud. On a positive note, cloud providers are continually working to strengthen security, and there is peace of mind knowing that hardware failure and loss of data are less likely in the cloud as a result of the seamless backup/data redundancy it provides.

The greatest obstacle to cloud computing is inadequate network bandwidth. In locations where Internet connectivity is limited, it may be difficult or even impossible to achieve desired results. Even short latencies can be challenging or annoying when attempting to visualize data through software installed on the cloud.

SAMPLE CLOUD INSTANCE

There are a few examples of successful cloud instances emerging in the atmospheric and oceanic sciences literature. For example, Molthan et al. (2015) tested the viability and feasibility of regional numerical forecast modeling in the cloud. They successfully configured and executed the Weather Research and Forecasting Environmental Modeling System (WRF EMS; http://strc.comet.ucar.edu/index.htm) with acceptable time and cost efficiencies using both private and public cloud-computing resources. However, examples of cloud-computing instances at *educational institutions* are just beginning to emerge in the literature.

In this section, we describe a successful implementation of cloud computing at Embry–Riddle Aeronautical University (ERAU) in Prescott, Arizona (AZ), and review lessons learned. To our knowledge, it is one of the first documented cloud-computing applications in the atmospheric sciences and the first university-hosted instance of the Advanced Weather Interactive Processing System II (AWIPS II) in the cloud (James et al., 2015). AWIPS II is the current interactive operational weather forecasting software system (http://www.raytheon.com/capabilities/products/awips/), developed for the National Weather Service (NWS). It is a Java application consisting of a back-end data server (Environmental Data Exchange [EDEX]) and a front-end graphical rendering client (Common AWIPS Visualization Environment [CAVE]). EDEX requires a minimum of four cores of CPU, 8 GB of RAM, 250 GB of storage, and CAVE requires a minimum of 4 GB of RAM and 2 GB of GPU with Open Graphics Library (OpenGL) support (http://www.unidata.ucar.edu/software/awips2/#sysreqs).

There are a number of factors that motivated ERAU's Applied Meteorology program to utilize cloud computing. One aim was to provide undergraduate meteorology students with invaluable experience using the Linux-based AWIPS II software. ERAU already had an excellent in-house meteorological analysis and display tool available to students called Meteorological Analysis & Diagnosis Software (MADS; Sinclair, 2015). MADS provides easy access to both current and past observational data and gridded model visualization. However, this tool was only accessible while connected to the campus network using the Microsoft Windows environment. Meteorology students were therefore well served to gain exposure to the Linux computing environment and to likewise have experience using the more sophisticated and powerful (albeit more complicated) operational AWIPS II package.

As AWIPS II was becoming available in 2014, the school was also in the process of implementing a new certificate program in Emergency Response Meteorology (ERM) to prepare selected students for careers as mission-critical forecasters in emergency management operations (Woodall et al., 2015). These students would need to have skills and experience analyzing gridded meteorological forecast model data and weather observations in AWIPS II. Moreover, the software would need to be accessible in real time to allow students to practice preparing weather forecast guidance in a timely fashion. Another important factor was that students would need access to AWIPS II via a laptop or mobile device from any location where mobile Internet service was available. The remote nature of emergency management operations required the ability to access data visualization from almost any location and with any device capable of running virtual machine software. Finally, having a cloud-based Linux server would help encourage undergraduates to become more comfortable using the Linux operating system and enable the installation of other useful forecasting tools such as hydrology models and dispersion models in collaboration with the NWS.

Commensurate with the rising educational software needs of the meteorology program were severe limitations in computing resources and a lack of Linux support on the campus. As a small, private university campus funded principally by student tuition, monies for specialized research and computing infrastructure were limited. Information Technology staff were dedicated to maintaining the Microsoft Windows operating system, and there were no dedicated Linux systems administrators on staff. Moreover, faculty did not have the expertise nor the time to configure the hardware and the necessary software chain to achieve real-time AWIPS II visualization. ERAU was not yet operating any clusters in the cloud, but it did have the desire to be able to share access to their AWIPS II server with the Emergency Management community. To provide an increasing number of people access to the server during times of crisis, the server would need to scale up its capacity to meet the growing demand. This type of event-driven demand is a very good use case for deploying in the cloud. Containerization and virtualization make great sense in this example, giving students and faculty access to AWIPS II from multiple operating systems and locations.

ERAU's Applied Meteorology program was therefore an excellent candidate for taking advantage of cloud-computing resources. Through financial support from an Unidata Equipment Grant and considerable technical assistance by Unidata staff, a cloud-computing instance was purchased. Fig. 7.1 gives a schematic illustration of the basic configuration of the cloud instance that was created using Amazon Web Services (AWS). By selecting a cloud server geographically collocated with Unidata's Local Data Manager (LDM) data stream, data have been supplied to ERAU's instance of AWIPS II at no additional cost. (LDM is a software component of the EDEX server that supplies data to the AWIPS II system.)

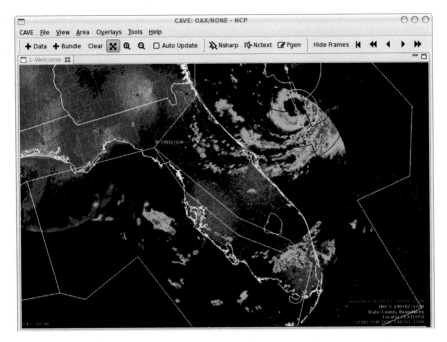

Figure 7.1 Data flow schematic for ERAU's instance of AWIPS II in the cloud.

We needed a high-end machine instance that met the minimum needs for AWIPS II and a fast network connection to help with the performance of the AWIPS II visualization client (CAVE) in a remote setting. HEC was required with many processors to accommodate the requirements of AWIPS II and additional software packages and software updates that would be installed on the system. We therefore selected a machine with the following specifications: 4×2.4-GHz Intel Xeon® vCPUs (virtual Central Processing Units), 16 GB of memory, and 750 GB of elastic SSD hard-drive storage. AWIPS II required Red Hat Enterprise Linux (RHEL) 6.5 to insure proper function. It also had specific demands for the graphics processing unit (GPU). Virtual instances do not have physical GPU's but do offer virtual GPU's (vGPU), and the back-end data server (EDEX Server) performed well in this virtual environment.

It has been difficult to nail down the costs of our cloud instance at ERAU, and changes in processing and data transfers out of the cloud have had to be monitored closely. Data flowing into the cloud is generally free, and moving it around the cloud can also be free if the data transfer remains within the same cloud zone. The greatest costs were incurred when bringing data out of the cloud. During times when the server was unused, the

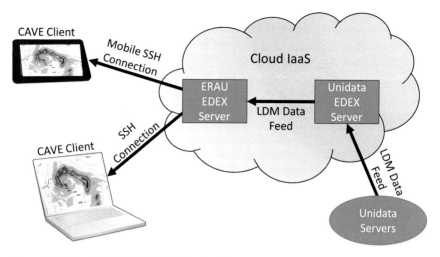

Figure 7.2 Sample AWIPS II CAVE display depicting Convective SIGMET outlines and radar reflectivity imagery around Tropical Storm Arthur off the coast of Florida on July 2, 2014.

image was saved and the machine was turned off to minimize expenses. On average, monthly charges have ranged from $200 to $250/month when school is in session. Compared to hiring part-time Linux support and obtaining the necessary hardware to host AWIPS II locally, this cost has been fairly reasonable.

Fig. 7.2 provides a sample AWIPS II display in CAVE, depicting radar data and Convective SIGMET (Significant Meteorological Information) boundaries surrounding Tropical Storm Arthur along the Southeast Coast on July 2, 2014. This figure demonstrates how AWIPS II enables meteorology students to overlay multiple datasets and imagery and serves as a powerful analysis and forecasting tool. Real-time surface and upper-air observations, remote sensing imagery, numerical forecast model output grids, NWS products, etc., are continually loaded onto the ERAU EDEX Server from Unidata's EDEX server. Using these data, the CAVE client renders two-dimensional maps, vertical cross sections, time–height cross sections, thermodynamic charts, and other output used in operational weather forecasting.

Using AWIPS II has been a pleasant experience thus far. By installing the Oracle Virtual Machine (VM) Virtual Box (www.virtualbox.org) software with a virtual RHEL 6.5 Linux machine on a Windows laptop or tablet, AWIPS II is easily accessible. This ability to achieve AWIPS II visualization from practically any location and on mobile devices supports the educational needs of the students pursing the ERM Certificate. One disadvantage that has been observed is a slight, noticeable lag compared to accessing an EDEX server on a local network, especially when there are multiple users or slow Internet connections. Small time lags up to 1-2 seconds have been observed depending on traffic and available bandwidth on the Internet connection. Nevertheless, the latencies have not deterred students from experimenting with this useful new visualization tool (as demonstrated in Fig. 7.2).

Security is a challenge, especially because users need to access the server from a variety of Internet providers. The dynamic nature of residential and mobile Internet Protocol (IP) addresses necessitates temporarily opening the firewall on the cloud server for any IP to enable Hypertext Transfer Protocol (HTTP) access for the CAVE client. Security problems of this nature require ongoing testing and research to avoid breaches in the future. For now, the only solution we have identified is to keep the firewall engaged as much as possible and not allow open access.

We are now beginning to test the utility of this AWIPS II instance for facilitating meteorology education. Instructions on how to access the CAVE client have been distributed to all stakeholders in the Applied Meteorology program, both students and faculty. In particular, students in an upper-division forecasting techniques course and other synoptic and dynamic meteorology courses in the curriculum are being encouraged to learn to use AWIPS II in preparation for future careers as forecasters. They are also beginning to incorporate the software in labs, discussions, briefings, and forecasts. Eventually, the advantages and disadvantages of cloud-hosting AWIPS II for student instruction will become evident. If the results are positive, it is anticipated that other meteorology schools will also want to provide their students with the same opportunity to access AWIPS II via the cloud.

In summary, educational institutions are seeing the benefits of cloud computing more and more. Learning Management Systems are the tip of the iceberg. As outlined above, there are so many benefits to hosting educational computing in the cloud. Of course there are legitimate security concerns preventing university campuses from moving their more valuable and mission-critical computing functions to the cloud. However, for the purpose of education, scientific computing, and data management, cloud computing is largely underutilized and can be a viable option. Especially for smaller schools with limited technical support, the cloud can be an affordable, flexible, and powerful computing solution in the atmospheric and oceanic sciences. With time, larger and more diverse datasets will be stored in private, open, and commercial cloud environments. The sharing of these data and the computational resources available in the cloud will allow previously prohibitive research and analysis to be achieved without the ongoing capital costs for increased hardware for the researcher.

REFERENCES

Allen, I.E., Seaman, J., 2013. Changing Course: Ten Years of Tracking Online Education in the United States. Babson Survey Research Group, Boston.

Bora, U.J., Ahmed, M., January 2013. E-learning using cloud computing. International Journal of Science and Modern Engineering 1 (2), 9–13.

Dong, B., Zheng, Q., Yang, J., Li, H., Qiao, M., 2009. An e-learning ecosystem based on cloud computing infrastructure. In: 9th IEEE International Conference on Advanced Learning Technologies, Riga, Latvia, pp. 125–127.

Dudley, J.T., Butte, A.J., 2010. In silico research in the era of cloud computing. Nature Biotechnology 28, 1181–1185.

Green, K.C., 2011/2012. The National Survey of Computing and Information Technology. The Campus Computing Project, Encino, CA.

James, C.N., Weber, J., Woodall, G.R., Klimowski, B.A., 2015. A cloud-based mobile weather server to support emergency response meteorology training and operations. In: 24th Symposium on Education. Amer. Meteor. Soc., TJ5.2, Phoenix, AZ.

Kats, Y., 2013. Learning management systems and instructional design: best practices in online education. IGI Global. http://dx.doi.org/10.4018/978-1-4666-3930-0 467 pp.

Molthan, A.L., Case, J.L., Venner, J., Schroeder, R., Checchi, M.R., Zavodsky, B.T., Limaye, A., O'Brien, R.G., 2015. Clouds in the cloud: weather forecasts and applications within cloud computing environments. Bulletin of the American Meteorological Society. http://dx.doi.org/10.1175/BAMS-D-14-00013.1.

Ostermann, S., Prodan, R., Schüller, F., Mayer, G.J., 2014. Meteorological applications utilizing grid and cloud computing. In: 3rd International Conference on Cloud Networking (CloudNet), Luxembourg, pp. 33–39. http://dx.doi.org/10.1109/CloudNet.2014.6968965.

Sinclair, M.R., 2015. Use of a simple graphics viewer for gridded data to improve data literacy. In: 24th Symposium on Education. Amer. Meteor. Soc., Phoenix, AZ.

Sultan, N., April 2010. Cloud computing for education: a new dawn? International Journal of Information Management 30 (2), 109–116.

Woodall, G.R., James, C.N., Klimowski, B.A., 2015. The "Emergency Response Meteorologist" Curriculum Development at Embry-Riddle Aeronautical University, Prescott, Arizona. In: 24th Symposium on Education. Amer. Meteor. Soc., TJ5.3, Phoenix, AZ.

CHAPTER 8

Cloud Computing for the Distribution of Numerical Weather Prediction Outputs

B. Raoult
European Centre for Medium-Range Weather Forecasts, Reading, UK; University of Reading, Reading, UK

R. Correa
European Centre for Medium-Range Weather Forecasts, Reading, UK

INTRODUCTION

This chapter considers various ways cloud computing could solve one of the greatest challenges being faced by numerical weather prediction systems: how to ensure that users can make use of the ever-increasing amount of data produced.

This chapter will explore two use cases:

- How to make very large amounts of data available to operational users in a timely fashion?
- How to make multi-Petabyte (PB) research datasets available to a large community of users?

It will be done in the context of a public cloud, i.e., a general-purpose cloud operated by an external company. The advantages of running a private cloud within a data center will also be considered.

PUSHING LARGE QUANTITIES OF DATA TO THE CLOUD UNDER TIME CONSTRAINTS

The first part of this chapter considers how the European Centre for Medium-Range Weather Forecasts' (ECMWF) dissemination system could benefit from cloud technology. One should only consider a change if the result is a better or cheaper service.

Cloud Computing in Ocean and Atmospheric Sciences
ISBN 978-0-12-803192-6
http://dx.doi.org/10.1016/B978-0-12-803192-6.00008-6

Copyright © 2016 Elsevier Inc.
All rights reserved.

ECMWF Real-Time Dissemination

The bulk of the products are generated by ECMWF's high-resolution forecast (HRES) (10 days, 137 levels, 16 km global grid) and its ensemble forecast (ENS) (51 members, 15 days, 91 levels, 32 km global grid). Each forecast produces respectively 260,000 global meteorological fields (1 Terabyte (TB) in total) and 5,400,000 (4.4TB in total) fields per run. There are two runs per day (00Z and 12Z). All model outputs are archived in ECMWF's Meteorological Archiving and Retrieval System (MARS (Raoult, 2001; Woods, 2006)).

Fig. 8.1 shows a graphical representation of two meteorological fields produced by ECMWF HRES. In the case of ECMWF's dissemination, users are provided with the binary representation of the fields, encoded using the GRIB format, a standard of the World Meteorological Organization, which allows them to make use of the actual values of the fields in their own applications.

To make use of the data to build their own products, users need access to the model's result as soon as possible. To achieve this, ECMWF has put in place a dissemination system that pushes the data to the user shortly after the model runs and according to an agreed schedule. All products of a given forecast run must be pushed to all users simultaneously in less than 1 h.

As it is not possible to push the entirety of the model outputs to all users within the allocated time, the ECMWF dissemination system allows its users to select which parameters, levels, forecast time steps, etc. they want delivered. It lets them specify subarea extraction as well as regridding, thus reducing greatly the size of the data to be transmitted. As a result, each user receives customized selections of fields, with very little overlap between the information requested by various users.

ECMWF also operates a private network (RMDCN) between itself and its Member and Cooperating States (Member States include Austria, Belgium, Croatia, Denmark, Finland, France, Germany, Greece, Iceland, Ireland, Italy, Luxembourg, the Netherlands, Norway, Portugal, Serbia, Slovenia, Spain, Sweden, Switzerland, Turkey and the United Kingdom; Cooperating States comprise Bulgaria, Czech Republic, Estonia, the former Yugoslav Republic of Macedonia, Hungary, Israel, Latvia, Lithuania, Montenegro, Morocco, Romania, and Slovakia). Member and Cooperating States may choose to have products delivered using either this network, which guarantees high availability via a service level agreement with a service provider, but has limited bandwidth, or the Internet that provides high transfer speed but with no guarantee of availability.

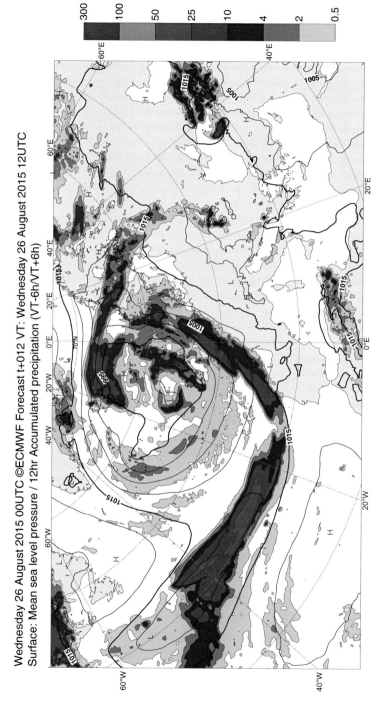

Wednesday 26 August 2015 00UTC ©ECMWF Forecast t+012 VT: Wednesday 26 August 2015 12UTC
Surface: Mean sea level pressure / 12hr Accumulated precipitation (VT-6h/VT+6h)

Figure 8.1 Two forecast fields (mean sea-level pressure and total precipitation) plotted on a map.

Improving Access to Ensemble Data

ECMWF has been pioneering ensemble predictions (Molteni et al., 1996), a system by which many weather predictions are run simultaneously based on slightly different initial conditions, to provide a measure of probability that certain events will arise, such as strong wind or large amounts of precipitation. Each weather forecast comprises 52 simulations of future weather, which increases the data volume by one order of magnitude (see Fig. 8.2).

Fig. 8.3 gives an example of a product that has been created using all 52 forecasts. These graphs (called "plumes") show the 10-day forecasts for Reading, United Kingdom, for three meteorological parameters (temperature at 850 hPa, total precipitation in 6-h periods, and geopotential at 500 hPa). These graphs also show the probabilities of a parameter value lying within a given range (shaded areas on the temperature and geopotential panels). These probabilities have been computed using the information from all 52 forecasts.

It is felt that products from the ensemble forecasts are not exploited as much as they could be because of the difficulties of handling the sheer volume of data involved and the limited time available to do so. Improving real-time access to these products would greatly increase their usefulness.

Preliminary Experiments

The Helix Nebula—The Science Cloud Initiative (HNI) pioneered the use of commercial cloud services for scientific research by organizations such as the European Organization for Nuclear Research (CERN), the European Space Agency (ESA), and the European Molecular Biology Laboratory (EMBL), as well as suppliers of cloud-based services, and national and European funding agencies (Helix Nebula). This initiative, supported by the European Commission, strives to bring coherence to a highly fragmented information technology services industry through the vision of a federated "Science Cloud" integrated with publicly funded scientific e-Infrastructures.

Helix Nebula, a science cross-domain initiative building on an active public–private partnership, is aiming to implement the concept of an open science commons (Open Science Commons) while using a cloud hybrid model (Jones, 2015) as the proposed implementation solution. This approach allows leveraging and merging of complementary data-intensive Earth Science disciplines (e.g., instrumentation and modeling), without introducing significant changes in the contributors' operational setup. Considering the seamless integration with life science (EMBL), scientific exploitation of

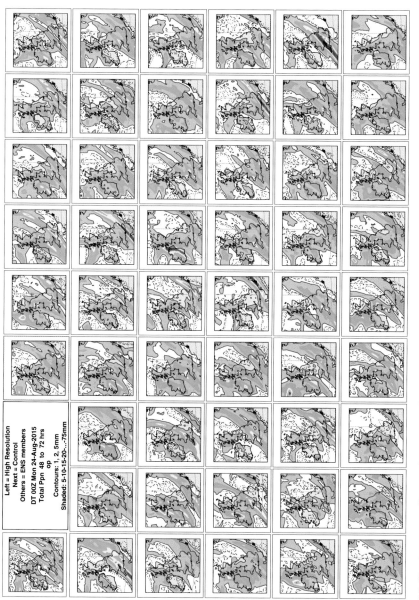

Left = High Resolution
Next = Control
Others = ENS members
DT 00Z Mon 24-Aug-2015
Total Ppn 48 to 72 hrs
op
Contours: 1, 2, 5mm
Shaded: 5-10-15-20-...75mm

Figure 8.2 "Postage stamp" maps showing 52 different 5-day forecasts total precipitations. All forecasts share the same reference time.

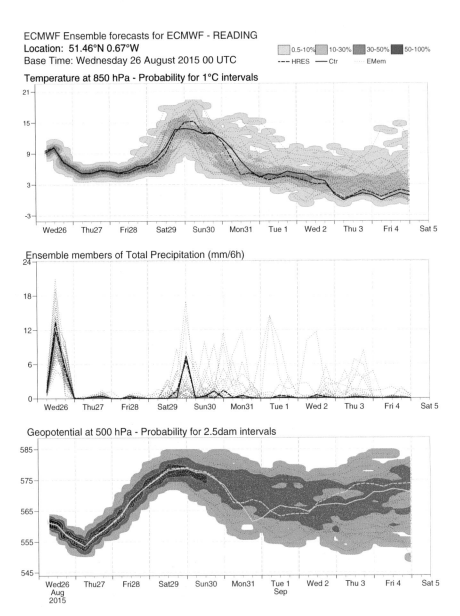

Figure 8.3 Plume plot for reading, United Kingdom.

meteorological, climate, and Earth observation data and models open an enormous potential for new *big data* science.

In the context of Helix Nebula, ECMWF collaborates with commercial cloud providers to investigate pushing its model output as quickly as possible into a public cloud infrastructure. To match the centralized dissemination schedule, the goal of this experiment is to push 2TB of model output in less than 1 h.

Bandwidth Tests

Initial bandwidth tests over the Internet were lower than expected with transfers being limited to around 2 Gbps (gigabits per seconds). The limitation seen on the initial tests seems to be related to the fact that acceptable levels of dropped TCP packets on typical Internet links (~0.5%) is enough to cause a significant drop in throughput when transferring high volumes of data over a low number of parallel TCP streams.

This was probably caused by congested links or peering limitations when traversing specific Internet Exchange Points on the network route taken between ECMWF's data center in the United Kingdom and the data center of a public cloud provider in Switzerland on the specific day the tests were carried out. Due to the dynamic nature of the Internet, this changes from day to day and the statement cannot therefore be generalized.

Further bandwidth tests have shown much more positive results by achieving up to 7 Gbps of real TCP/IP traffic, with a latency of just 29 ms, when the traffic happens to be routed solely within GÉANT (*GÉANT Network*) and the National Research and Education Networks (NRENs). This was achieved using just five parallel TCP streams, from five different physical hosts at ECMWF to five different virtual machines at the cloud provider.

GÉANT is a pan-European research and education network that interconnects Europe's NRENs. It has extensive links to networks in other world regions, such as North America or the Asia–Pacific Region, reaching over 50 NRENs and is partly funded by the European Commission. The main benefit of using GÉANT rather than commodity Internet is the superior performance of TCP, due to the lower congestion (fewer dropped packets).

Further Tests Are Needed

It should be noted that these tests only measure network bandwidth, and the data was discarded by the target machines; further tests will have to consider the time required to transfer the data to the cloud provider's storage, but no problems are anticipated as physical hosts can typically write to disk at 12 Gbps over dual controllers. Assuming that the cloud provider assigns 2 Gbps per physical target and that the target virtual machines (VMs) are on different physical machines, it should be possible to write the data to disk faster than it is transferred over the network.

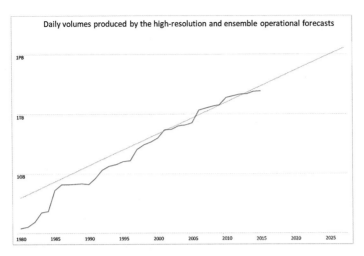

Figure 8.4 Growth of the average daily volumes produced by ECMWF main forecasts (log scale). The daily volume end of 2015 is around 13 TB. If the current growth rate remains the same, 1 PB will be reached in 12 years.

These results of the initial bandwidth tests are very encouraging, and further experimentation involving possible users of the data could be considered.

In should be noted that the volume of products from ECMWF high-resolution and ensemble forecasts has grown at a rate of between 30 and 40% per annum during the last decade (Fig. 8.4). Such a service would only be sustainable if the infrastructure (network, disks, and virtual machines) can accommodate this growth at an affordable cost.

The Cost of Such a Service

The cost of high-performance cloud services can be broken down as:
- Cost to transfer the data into the cloud (cost per volume and/or transfer requests)
- Cost to transfer the data out of the cloud (cost per volume and/or transfer requests)
- Cost to move the data between two data centers of the same provider
- Cost to use the data within the cloud (central processing unit, memory)

These costs are typically borne by the end users. Users will only consider such a service if it provides clear benefits to them, such as faster access to the products, access to more products, or collocation with data from other scientific disciplines. In any case, users will need to ensure that their applications are run as close as possible to the data, e.g., that their virtual machines are deployed in the same data center that holds the products; this is a service offered by cloud providers at extra cost.

At present it is premature to assess the ratio between costs and benefits. This will be evaluated during further testing involving actual users.

MAKING A MULTI-PB DATASET AVAILABLE IN THE CLOUD

The second part of this chapter considers how ECMWF's own data portals could benefit from cloud technology.

Making Research Data Available to the Community

Research is another of ECMWF's objectives. ECMWF has produced several reanalysis (Dee et al., 2011, 2014; Uppala et al., 2005) datasets and has made them available to the public, including research and commercial users, via several generations of data portals. These datasets have proven very popular, with thousands of registered users worldwide.

Unlike most data portals, ECMWF abstracts files and provides access to any user-defined collection of fields (Raoult, 2005; Raoult, 2003). It is common to see requests for a single variable for a very long time series (e.g., 100 years), whereas other users will retrieve all variables, at all levels, for a given month.

Because of the size of the datasets offered, it would be too costly to hold them all on disk. Consequently, for each request, the data portal accesses data from ECMWF's meteorological archive (MARS).

MARS holds its data on an automatic tape library. It also implements field-level caching so popular fields are always online, whereas infrequently accessed parameters are retrieved from tape. A series of queues, and the use of priorities and limits, protect the MARS system from unreasonable requests.

Users are given the ability to ask for subarea extraction, regridding, and format conversion. This leads to a great reduction of the volumes of data transferred over the Internet.

This paper now considers how ECMWF's next-generation reanalysis (ERA5) could be provided by means of the cloud. ERA5 is a new atmospheric reanalysis that will be produced at ECMWF and will be a flagship deliverable of the newly established Copernicus Climate Change Service (Copernicus Climate Change Service). This dataset is expected to be complete by the end of 2017, and its size is estimated to be over 5 PB, for 14 billion fields. The fields will be encoded in GRIB, which enables a volume reduction of a factor of about four compared to the raw model outputs.

Data Volume: How Big Is 5 PB?

People find it difficult to grasp how big 5 PB are. To put it into context they can be asked: "*imagine you were to copy this data to your PC's hard disk (assuming it is big enough)*", then presented with the following argument:

USB 3.0 can achieve around 640 MB/s (megabytes per second) (5 Gbps). Writing 5 PB would take about 100 days, which is over 3 months. This assumes that the underlying disk can sustain this throughput. Typical desktop PC hard disks usually range between 100 MB/s and 200 MB/s, therefore copying the data would take between a year and 18 months.

The Cost of Pushing the Data

The first challenge would be transferring the data to the cloud provider.

Sending Tapes

Some authors argue that is still cheaper to send very large amounts of data on physical media (e.g., USB disks) to the cloud provider (Armbrust et al., 2010). One should also consider the time it takes to write these disks in the first place, and to read them at the other side (see section: Data Volume: How Big Is 5 PB?). Furthermore, such a process requires a large degree of human intervention to mount and unmount the disks.

In 2003, ECMWF transferred its ERA-40 reanalysis (Uppala et al., 2005) to the National Center for Atmospheric Research (NCAR). Thirty-three TB of data were sent using 178 LTO2 tapes, of which two were unreadable and needed to be resent (for a detailed inventory of each of the tapes was kept). It took 6 weeks to write the tapes and it took NCAR around 4 months to read and ingest the data into their mass storage system. Including the preliminary discussions, purchasing the automatic tape library, testing with some samples, the whole process took around 8 months.

Today, tapes like T10KD can hold around 10 TB, which means that 5 PB would require over 500 tapes. It would take, based on our previous experience, about 18 months to transfer the data via physical media, and would require some costly human resources.

Network Transfers

As seen in section Bandwidth Tests, data can be transferred over the GÉANT network at around 7 Gbps, which is comparable to 5 Gbps of USB3.0 mentioned in section Data Volume: How Big Is 5 PB?, meaning that the order of

magnitude of transferring 5 PB over to the cloud provider would theoretically take around 3 months. In practice, this will take much longer:
- the data is originally on tapes;
- it will have to be routed via internal routers and across firewalls;
- components will fail or be subject to scheduled maintenance;
- the data will have to be transferred in files of manageable size, leading to the creation of a very large number of files, which comes with its own overheads;
- a sustained bandwidth rate of 5 Gbps is likely to occasionally suffer from bad performances during such a long period.

The copying process will have to be carefully planned and will have to be supervised so that failures are immediately remedied.

The Cost of Holding and Serving the Data

The cost of holding 5 PB in a public cloud, based on the price list of the most popular cloud provider, June 2015, is above $1.5 million per year.

As a comparison, offering such a dataset using ECMWF existing disk/tape-based infrastructure is estimated to be one order of magnitude cheaper. Such a service would be of course much slower.

ECMWF currently serves around 1 TB/day from its public data servers, in about 150,000 requests. The cost of serving 30 TB/month via the cloud will be around $30,000 per year (free if used from the same data center, four times cheaper to other data centers of the same provider).

It is expected that having this dataset in the cloud will increase its popularity and that the amount of data served will grow, and consequently so will the cost of transfers. On the other hand, the costs of transfers in and out of the cloud are clearly negligible compared to the cost of holding the data.

The costs of providing access to a multi-PB dataset in the cloud are currently very high. They could be lowered if the dataset becomes very popular and the cloud provider can recover these costs through the users' payments of the cloud services.

The Need for a Catalog

Very large datasets often mean very large numbers of files, or large files with a very large number of "data items"; in the case of ERA5, these are 2D global meteorological fields. There will be 1.4×10^{10} fields, averaging 400 KB in size on average.

It is not possible to create one file per field (there would be too many files). Fields need to be grouped and aggregated into large files, but users should not have to scan files to retrieve the data they want. They should also not be aware of any file-naming convention.

A possible solution is to reuse the architecture of ECMWF's archival system (MARS). MARS provides a way to map a user's request expressed in meteorological terms (e.g., air temperature and geopotential at all pressure levels for the years 2000–2015) into data locations, wherever the data may be. For that, MARS implements a very efficient indexing system that keeps track of the location of every field: this index can, for any field, return the filename, offset, and length of the field in the file. This system is very scalable, and MARS currently provides direct and constant-time access to more than 2×10^{11} items (200 billion), for a total volume of 64 PB.

Users first query the index and a list of (file, offset, index) tuples is returned which they can then iterate. In a distributed system, "file" can be substituted by URL. This scheme will work with any system/protocol that supports partial access to a resource. POSIX file systems do (via the C library calls `fopen()`, `fseek()`, `fread()`, and `fclose()`), as does HTTP/1.1 with the `Range` header (*RFC 2116*). For example, Amazon Simple Storage Service (S3) REST API supports the `Range` header.

Data Access Models

Cloud computing introduces the following *service models* (Sakr et al., 2011; Demchenko et al., 2013; Esteves, 2011):

- Infrastructure as a service (IaaS): This is typically the provision resources such as storage or virtual machines (e.g., Amazon S3)
- Platform as a service (PaaS): This is the provision of a software stack with which users can develop applications (e.g., Google App Engine)
- Software as a service (SaaS): These are applications that rely on the cloud (GMail, DropBox, etc.)
- Information as a Service (InfoaaS): When data is presented in a format that is ready to be consumed by the end user (e.g., Google Finance)
  Access to the data could be provided for each of the service models:

Infrastructure as a Service

In this model, users will instantiate VMs using the same cloud provider, and will directly access the products by "mounting" the file systems containing the data and indexes. Users would then directly access the data and indexes, provided they install the client and libraries that will be made available for download.

This model would provide the best throughput to the data assuming the file systems are mounted from disk storage from the same data center.

Platform as a Service

In this model, users will be able to instantiate VMs with a predefined software stack that will abstract the access to the data and indexes, for example based on ECMWF's data manipulation and visualization software Metview (Woods, 2006; Daabeck et al., 1994).

MapReduce-based tools (Dean and Ghemawat, 2008) using Hadoop (White, 2012) should be made available. ECMWF has been working on a prototype that extends Hadoop to directly access GRIB data, unpacking the values only when necessary.

Software as a Service

In this model, ECMWF would run part of its MARS system in the cloud to index and serve the data as well as redeploy its existing data portal to provide it with a Web-based user interface.

From an end-user point of view, there would not be any difference in the services already provided by ECMWF. From the infrastructure point of view, these services will benefit from the elasticity of the cloud, scaling out as the load increases.

Information as a Service

In this model, the raw data provided by ECMWF and possibly other data sources would be postprocessed by third parties to offer a variety of ready-to-consume value-added information services to other entities. This model creates a value chain that could be of great interest for the wider community.

PRIVATE CLOUD

A potentially more cost-effective solution to the "bring the user to the data" problem would be the establishment of a private cloud infrastructure on the ECMWF premises.

ECMWF is already running a data center and has the necessary infrastructure to deploy and host a private cloud. This could be done with in-house skills or by means of a public–private partnership with a commercial cloud provider.

Users will then be able to create VMs on demand on that infrastructure, running their own application next to the data, and downloading results that would be of smaller orders of magnitude. The service models offered would be the same as described in data access models.

A hybrid model that would allow the private cloud to scale during periods of high demand by automatically allocating more resources in the public cloud could also be considered (Sakr et al., 2011). In this scenario, users would need to consider that scaling to the public cloud would slow down their applications as the resources could once again not be collocated with the data.

A private cloud scenario could also prevent the collocation of data from different sources other than ECMWF.

CONCLUSION

The cost of holding a very large amount of data in a public cloud is still prohibitive, especially for an organization like ECMWF that is already running a data center and has an infrastructure in place. This could be alleviated if the hosting of the dataset is funded by its users, for example by having the cloud provider charge for access to the data.

Already having a copy of the data, it could be more cost-effective to run a private cloud at ECMWF and only provide on-demand computing power to users. This could prove a lot cheaper overall but would restrict users from accessing data from multiple sources and could limit the scaling of their applications even with a hybrid cloud model.

As far as using a cloud provider for the distribution of real-time model output is concerned, preliminary results are very promising. In this case, the transient nature of the data will mean that no long-term data storage is required from the cloud provider, only data transfers, which are a lot cheaper. Furthermore, the value of real-time weather forecasts is such that many users are already paying for faster, more reliable access to them, and they might be ready to move to a cloud-based distribution system if this gives them a better service.

REFERENCES

Armbrust, M., et al., 2010. A view of cloud computing. Communications of the ACM 53 (4), 50–58.
Raoult, B., 2005. On-line provision of data at ECMWF. In: 21st International Conference on Interactive Information Processing Systems.
Copernicus Climate Change Service, . Available from: http://www.copernicus-climate.eu.
Dee, D., et al., 2014. Toward a consistent reanalysis of the climate system. Bulletin of the American Meteorological Society 95 (8), 1235–1248.
Dee, D., et al., 2011. The ERA–interim reanalysis: configuration and performance of the data assimilation system. Quarterly Journal of the Royal Meteorological Society 137 (656), 553–597.

Demchenko, Y., Ngo, C., Membrey, P., 2013. Architecture framework and components for the big data ecosystem. Journal of System and Network Engineering 1–31.

Daabeck, J., Norris, B., Raoult, B., 1994. Metview-interactive access, manipulation and visualisation of meteorological data on UNIX workstations. ECMWF Newsletter 68, 9–28.

Dean, J., Ghemawat, S., 2008. MapReduce: simplified data processing on large clusters. Communications of the ACM 51 (1), 107–113.

Esteves, R., 2011. A taxonomic analysis of cloud computing. In: 1st Doctoral Workshop in Complexity Sciences ISCTE-IUL/FCUL.

GÉANT Network. Available from: http://www.geant.org/geantproject/About/Pages/Home.aspx.

Helix Nebula - The Science Cloud. Available from: http://www.helix-nebula.eu.

Jones, B., 2015. Towards the European Open Science Cloud. Available from: http://dx.doi.org/10.5281/zenodo.16001.

Molteni, F., et al., 1996. The ECMWF ensemble prediction system: methodology and validation. Quarterly Journal of the Royal Meteorological Society 122 (529), 73–119.

Open Science Commons. Available from: http://www.egi.eu/news-and-media/publications/OpenScienceCommons_v3.pdf.

Raoult, B., 2001. MARS on the web: a virtual tour. ECMWF Newsletter 90, 9–17.

Raoult, B., 2003. The ECMWF public data server. ECMWF Newsletter 99.

RFC 2116 - Byte Ranges. Available from: https://tools.ietf.org/html/rfc2616\l"section-14.35".

RMDCN. Available from: http://www.ecmwf.int/en/computing/our-facilities/rmdcn.

Sakr, S., et al., 2011. A survey of large scale data management approaches in cloud environments. Communications Surveys & Tutorials, IEEE 13 (3), 311–336.

Uppala, S.M., et al., 2005. The ERA-40 re-analysis. Quarterly Journal of the Royal Meteorological Society 131 (612), 2961–3012.

Woods, A., 2006. Archives and Graphics: Towards MARS, MAGICS and Metview. Medium-Range Weather Prediction: The European Approach, pp. 183–193.

White, T., 2012. Hadoop: The Definitive Guide. O'Reilly Media, Inc.

CHAPTER 9

A2CI: A Cloud-Based, Service-Oriented Geospatial Cyberinfrastructure to Support Atmospheric Research

W. Li, H. Shao, S. Wang, X. Zhou, S. Wu
School of Geographical Sciences and Urban Planning, Arizona State University, Tempe, AZ, USA

INTRODUCTION

It is generally agreed that an atmospheric process refers to a systematic and dynamic energy change that involves physical, chemical, and biological mechanisms. In recent years, atmospheric research has received increasing attention from environmental experts and the public because atmospheric phenomena such as El Niño, global warming, ozone depletion, and drought that may have negative effects on the Earth's climate and ecosystem are occurring more often (Walther et al., 2002; Karl and Trenberth, 2003; Trenberth et al., 2014). To model the status quo and predict the trend of atmospheric phenomena and events, researchers need to retrieve data from various relevant domains, such as chemical components of aerosol and gas, the terrestrial surface, energy consumption, the hydrosphere, the biosphere, etc. (Schneider, 2006; Fowler et al., 2009; Guilyardi et al., 2009; Ramanathan et al., 2001; Katul et al., 2012). In complex Earth system modeling, the data and services for atmospheric study present characteristics of being distributed, collaborative, and adaptive (Plale et al., 2006). The massive volume, rapid velocity, and wide variety of data has led to a new era of atmospheric research that consists of accessing and integrating big data from distributed sources, conducting collaborative analysis in an interactive way, providing intelligent services for data management, and integration and visualization to foster discovery of hidden or new knowledge. One of the most important ways to support these activities is to establish a national or international spatial data infrastructure and geospatial cyberinfrastructure on which the data and computational resources can be easily shared, the spatial analysis tool can be executed on-the-fly, and the scientific results can be effectively visualized (Yang et al., 2008; Li et al., 2011).

Cloud Computing in Ocean and Atmospheric Sciences
ISBN 978-0-12-803192-6
http://dx.doi.org/10.1016/B978-0-12-803192-6.00009-8

Copyright © 2016 Elsevier Inc.
All rights reserved.

Technically, a geospatial cyberinfrastructure (GCI) is an architecture that effectively utilizes georeferenced data to connect people, information, and computers based on the standardized data-access protocols, high-speed Internet, high-performance computing facilities (HPC), and service-oriented data management (Yang et al., 2010). Since the concept's official introduction by the National Science Foundation (NSF) in its 2003 blue ribbon report, cyberinfrastructure research has attracted much attention from the atmospheric science domain because of its promise of bringing paradigm change for future atmospheric research. As a result, several GCI portals to support atmospheric data analysis and integration have been developed:

- CASA (Collaborative Adaptive Sensing of the Atmosphere; http://www.casa.umass.edu), a real-time sensor network for acquiring and processing atmospheric data to predict hazardous weather events;
- LEAD (Linked Environments for Atmospheric Discovery; http://lead.ou.edu; Droegemeier, 2009), a service-oriented cyberinfrastructure that establishes a problem-solving environment which hides complex computation details by providing efficient workflow management and data orchestration mechanisms;
- Polar Cyberinfrastructure (http://polar.geodacenter.org/gci; Li et al., 2015), the cross-boundary GCI that supports sustained polar science in terms of atmospheric, environmental, and biological change.
- RCMED (Regional Climate Model Evaluation System; Mattmann et al., 2014), a cyberinfrastructure platform that supports the extraction, synthesis, and archive of disparate remote sensing data in various formats, including hierarchical data format (HDF) (Fortner, 1998), Network Common Data Format (NetCDF) (Rew and Davis, 1990), and Climate Forecast (CF) (Eaton et al., 2003), etc.

These GCI solutions share similar goals and characteristics. They (1) provide precise and effective access to distributed geospatial data; (2) collaboratively analyze large-scale data through remote communication and sharing; (3) use an intelligent, systematic, and user-friendly interface to improve user experience; and, more importantly, (4) support on–demand data integration, modeling, and analysis. Achieving these goals simultaneously requires a new computing platform that supports scalable, elastic, and service-oriented data analysis. Cloud-computing platforms, known for their capability to provide a converged infrastructure and shared service to maximize the efficiency in data sharing, storage, and computation, are a promising solution. First, the three levels of resource-sharing strategies, Data as a Service (DaaS), Software as a Service (SaaS), and Platform as a Service (PaaS), provide strong architectural and technical support that will enable the establishment of a service-oriented GCI for atmospheric research. Second, cloud computing can broaden the utilization of GCI by providing an open, collaborative, Web-based environment for decision-making (Li et al., 2013). Finally, cloud-computing techniques can provide strong support for efficient processing in a cyberinfrastructure environment in which many data/processing requests from distributed users may arrive simultaneously because of the cloud's ability to utilize the spare resources in a network of distributed computers (Yang et al., 2008, 2011). We believe that the integration of cloud-based GCI represents a new frontier for atmospheric research and that the marriage of the technology will foster the establishment of a software infrastructure that enables deep computing in atmospheric analysis and extensive use of cybertechnology across different communities and organizations.

In this paper, we describe our research in developing a cloud-based, service-oriented Atmosphere Analysis Cyberinfrastructure (A2CI) to support the effective discovery of distributed resources.

LITERATURE REVIEW

In the literature, cloud-computing has been widely used as the computing infrastructure to host various online decision-making tools (Yang et al., 2011; Li et al., 2015), because of its advantages of being convenient, easy to manage, and having an elastic capacity to handle various workloads over time. This popular service is also known as Infrastructure as a Service (IaaS), one of the four major types of cloud service defined by the National Institute of Standards and Technology (NIST). In comparison to IaaS, the other three types of cloud services—DaaS, SaaS, and PaaS, require a deeper level of understanding of the domain knowledge, including the distribution pattern of data, the computational characteristics of the analytical tools, and the extendibility of the cyberinfrastructure platform. We first provide a general review of the DaaS, SaaS, and PaaS principles, then we describe the design and development of cloud services following these principles to develop next-generation cyberinfrastructure for atmospheric studies.

Data as a Service

DaaS provides an effective way to make massive data readily available for clients and represents a great improvement over traditional approaches. In conventional methods, accessing data to support atmospheric research requires the development of a stand-alone tool for each data repository.

In the big data era, however, the countless number of data repositories, data structures, data formats, and other factors make it a significant challenge to build and integrate all kinds of customized tools for integrative data analysis. DaaS, which focuses on discovering needed data associated with a specific analysis task and delivering data to researchers without considering the physical location and sources of the data through the provision of services, becomes a promising strategy to address the aforementioned challenges. For data users, DaaS in GCI enables users from a wide range of communities to discover and access needed data by appropriate protocol. For data providers, DaaS reduces the cost of data storage and data management by moving the data server onto the cloud and increases data reuse and management efficiency through the provision of standardized data services.

Software as a Service

SaaS provides on-demand tools and applications for performing specific tasks in the first layer on the top of a cloud server infrastructure. This elasticity enables atmosphere researchers to perform computing tasks without installing a model or analysis tool locally. For example, various analysis functions that monitor and predict atmospheric events can be deployed in a cloud environment and made available to online users. These functions can be deployed as Open Geospatial Consortium (OGC)-compliant processing services, such as Web Processing Service (WPS), or Web Coverage Processing Service (WCPS), if a GCI primarily deals with raster data processing. Researchers access the online portal and send commands through the SaaS to run model simulations in the cloud. Some other services, such as data discovery service or data visualization service can also be encapsulated as services for easy integration into GCI solutions. The benefits of SaaS include eliminating local constraints of limited computing resources and software tools.

Platform as a Service

PaaS enables virtualized computing between data providers and data users. In a cloud-based GCI, data and computing resources to satisfy the demands of atmospheric research can be provided using the mechanism of virtual space and virtual storage. Furthermore, the scalable data and computing resources defined on top of a cloud platform can be dynamically configured and scaled depending on the demand of atmospheric analysis request. In a PaaS model, it is not just the elements of CI, such as data or software provided, it is actually the entire CI problem-solving environment that is provided. In such an online environment, multidisciplinary data is accessible through DaaS, the various software tools for atmospheric analysis are made available through SaaS, and the whole CI software platform is provided to support online data analysis and decision–making.

The next sections describe the proposed CI framework for sharing, analyzing, and visualizing atmospheric data in a cloud environment.

CLOUD-BASED CI FRAMEWORK FOR ATMOSPHERIC RESEARCH

Our proposed Atmosphere Analysis CyberInfrastructure (A2CI) was designed following the principle of service orientation (DaaS, SaaS, and PaaS) to integrate as many data sources as possible and provide the best user experience. Fig. 9.1 demonstrates the architecture of the A2CI.

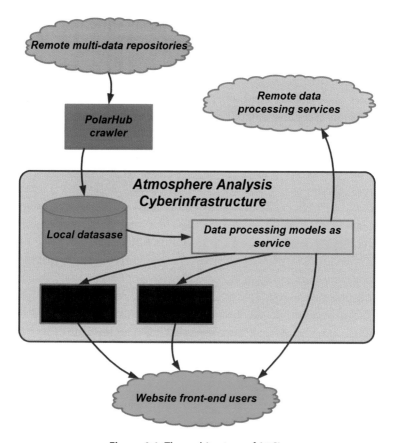

Figure 9.1 The architecture of A2CI.

There are five types of services being managed in the A2CI framework: data service, data discovery, data search, data processing, and data rendering and visualization. Atmospheric data is location specific, and, therefore, a typical type of geospatial data. These data are diverse and heterogeneous in nature. They have various encoding structures, i.e., vector or raster, due to different data providers' conventions. They may also have different spatio-temporal scales, spatial reference systems, and accuracy, etc.

To ensure that data from various sources are sharable and interoperable, our A2CI implements *a data service module* that can handle disparate data according to OGC standards. The OGC is an international industry consortium of over 500 universities, companies, and government agencies participating in a consensus process to standardize publicly available Web interfaces. This standard has been widely adopted by the atmospheric research community for sharing data and other resources. The data service module also provides the capability to publish raw datasets into services. To enrich the data repository in the A2CI system and support researchers' needs, a *data discovery service* is provided. This data discovery service is implemented through PolarHub, a large-scale Web crawler that has the ability to identify distributed geospatial data resources on the Internet.

Besides managing data resources, a key feature of the A2CI framework is support for online spatial analysis of datasets. All the processing functions are developed as processing services such that they are easily transplanted and reused. Two types of processing services are enabled: *local data processing services* and *remote data processing services*. The local data processing services are the locally deployed spatial analytical tools and models as services. The remote data processing services refer to the analytical capabilities provided by another CI system with standard service interface. The capabilities of these services and their invocation protocol are stored in the A2CI database. Once needed, the high-level processing service will compose an analytical workflow that orchestrates the data and local or remote processing services for accomplishing an analysis task. Because of its ability to glue together other components and at the same time provide an array of useful functions to users, the "data processing models as service" becomes an A2CI core service.

The results generated, either in the form of a map or statistical chart, will be developed through the *rendering and visualization services*. The rendering service is mainly responsible for defining the symbology and style for rendering different types of geodata. It is actually part of the *visualization service*, which provides a platform that is either two-dimensional (2D) or multidimensional for intuitive presentation of the data and analysis results.

The details of geodata rendering and visualization will be discussed further in the section "Multidimensional Visualization Service".

All of these service components are coordinated by A2CI middleware to support a complete question-answering cycle from searching for data, integrating, and mapping data, analyzing the data, and visualizing the analytical results. These services are available as cloud services and are remotely programmable; hence, they can be easily integrated and reused by other CI solutions to reduce code duplication and to accelerate the knowledge discovery process. The next section describes the service modules in detail.

COMPONENTS

PolarHub: A Large-Scale Web Crawler to Provide a Data Discovery Service

Big data, which is dominating the Internet, has played an important role in improving our understanding of the Earth and supporting scientific analysis. There is an ever-increasing amount of Earth observation data being recorded and made available online, including past and current atmospheric data. These data are widely dispersed on the Internet, making effective discovery a significant challenge. Various search engines, such as Google and Bing, are convenient for the public's daily information discovery needs. However, they are not suitable to search for the geospatial and GIS data that helps the ocean and atmospheric research. To bridge the gap, we have developed a Web crawler, the PolarHub, and provided a data discovery service as part of A2CI to support scientific research and effective and accurate access to the desired data.

PolarHub starts by visiting the Web from a set of predefined Universal Resource Locators (URLs), then gradually follows the hyperlinks within these Web pages to expand the search scope until the entire Web or most of the interlinked Web has been visited. While PolarHub is scanning Web pages, it extracts any relevant geospatial data. As the crawling process continues, it identifies and downloads the metadata describing the content and the permanent link of the dataset. To efficiently accomplish the task, two questions need to be answered— (1) Where do we start to visit the Web? (2) What kind of geospatial data is our target?

For the first question, we rely on search results from popular search engines as the crawling seeds. This is because the Web is like an ocean of data; but what we have is no more than a big fish net. Looking for data on the Web requires huge computing power to continuously crawl the Web pages in it. Although general search engines such as Google and

Bing are not effective enough to look for geospatial data, they are still utilizable here because they provide a great initial start for our spatial data-oriented crawling due to the large amount of Web pages they have indexed.

For the second, the OGC Web Services (OWS) become the main data source for discovery because of the service-oriented principle adopted in our A2CI framework. Upon finding an OWS, PolarHub will invoke its semantic analysis module to determine whether a service is a related atmospheric dataset. The knowledge used in the semantic analysis is a bag of words imported from National Aeronautics and Space Administration (NASA) Global Change Master Directory (GCMD) science keywords. These words are scientific terms used to measure the atmospheric condition, such as "Albedo" or "Atmospheric pressure measurement". The services not related to atmospheric science are disregarded by PolarHub. Fig. 9.2

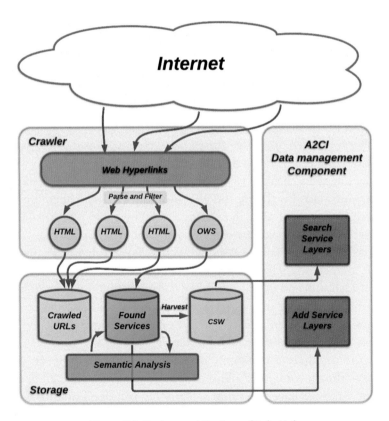

Figure 9.2 System architecture of PolarHub.

describes the software framework of PolarHub, which is composed of the crawler component and the storage component and interacts with the A2CI data management component.

The crawler component is responsible for crawling the Internet and finding relevant data. A typical crawling task starts with keywords that indicate the theme of the desired geospatial data. Taking the search for atmospheric data as an example, a user may enter "World Sea Surface Temperature" or simply "World SST" as a start. Once it receives the keywords, PolarHub will redirect the query to a general search engine such as Google or Bing to retrieve a set of relevant Web pages as crawling seeds. All of these seeds will be put into a seed pool that holds the URLs waiting for processing. There are mainly two kinds of URL processing. One is for URLs that link to an Hypertext Markup Language (HTML) Web page. The other is for the Extensible Markup Language (XML) documents, in which the OWS metadata are encoded.

At the beginning of processing, the crawling engine retrieves a URL from the seed pool then analyzes the file type of the URL, i.e., whether it is an HTML Web page or an XML metadata file. The extracted URLs linking to a dynamic HTML Web page are then inserted into another seed pool. At the same time, these URLs are checked to determine whether they belong to a specific type of OWS. If so, the link information together with all the metadata associated with the data or processing service will be sent to the local data repository for caching.

To foster cross-CI data discovery, namely to make PolarHub discovered data available to other CI portals, all the local data will be harvested into an OGC-compliant online catalog with a Catalog Service for the Web (CSW) interface openly accessible by a Web client or other portals. This catalog supports full-text searching through a structured query language, Contextual Query Language (CQL), and interfaces with A2CI middleware's data management component. This provides an important data source for atmospheric data search and analysis.

Taking advantage of the data discovery service provided by PolarHub, A2CI has successfully discovered 1816 atm-related datasets. These datasets are distributed across 32 countries including the US, Canada, Australia, and several EU countries. Fig. 9.3 demonstrates the spatial distribution of the Web servers that host these datasets.

The labeled countries on the main map are those providing more than 62 atmospheric data services (top three classes per natural break classification). The inset map shows a closer look at the service providers located in

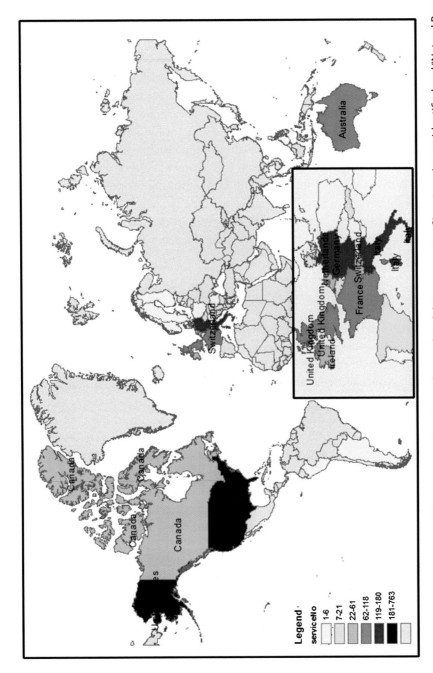

Figure 9.3 Spatial distribution of Web-server hosting atmospheric-related data per country. Six categories are identified and "Natural Break" is used as the classification method. Gray color means no data.

Europe. Among all these countries, the US provides almost half of the total datasets in the OWS format. European countries such as Germany, Italy, the United Kingdom, as well as Canada and Australia are all main providers for the atmospheric datasets. None of the relevant data are found in Asian countries, such as China, or many of the African countries, such as Libya and Sudan. This pattern reflects the fact that Western developed countries are more active in sharing data as services for reuse and integration. We did not find any OGC-related data from Asian developed countries such as Japan or Korea. This may due to the relatively less active role of these countries in the OGC community.

In the US, Federal government agencies that focus on Earth and space research, including the US Geological Survey (USGS), National Oceanic and Atmospheric Administration (NOAA), NASA, and NationalAtlas.gov, are the main contributors for atmospheric data services. Besides these government agencies, several US universities, and national data centers, such as the University of Hawaii and National Snow and Ice Data Center (NSIDC), are also contributing to the sharing of atmospheric data. Fig. 9.4 lists the top 10 atmospheric data providers in the US as determined through a further analysis on the atmospheric data holdings existing in the A2CI data portal. The available datasets provide us rich resources to conduct atmospheric

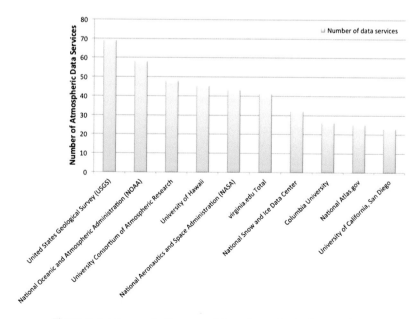

Figure 9.4 US organizations providing atmospheric data services.

analysis. At the same time, however, an effective data/service management mechanism is needed for better coordination of these distributed data resources. A2CI aims at providing a flexible and easy-to-use platform to support this coordination.

Data and Service Management

The A2CI platform is user centric as different users can work on different applications or analyze multiple data of interest. All data used for a specific application are organized into a workspace. One user can create multiple workspaces, meaning that they can start multiple research topics on our A2CI platform. All user-created data and intermediate results are saved in the A2CI database to facilitate multiphase scientific analysis. To support this, we developed a data and service management module to manage user data, workspace data, and atmospheric data collected by PolarHub, as well as other analysis-related information. Fig. 9.5 demonstrates the Unified Modeling Language (UML) diagram of the database schema.

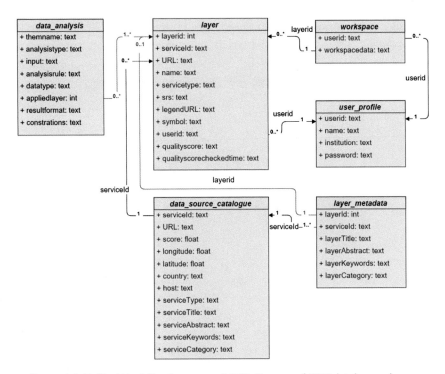

Figure 9.5 Unified Modeling Language (UML) diagram of A2CI database schema.

The *user_profile* table records users' email addresses as their unique ids and other profile information such as full name, institution, etc. This information is obtained during the registration process. User's passwords are stored as an encrypted string. The user id also plays a key role for linking different records together among the "*user_profile*", "*layer*," and "*workspace*" schemas. Any end user who wants to conduct analysis on the A2CI portal must register in the system and maintain a user profile. This way, the user's data viewing and analysis history can be retained and users' working results can be saved for multilogon processing.

The *workspace* table contains all of the essential attributes of a user-created workspace, including the workspace name, the default spatial reference system used, all data layers in a workspace, the layers' display order, etc. This information is formatted as a JavaScript Object Notation (JSON) string and compressed into one single data column titled *workspacedata*. Each time a user logs into the A2CI portal, the workspace information will be directly loaded into the Web front-end and processed by JavaScript. Because JSON strings were originally designed for the JavaScript language, saving workspace data into JSON makes it much easier for the front-end Web portal to present and visualize workspace data. The *workspace* table is linked to the *user_profile* table by the unique user id. The data layers loaded in each workspace are linked to the layer table by unique layer ids.

Different workspaces can load the same data layer, as long as this information is explicitly listed in the *workspacedata* field in the *workspace* table. When a user deletes a data layer in a workspace, the layer record will not be deleted in the layer table, instead only the record linking the workspace and the data layer will be removed. This way, different users' or a single-user's operations on the same data layer in different workspaces will not be affected. Moreover, because one data layer only has one permanent record in the database no matter how many times it is used by various workspaces, data redundancy can be avoided.

The layer information is stored in two data tables: the *layer* table for storing visualization and analysis-related data used in the A2CI system and the *layer_metadata* table for storing the static layer information associated with the data service it belongs to. In the *layer* table, essential information such as layer name, its Web URL, and spatial reference information are saved. When a data layer is requested by a user in the A2CI portal, this information will be encoded into a JSON object and pushed to an OpenLayers' Web client to generate a data retrieval request, download remote data, and display it in the map container. The *userid* records the user that the layer belongs to.

The *symbol* column records the customized symbol schema for each layer, such as opacity, size, and color of the layer's features. The *quality score* of each layer indicates the service's performance of each layer, such as the data transmission rate. This *layer* table is linked through the primary key *layerId* with the *layer_metadata* table, which stores additional thematic information about the layer's content, such as the title, abstract, and keywords. Though this information will not be used for data retrieval, they are important in supporting the discovery of relevant datasets by matching the thematic data.

These two-layer tables are mainly responsible for managing data layers that are the atomic unit for visualization. To manage the data resources at a higher level—the data service level, we designed another table, the *data_source_catalog*, as the data layers come from different data services. In fact, PolarHub uses this data table to manage all data services crawled. The crawled data service information includes the URL, which links to the original data provider, and the physical locations of the service hosts in the *latitude, longitude,* and *country* columns. The *score* is a quantitative measure to indicate the quality of the data service. The service table and the two-layer tables are linked by the unique *serviceId*. A layer discovering service is designed and implemented to help users efficiently find appropriate data resources on a layer level by searching from the *layer_metadata* table or at a service level by redirecting the search to the *data_source_category* table.

Besides the mechanism to manage data resources, another database table, *data_analysis*, is also included. Many data analysis models are implemented as services in A2CI's front end and back end. The profile information of these analysis services, including service name, service URL, the input/output of each service model, data type that each service takes and generates, analysis rule description, and the special constraints of each model, as well as the data layer in the workspace that binds and presents the results, are all stored in this table. Using this information, the A2CI front end can generate a scientific workflow automatically by linking the data and analytical components, significantly accelerating the process of data analysis and knowledge discovery.

Service-Oriented Data Integration

The conventional method of data integration accesses all atmospheric study-related data sources and stores them in a local data repository. The countless number of data repositories, data structures, data formats, and other factors in the big data era make it impossible to build a local database and integrate all the independent data repositories together. Therefore, a service-oriented data integration strategy is adopted in A2CI. It was designed following the principle of DaaS to provide an effective way of making massive data available for clients without the need for a local database. To save storage space and reduce management efforts, only the metadata about a distributed dataset is recorded. When users require the actual data for viewing or analyzing, the data will be directly retrieved from the original data provider and transferred to the end users. This data integration strategy significantly reduces the data-loading pressure on the server end of the A2CI platform. By crawling as many atmospheric study-related data sources as possible and providing a comprehensive data search interface to users, A2CI can also increase users' data discoverability and usability. Fig. 9.6 represents the workflow of the service-oriented data integration in A2CI.

The first level of integration occurs for presenting and overlaying disparate data sources. As discussed in the previous section, these datasets could come from remote data repository hosted by government agencies such as NASA, NOAA, USGS, or from some large data centers such as NSIDC. Through a data service layer, these data could be encoded into different OWS formats according to different application needs. For instance, remote sensing images in raster format could be published into the OGC Web Map Service (WMS) for visualization. Using this service standard, the back-end datasets, regardless whether they are in GeoTiff or *Global Resource Information Database* (GRID) formats, will be rendered into a static map covering the requested geographical region. The static map if requested and returned in Portable Network Graphics (PNG) format, can be tailored to set the transparency level at the client side to make a better visual effect when overlaying with other data layers.

Raster-based atmospheric data can also be shared in the OGC Web Coverage Service (WCS), which returns actual data in comparison to the static image given by a WMS. The disadvantage of a WCS is that the data file is much larger than the size of the static image. Vector-based data, in comparison, can be shared through the OGC Web Feature Service (WFS), which serializes the raw data formats, in an Environmental Systems Research Institute (ESRI) shapefile for example, to a standardized Geography Markup Language (GML) or Geographic JavaScript Object Notation (GeoJSON) for client visualization. These feature data can be customized with different display styles, including the use of different color ramp, point, or line size, and color to enhance the visual analysis in combination with raster datasets, etc.

Note that a successful integration of disparate data for conducting atmospheric analysis requires an intelligent method to search for the most

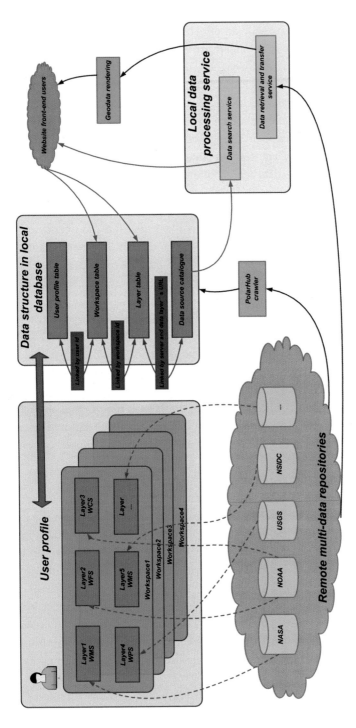

Figure 9.6 The workflow of service-oriented data integration.

suitable dataset from a set of data holdings. To accomplish this, we developed a data search service. Different from the data discovery service, which focuses on locating the distributed resources from the Internet to build a large atmospheric data clearinghouse, the data search service deals with finding relevant dataset that satisfies a user's request from the clearinghouse. This service provides a comprehensive search interface for users to find the data resources meeting certain conditions such as keywords, period, formats, region of interest, and spatial reference system (SRS). The search results will be presented to users as a formatted table. Additional functions are also provided to help the user better understand the data source. For example, a data quality score regarding the server response time, degree of metadata completeness, and other factors will be calculated and reported to the user. A preview thumbnail image of the data will also be provided if standard OGC WMS or WFS protocol publishes the data. Once users select a data source, the data source will be stored in the user database as a record associated with their user profile. This way, even if the data sources are distributed in many different remote data repositories and vary a lot, they can be uniformly integrated and stored in A2CI's local database.

The data search and integration strategies facilitate the integrated analysis of data from multiple sources. This integration is at a more static level, which means visual analysis of atmospheric phenomena is based upon the information that already exists in the datasets. To realize integration at a higher knowledge level, analytical tools are enabled in the A2CI platform as OGC services—the OGC Web Processing Services (WPSs). These WPSs take the static data as input and create new information and knowledge through different analysis methods. This type of service orientation belongs to a specific type of SaaS—Spatial Analysis as a Service (SAaaS) in the cloud-computing paradigm. The advantage of publishing various spatial analysis tools, such as zonal statistics, into Web services, is to increase their reusability and enhance A2CI's analytical capability by fusing existing data and creating unknown knowledge to further advance science.

Multidimensional Visualization Service

Once the data and analytical results are generated through the cloud data-processing services, they are sent to the A2CI's visualization service for mapping and rendering. Both 2D and three-dimensional (3D) visualization mechanisms are provided. The layer data management module controls the visualization mode and all the visualization-related data, including all the layer information, their display orders, as well as the uniform projection

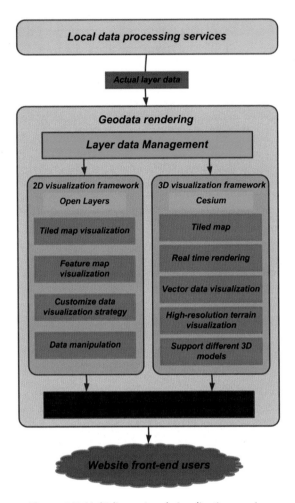

Figure 9.7 Multidimensional visualization service.

data. When a user chooses to switch between the 2D and 3D visualization, all the layer information is sent to the corresponding visualization engine for result rendering. Fig. 9.7 presents the data visualization framework.

2D VISUALIZATION SERVICE

The 2D data-rendering service is built based on OpenLayers, an open-source geodata visualization and manipulation library in JavaScript. The 2D mapping module includes the following main functions to visualize atmospheric dataset:

1. Use of OpenLayers in A2CI portal for image and feature map visualization: OpenLayers supports standard WMS protocol and is able to import different kind of image formats such as Tagged Image File Format (TIFF), Joint Photographic Experts Group (JPEG) File Format (.jpeg and .jpg file), and PNG into its map panel. In addition, OpenLayers also provides a built-in interface for integrating base maps provided by commercial companies, such as Google, ESRI, Open Street, and Bing, which helps users better understand the context of their study region. OpenLayers is also capable of importing feature layers in various formats, such as ESRI JSON, GeoJSON, GML, Keyhole Markup Language (KML), WFS, and XML files. What's more, OpenLayers also supports basic geofeature reprojection between different Spatial Reference Systems (SRSs), which means OpenLayers makes users' available layer source much richer.

2. Tiled map visualization: one of the main data sources in A2CI is OGC WMS, which returns static image from the remote server. When the requested extent becomes too large, loading and visualizing the entire dataset as a single piece may become very slow. Tile-based visualization resolves this issue by partitioning big datasets into smaller pieces with each piece's data individually and concurrently requested, retrieved, and displayed. This parallel processing strategy helps improve system performance and user experience.

3. Customize data visualization strategy: In the A2CI portal, users can define the opacity of tiled and feature maps. For the feature maps, users can also customize the symbols of layers, such as fill color, stroke color of polygons, and symbol size of points, presenting rich layer attributes to users.

4. Data manipulation: Through the powerful interaction interface provided by OpenLayers and A2CI visualization service, users can draw their own features or revise existing feature data on the 2D map.

3D VISUALIZATION SERVICE

Besides the 2D map container, A2CI also supports the visualization of atmospheric data in a 3D globe. As we know, 2D maps allow us to view the data on the entire Earth in one scene, but 3D maps usually cannot. However, 3D visualization has the advantage of introducing none or very little distortion in the original data, in comparison to a 2D map. Therefore, both visualization strategies are supported as they can complement each other to realize the best visualization effect for presenting atmospheric data in the A2CI portal.

The embedded 3D virtual globe is built upon Cesium, an open-source virtual globe made with WebGL technology. This technique utilizes graphic resources at the client side by using JavaScript-based library and Web Graphics Library (WebGL) to accelerate client-side visualization. The virtual globe has the capability of representing many different views of the geospatial features on the surface or in the atmosphere of the Earth, and can support the exploration of a variety of geospatial data. By running on a Web browser and integrating distributed geospatial services worldwide, the virtual globe provides an effective channel to find interesting meteorological phenomena or the correlation between heterogeneous datasets (Hecher et al., 2015).

As part of the visualization service, the virtual globe can dynamically load and visualize different kinds of geospatial data, including tiled maps, raster maps, vector data, high-resolution worldwide terrain data, and 3D models. In the following, we list the main functions supported by the 3D visualization module:

1. The world bathymetry and other terrain data is georeferenced and pre-rendered to serve as the base map of the virtual globe. All of the tiled map services, such as Web Map Tile Service (WMTS) specification developed by the OGC, Tile Map Service specification developed by the Open Source Geospatial Foundation, ESRI ArcGIS Map Server imagery service, OpenStreetMap, MapBox, and Bing maps, can be easily loaded into the virtual globe as base map.

2. Besides the base map, the virtual globe can accept real-time rendered WMS map services and georeferenced static raster maps as imager layers to overlap on the base map. All the imagery layers can be ordered and blended together. Each layer's brightness, contrast, gamma, hue, and saturation can be customized by the end user and dynamically changed.

3. The virtual globe also provides a library of primitives, such as point, line, polygon, rectangle, ellipsoid, etc., for vector data visualization. To improve the transmission speed of big atmospheric data over the Internet (Li et al., 2015), open specifications such as GeoJSON and TopoJSON (Topological Geospatial Data Interchange Format based on GeoJSON) are adopted to encode collections of simple spatial features along with their nonspatial features. Besides these primitives, the virtual globe visualization module also supports complex spatial entities by aggregating related visualization and data information into a unified data structure.

The A2CI will provide uniform layer control and interaction interface on both 2D and 3D visualization frameworks to help users efficiently manipulate and customize their data layer and to conduct experiments.

GRAPHICAL USER INTERFACE OF A2CI

Fig. 9.8 demonstrates the Graphical User Interface (GUI) of the A2CI portal, which as a whole can be considered as a PaaS—Platform as a Service, and is deployed onto an Amazon Elastic Cloud Compute (EC2) cloud infrastructure.

Box 1 (top left navigation) lists the layer management module, in which a list of data layers is included. These data are loaded from the database once the user is recognized in the system. Box 2 (bottom left navigation) is the legend component that displays the information of all visible layers. This data is loaded in real time from a remote data server at the same time the map data is requested. Box 3 (top right navigation) is the data search service, supported by the OGC Catalog Service for Web (CSW) 2.0.2. The small box at the bottom shows the results returned by searching data related to Sea Surface Temperature (SST). An abstract of each resulting data layer is provided. Users can use this information to decide whether the data is of interest. If so, the user can click the "Add WMS to Map" button to add the service to the map panel for visualization. Box 4 (right header) is the module for various spatial analyses. For instance, zonal statistics of the SST can be conducted on a state basis to obtain a long-term mean temperature for different areas. Box 5 (middle) is the map service. In the example, four data layers from three remote servers (the nationalatlas.gov, ciesin.columbia.edu, nsidc.org) are integrated for atmospheric analysis. The line data shows the trajectories of all tropical storms in the 2000s in the Pacific and Atlantic Oceans, respectively. Different colors represent different frequencies on a particular trajectory. The blue–red image data shows the SST data in 2008. The country boundary data from NSIDC is also loaded.

The same data can also be loaded onto and visualized in the 3D globe, shown in Fig. 9.9. From the integrated map, we can see that tropical storms originating in the Atlantic Ocean travel longer distances and have a higher intensity than those developed in the Pacific Ocean. This means that Atlantic Ocean storms carry more energy than Pacific Ocean storms as they travel. The energy could potentially come from the warmer weather in the same latitude in the Atlantic Ocean than the Pacific Ocean, reflected in the SST data in the 2D map. Further analysis on the intensity can reveal more quantitative evidence on atmospheric conditions and the formation of tropical storms.

CONCLUSION AND DISCUSSION

This paper introduces a service-oriented cyberinfrastructure, the A2CI, to support atmospheric research. This CI framework introduces and integrates

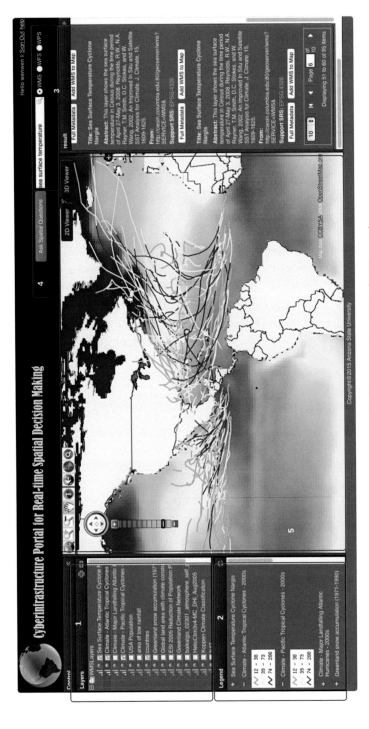

Figure 9.8 Graphic User Interface (GUI) of A2CI portal.

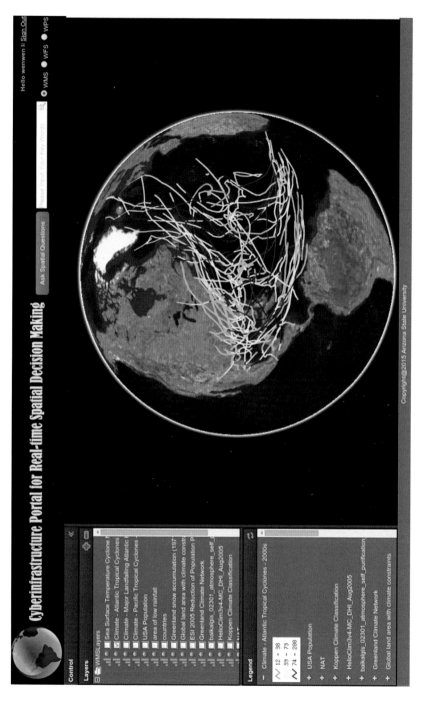

Figure 9.9 A2CI 3D visualization.

four primary principles—the DaaS, SaaS, PaaS, and IaaS—in the cloud-computing paradigm. There are a number of advantages in adopting these design principles. First, the use of cloud-computing platforms, such as Amazon EC2, to deploy the whole A2CI platform (the IaaS principle) can eliminate the hurdle of managing Web servers and at the same time achieve high stability of the operational system. Second, the highly modularized data service and software service-oriented design (following the DaaS and SaaS principles) makes the software system and its data and analysis resources highly extendable, sharable, and reusable. Third, as Web-based analysis is becoming more common, sharing the A2CI framework and making it easily accessible for physically distributed users (PaaS principle) will significantly reduce scientists' time in collecting and preprocessing data, which always takes over 80% of the scientific analysis. This will greatly speed up the knowledge discovery process. To ensure that the A2CI platform stays relevant, we will continue to enhance the system and make it more intelligent in terms of understanding users' search and analysis behavior to suggest the most suitable data and tools to conduct online analysis (Li et al., 2014). We will also closely collaborate with atmospheric scientists to integrate more simulation models and analysis tools to benefit a broader user community.

ACKNOWLEDGMENT

This work is supported in part by the National Science Foundation under Grants PLR-1349259, BCS-1455349 and PLR-1504432. Any opinions, findings, and conclusions or recommendations expressed in this material are those of the author and do not necessarily reflect the views of the National Science Foundation.

REFERENCES

Droegemeier, K.K., 2009. "Transforming the sensing and numerical prediction of high-impact local weather through dynamic adaptation." Philosophical Transactions of the Royal Society of London A: Mathematical, Physical and Engineering Sciences 367, no. 1890: 885–904.
Eaton, B., Gregory, J., Drach, B., Taylor, K., Hankin, S., Caron, J., Juckes, M., 2003. NetCDF Climate and Forecast (CF) Metadata Conventions. http://cfconventions.org/ (last accessed on 11.09.15.).
Fortner, B., 1998. HDF: The Hierarchical Data Format. http://www.hdfgroup.org/HDF5/ (last accessed on 11.09.15.).
Fowler, D., Pilegaard, K., Sutton, M.A., Ambus, P., Raivonen, M., Duyzer, J., Zechmeister-Boltenstern, S., 2009. Atmospheric composition change: ecosystems–atmosphere interactions. Atmospheric Environment 43 (33), 5193–5267.
Guilyardi, E., Wittenberg, A., Fedorov, A., Collins, M., Wang, C., Capotondi, A., Stockdale, T., 2009. Understanding El Niño in ocean-atmosphere general circulation models: Progress and challenges. Bulletin of the American Meteorological Society 90 (3), 325–340.

Hecher, M., Traxler, C., Hesina, G. and Fuhrmann, A., 2015. Web-based Visualization Platform for Geospatial Data. In Proceedings of the 6th International Conference on Information Visualization Theory and Applications (VISIGRAPP 2015), pages 311–316. http://dx.doi.org.10.5220/0005359503110316, ISBN 978-989-758-088-8.

Katul, G.G., Oren, R., Manzoni, S., Higgins, C., Parlange, M.B., 2012. Evapotranspiration: a process driving mass transport and energy exchange in the soil–plant–atmosphere–climate system. Reviews of Geophysics 50 (3).

Karl, T.R., Trenberth, K.E., 2003. Modern global climate change. Science 302 (5651), 1719–1723.

Li, W., Yang, C., Nebert, D., Raskin, R., Houser, P., Wu, H., Li, Z., 2011. Semantic-based web service discovery and chaining for building an Arctic spatial data infrastructure. Computers & Geosciences 37 (11), 1752–1762.

Li, W., Goodchild, L., Anselin., Weber., 2013. A service-oriented smart cybergis framework for data-intensive geospatial problems. CyberGIS Fostering a New Wave of Geospatial Discovery and Innovation 107–123.

Li, W., Bhatia, V., Cao, K., 2014. Intelligent polar cyberinfrastructure: enabling semantic search in geospatial metadata catalogue to support polar data discovery. Earth Science Informatics 8 (1), 111–123.

Li, W., Song, M., Zhou, B., Cao, K., Gao, S., 2015. Performance improvement techniques for geospatial web services in a cyberinfrastructure environment–A case study with a disaster management portal. Computers, Environment and Urban Systems. http://dx.doi.org/10.1016/j.compenvurbsys.2015.04.003.

Mattmann, C.A., Waliser, D., Kim, J., Goodale, C., Hart, A., Ramirez, P., Hewitson, B., 2014. Cloud computing and virtualization within the regional climate model and evaluation system. Earth Science Informatics 7 (1), 1–12.

Plale, B., Gannon, D., Brotzge, J., Droegemeier, K., Kurose, J., McLaughlin, D., Chandrasekar, V., 2006. Casa and lead: adaptive cyberinfrastructure for real-time multiscale weather forecasting. IEEE Computer 39 (11), 56–64.

Ramanathan, V.C.P.J., Crutzen, P.J., Kiehl, J.T., Rosenfeld, D., 2001. Aerosols, climate, and the hydrological cycle. Science 294 (5549), 2119–2124.

Rew, R., Davis, G., 1990. NetCDF: an interface for scientific data access. Computer Graphics and Applications, IEEE 10 (4), 76–82.

Schneider, T., 2006. The general circulation of the atmosphere. Annual Review of Earth and Planetary Sciences 34, 655–688.

Trenberth, K.E., Dai, A., van der Schrier, G., Jones, P.D., Barichivich, J., Briffa, K.R., Sheffield, J., 2014. Global warming and changes in drought. Nature Climate Change 4 (1), 17–22.

Walther, G.R., Post, E., Convey, P., Menzel, A., Parmesan, C., Beebee, T.J., Bairlein, F., 2002. Ecological responses to recent climate change. Nature 416 (6879), 389–395.

Wang, L., Von Laszewski, G., Younge, A., He, X., Kunze, M., Tao, J., Fu, C., 2010. Cloud computing: a perspective study. New Generation Computing 28 (2), 137–146.

Yang, C., Li, W., Xie, J., Zhou, B., 2008. Distributed geospatial information processing: sharing distributed geospatial resources to support Digital Earth. International Journal of Digital Earth 1 (3), 259–278.

Yang, C., Nebert, D., Taylor, D.F., 2011. Establishing a sustainable and cross-boundary geospatial cyberinfrastructure to enable polar research. Computers & Geosciences 37 (11), 1721–1726.

Yang, C., Raskin, R., Goodchild, M., Gahegan, M., 2010. Geospatial cyberinfrastructure: past, present and future. Computers, Environment and Urban Systems 34 (4), 264–277.

CHAPTER 10

Polar CI Portal: A Cloud-Based Polar Resource Discovery Engine

Y. Jiang, C. Yang, J. Xia, K. Liu
George Mason University, Fairfax, VA, USA

BACKGROUND AND CHALLENGES

The Polar Regions have received increasing attention as places for (1) natural resources, (2) sensitive indicators of human activities and global, environmental, and climatic changes, (3) preserving histories of the Earth and biological evolution, (4) space–Earth interactions, and (5) seeking answers to many other 21st century challenges (Yang et al., 2011). Recent international collaborations for helping explore and preserve the Polar Regions have been conducted in both scientific research and policymaking. For example, the Third International Polar Year (IPY) 2007–2009 helped raise the public awareness of the importance of Polar Regions, plan more observational networks, and advocate more engagements that are international. To effectively address the challenges and efficiently use available resources, a new and novel framework for long-term cooperation between stakeholders with a mandate and interest in the Polar Regions, entitled the "International Polar Initiative" (IPI) (World Meteorological Organization, 2012), has been proposed by the World Meteorological Organization.

Increased deployment of sophisticated sensors in both the Arctic and Antarctic regions combined with enhanced computational power enable scientists to observe and describe the present state of the Polar Regions, unveil the past trends, and predict future climate and environmental changes. These efforts often require extremely sophisticated integration of theoretical, experimental, observational, and modeling results as well as virtual networks for sharing data, information, and knowledge. The recent increase in volume and complexity of data and technologies, which support scientific discovery, demand a transformed infrastructure (National Science Foundation (NSF), 2011).

Cloud Computing in Ocean and Atmospheric Sciences
ISBN 978-0-12-803192-6
http://dx.doi.org/10.1016/B978-0-12-803192-6.00010-4
Copyright © 2016 Elsevier Inc.
All rights reserved.
163

For the purpose of better studying Polar Regions, scientists need to find the right data in a fast and accurate fashion. To facilitate the research, exploration, and development for better understanding, utilizing, and protecting the Polar Regions, a Geoscience Cyberinfrastructure (GCI) is needed to help collect data, integrate information gathered or data in real time from in situ and satellite sensors, and model the geophysical, biological, ecological, and social phenomena better decision support information.

Challenges

Polar geospatial resource discovery is a critical step for conducting polar science research. With the increasing number of geospatial resources available online, many Spatial Data Infrastructure (SDI) components (e.g., catalogs and portals) have been developed to help manage and discover geospatial resources. However, efficient and accurate geospatial resource discovery is still a big challenge in several respects:

1. Cross-domain barriers. Existing SDIs are usually established to enable organizations and users in specific geoscience domains (e.g., ocean, atmospheric, ecosystem, and demographics) to discover resources with defined types (e.g., dataset, Web services, applications, and documents). To find appropriate resources, users typically must visit/search multiple portals serving different communities (Gui et al., 2013).

2. Specialized technological knowledge is required. Different SDIs adopt diverse protocols/Application Programming Interfaces (APIs) (e.g., Open Geospatial Consortium (OGC) Catalog Service for the Web (CSW), Z39.50, OpenSearch, Really Simple Syndication, Atom, Web Accessible Folders, Open Archives Initiative Protocol for Metadata Harvesting (OAIPMH)) and work flows to access resources. Given the various metadata formats, specifications (e.g., International Organization for Standardization (ISO) 19139, National Aeronautics and Space Administration (NASA) Global Change Master Directory, Directory Interchange Format, and Service Entry Resource Format, Organization for the Advancement of Structured Information Standards' Electronic business Registry Information Model), Web User Interfaces (UIs), and the representations of metadata also vary significantly. These issues dramatically increase the difficulty of querying and integrating heterogeneous metadata from dispersed sources.

3. Implementing distributed search and federation at the SDI component level is costly. Many registries adopt a periodic harvesting mechanism to retrieve metadata from other federated catalogs. These time-consuming processes lead to network and storage burdens, data redundancy, and the overhead of maintaining data consistency.

4. Keyword-based search technologies are widely adopted in operational SDIs. However, because keyword search uses string matching rather than semantic matching, precision and recall, the two important measurements for search relevance, are hard to guarantee for the following reasons (Tran et al., 2007; Lei et al., 2006). The terminologies and vocabularies representing the same concept may vary with domains and languages. The search content cannot be expressed using several keywords explicitly (such as the subconcepts of a concept). The search keywords may appear in the metadata description but do not accurately reflect the content of the metadata (Bergamaschi et al., 2010). By contrast, semantic search offers a solution to address these issues. However, integrating semantics into existing SDIs is challenging due to the limited expandability of metadata templates, frameworks, and other issues.

5. Current representation and visualization methods for search results lack information and functionalities for assisting users' selection of polar resources. Plain text-based content presentation is not an effective way to help users distinguish records to make a final selection (Schwarz and Morris, 2011). In addition, polar data and services for the Arctic and Antarctic regions are distorted in widely used world projections (Gui et al., 2013).

6. Although various technologies have been adopted to improve the performance of resource discovery engines (Foster et al., 2008), enabling high-speed resource access for polar scientists remains challenging because of the computing complexity and concurrent intensity that comes with massive data query requests by end users (resource consumers) (Xia et al., 2014; Buyya et al., 2009).

The first issue increases entry barriers/learning curves (Craglia et al., 2011; Mazzetti and Nativi, 2011) for both end users and providers. The second issue causes complexity and high maintenance cost for SDIs. The last three issues are critical for effective and efficient polar resources discovery.

Objectives

To facilitate research, exploration, and development for better understanding, utilizing, and protecting the Polar Regions, the NSF launched the Polar CI program initiative (Tedesco, 2013). Funded by the program, we prototyped a cloud-based polar resource discovery engine—Polar CI Portal—that addresses these issues. The Polar CI Portal is a one-stop portal that makes it easy for users to discover and access polar-related geospatial resources across different online environments. The Portal is characterized as follows: (1) a lightweight Web engine framework is developed to search geospatial resources; (2) a data warehouse is built to harvest, store, search, and distribute geospatial information; (3) harvesting middleware is established to harvest data from different online environments and convert them into formats that the data warehouse supports; (4) a semantic-based query statement refinement is proposed to improve recall level and precision; (5) sophisticated functionalities (e.g., service quality monitoring and polar viewer) are used as means of improving user experience and assisting decision making; (6) a Quality of Service (QoS) engine is used to provide users with service quality information; and (7) virtualized resources within Amazon Elastic Cloud Compute (EC2) are leveraged to support a high-performance search experience.

This chapter introduces this portal in detail: section System Architecture introduces the architecture of the polar resource discovery engine. Section Implementation and Methodology describes the components and methodologies. Section Status concludes with status of each key component. Lastly, section Conclusions and Discussion discusses future research.

SYSTEM ARCHITECTURE

The search engine of our Polar CI Portal system searches against the dispersed heterogeneous SDIs (catalogs and portals) and consists of six loosely coupled components: Search broker, Data warehouse, Data harvesting middleware, QoS engine, Semantic engine, and Visualization tool (Fig. 10.1). The search broker is the key component that drives the entire system. It connects directly to other components and processes, responds to users' search requests, and plots search results. The primary function of the data warehouse is to

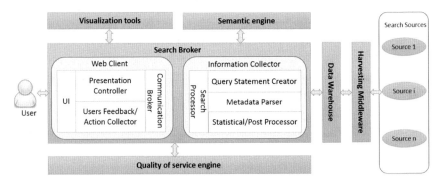

Figure 10.1 Architecture of the Polar CI Portal.

store data securely, supporting best practices and allowing for retrieval upon a request from search broker. As a component connecting the data warehouse and different sources, the data-harvesting middleware is developed to collect data using different techniques and convert data formats from different sources into formats supported by the data warehouse. The QoS engine is employed to monitor service performance, collect subjective evaluations from users, and calculate comprehensive (overall) scores. The semantic engine translates search input into query rules and calculates relevance indices between the search subject and the retrieved records. The visualization tool is used to visualize search results, display quality and statistical information, and help the user to make a selection that is more informed. The entire system is deployed onto the Amazon EC2 cloud-computing platform.

IMPLEMENTATION AND METHODOLOGY

Implementation

The search broker is developed by using Microsoft.Net Framework 4. The Web client adopts Silverlight to provide cross-browser uniform graphical user interface (GUI) and rich interactive capacity. Silverlight is a powerful development tool of Microsoft for creating engaging and interactive user experiences for Web and mobile applications. The choice of Microsoft-centric solutions rather than open-source techniques helped us best leverage the Microsoft Azure cloud-computing platform and using the Microsoft Web application framework-relevant techniques, such as Silverlight, enabled us to quickly develop user-friendly GUIs. Interactive maps used in our system are developed by using Bing Maps and OpenLayers APIs. The information collector is based on ASP.NET, which is part of the .Net Framework and allows programmers to build dynamic Web sites, Web applications, and services. The communications between the search broker and other components (Semantic engine, QoS engine, and visualization viewer), and also search-broker internal components, are implemented by following standard Simple Object Access Protocol (SOAP)/ Representational State Transfer (REST)ful Web-service interfaces. Asynchronous communication between client and server are achieved by leveraging the "Silverlight-enabled Windows Communication Foundation" (WCF) services (Klein, 2007).

The data warehouse adopts GeoNetwork, an open source catalog application, to manage spatially referenced resources. GeoNetwork has been used in a number of Spatial Data Infrastructure initiatives (Liu et al., 2011). By incorporating Apache Lucene and PostGIS, it also provides high-performance search capability through spatial and nonspatial indexing. Data harvesting middleware harvests data from other sources through CSW, REST search interface, and OpenSearch techniques. Once all these components are configured, the entire system is deployed to Amazon EC2. In the end, all these technologies work together to create a user-friendly experience and high-performance searching capabilities.

Search Broker

The search broker is made up of two tiers, a Web client and an information collector at the server side (Fig. 10.1). The Web client provides a Web-based GUI and all interactive functions. (1) A UI involves a number of Web components or pages, including basic/advanced search pages, multiple record views (Paged list viewer, Bing Maps Viewer, Pivot Viewer), a quality viewer, a metadata detail viewer, and others. (2) A Presentation Controller initializes Web components or invokes external tools to show search results, data, or quality information and controls user interactions through the GUI components that change with user context. (3) A Communication Broker manages the communication between client and server.

The Information Collector retrieves metadata records and additional information and responds to search requests from clients (Fig. 10.1). (1) A Search Processor handles search requests from Web clients and collaborates with other components as a coordinator. (2) A Query Statement Creator constructs the query statement based on query rules. (3) A Metadata parser retrieves metadata abstract/detail from the data warehouse. (4) A Statistical/ Post Processor acquires, calculates, and associates additional information (e.g., overall quality, relevance index) with retrieved records.

Data Warehouse

The data warehouse is the central component designed to harvest, store, search, and distribute polar resource information. It is built upon GeoNetwork, an open-source catalog application for managing spatially referenced resources that has been widely used in many SDI developments. The data warehouse supports numerous SDIs' protocols/APIs (e.g., GeoNetwork, CSW, Z39.50, OGC Windows XML Setup file (WxS), Web Distributed Authoring and Versioning (WebDAV), Thematic Real-Time Environmental Distributed Data Services (THREDDS), Local file system, and OAIPMH) and metadata format standards (e.g., ISO19115/ISO19139, Federal Geographic Data Committee (FGDC), and Dublin Core), and provides scheduled harvesting and synchronization of metadata between distributed catalogs. To speed up the search process, Apache Lucene is used for text indexing, and a spatial index is built using PostGIS. Apache Lucene is a high-performance, full-featured text search engine library written entirely in Java. It is a technology suitable for nearly any application that requires full-text search, especially cross-platform. The warehouse uses PostGIS, a spatial database extender for the PostgreSQL object-relational database. PostGIS adds support for geographic objects allowing location queries to be run in Structured Query Language (SQL). Additionally, GeoNetwork includes online metadata editing tools that allows users to publish geospatial resources manually by themselves.

Data Harvesting Middleware

Although the data warehouse provides a built-in data harvesting function, it also supports other search techniques such as the REST search interface which is adopted by many SDIs. Another issue is that the data warehouse requires data sources to strictly follow the GeoNetwork built-in standards of service request and metadata format. However, due to the unique needs of different domains, some SDIs have made slight changes to the unified standards and metadata formats. These changes make it difficult for GeoNetwork to harvest data from these SDIs. Therefore, a middleware is needed to perform customized data harvesting for each of the SDIs that cannot work well with GeoNetwork. To deal with heterogeneities (e.g., protocol, format, and description) among various search sources, a Web application was developed to customize search requests for each SDI, and to convert their responses into formats that are recognizable to GeoNetwork. These data can then be imported into the data warehouse and made available to users.

CSW, OpenSearch, and REST search interface are three primary techniques used by our data-harvesting middleware (Fig. 10.2), based on the search interfaces provided by our target sources, such as National Oceanographic Data Center (NODC), Earth Observing System (EOS), EOS Clearing House (ECHO), and National Snow and Ice Data Center (NSIDC). Catalog Service for the Web (CSW) (http://www.opengeospatial.org/standards/cat), sometimes referred to as Catalog Service–Web, is a standard for exposing a catalog of geospatial records in Extensible Markup Language (XML) on the Internet (over Hypertext Transfer Protocol (HTTP)). OpenSearch (http://www.opensearch.org/Home) is a collection of technologies that allow publishing of search results in a format suitable for syndication and aggregation. It is a way for Web sites and search engines to publish search results in a standard and accessible format. REST (http://www.restapitutorial.com/lessons/whatisrest.html) is a software architecture

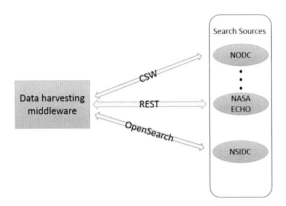

Figure 10.2 Search techniques for different sources of data-harvesting middleware. *CSW*, Catalog Service for the Web; *REST*, Representational State Transfer; *NODC*, National Oceanographic Data Center; *NASA*, National Aeronautics and Space Administration; *ECHO*, Earth Observing System Clearing House; *NSIDC*, National Snow and Ice Data Center.

style consisting of guidelines and best practices for creating scalable Web services. It is a coordinated set of constraints applied to the design of components in a distributed hypermedia system that can lead to better performance and easier maintenance of the architecture. The utilization of REST helps us leverage the characteristics of uniform interface, stateless, cachable, client–server layered system, and code on demand.

Semantic Engine

The purpose of the semantic engine is to improve the recall accuracy by means of expanding the search query to match the entire cluster (e.g., different expressions, terms, and subconcepts of the same concepts) rather than one or several terms. In most operational catalogs and portals, searching is based on keyword matching between the search input and the content of metadata fields (Nasraoui and Zhuhadar, 2010). However, it is inevitable that both user inputs and metadata descriptions include ambiguity, inaccuracy, and domain/culture limitations (Lutz and Klien, 2006). In addition to that, semantic similarity evaluation provides a way to sort and filter search results by relevance (Nasraoui and Zhuhadar, 2010). These issues drive us to employ semantic techniques to improve search accuracy and better understand users' search intents. To implement semantic-aided search with minimum reconstruction costs by utilizing existing GCI components, a semantic engine has been developed by extending the Earth Science Information Partners (ESIP) Semantic Testbed (http://wiki.esipfed.org/index.php/

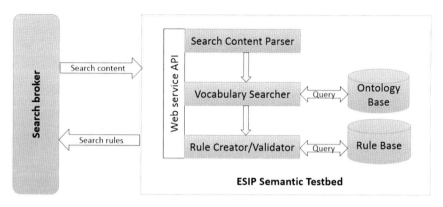

Figure 10.3 Interoperation between search broker and ESIP Semantic Testbed.

Semantic_Web_Testbed_Implementation) based on Semantic Web for Earth and Environmental Terminology (SWEET) (https://sweet.jpl.nasa. gov/) ontologies. This system translates users' search input into a set of search rules that consist of a series of constructs and related vocabularies. Fig. 10.3 illustrates the interpretation process.

The following example shows how the semantic engine works. When a user inputs "USGS ecosystem WMS" as a search content, the search broker sends it to the semantic engine in which the search content is compared sequentially with the terminologies about the providers and resource types in the established ontology base and is then decomposed into several segments. If a given segment is neither provider nor resource type, it is considered as search subjects. In this case, "USGS," "ecosystem," and "WMS" are recognized as constraints of provider, search subject, and resource type, respectively. After this, the subject "ecosystem"-related terminologies are queried. Based on the interpretation result, the semantic support system will wrap the following rules in an XML payload and send it back to the search broker:

- Resource provider: USGS | US Geological Survey | {…subsectors…
- Resource Type: OGC WMS (Web Map Service)
- Subject: ecosystem | {biome, ecosys, and terminologies (subdomain/multilingual)…}

Using the previous query rules, the query statement creator will construct statements for different data sources. For Global Earth Observation System of Systems (GEOSS) Clearinghouse, the following OGC filter fragment will be created to create CSW GetRecords request (Table 10.1).

Because we still use keywords to search against each source, irrelevant record retrievals cannot be avoided. To help users identify them, we

Table 10.1 OGC filter fragment sample for CSW GetRecords request

```
<ogc:Or>
    <!--provider constraints-->
    <ogc:PropertyIsLike escapeChar="\" singleChar="?" wildCard="*">
        <ogc:PropertyName>OrganizationName</ogc:PropertyName>
        <ogc:Literal>USGS</ogc:Literal>
    </ogc:PropertyIsLike>
    <ogc:PropertyIsLike>. . .</ogc:PropertyIsLike>
</ogc:Or>
<ogc:Or>
    <!--resourceType constraints-->
    <ogc:PropertyIsEqualTo matchAction="One">
        <ogc:PropertyName>ServiceType</ogc:PropertyName>
        <ogc:Literal>OGC:WMS</ogc:Literal>
    </ogc:PropertyIsEqualTo>
</ogc:Or>
<ogc:Or>
    <!--keywords or full text constraints-->
    <ogc:PropertyIsLike escapeChar="\" singleChar="?" wildCard="*">
        <ogc:PropertyName>Subject</ogc:PropertyName>
        <ogc:Literal>ecosystem</ogc:Literal>
    </ogc:PropertyIsLike>
    <ogc:PropertyIsLike>. . .</ogc:PropertyIsLike>
</ogc:Or>
```

introduced a relevance index from semantic similarity evaluation to measure the proximities between the search results and the user's input. We
adopted an edge-based similarity evaluation approach (Jiang and Conrath,
1997; Lutz, 2007): (1) predefine feature set $F = \{f_i\}$ (metadata fields which
participate in the calculation) and their contribution (weight set $W = w_i$) to
the relevance; (2) calculate the similarity between each feature f_i and corresponding query rule r_i in rule set R by measuring the distance in the
semantic ontology (Eq. [10.1]); (3) summarize the similarity of every feature
with its weight w_i into a single value Re (relevance index, Eq. [10.2]).

$$\text{sim}(f_i, r_i) = \frac{e}{\text{dis}(f_i, r_i) + e}$$ [10.1]

$$Re = \sum w_i * \text{sim}(f_i, r_i)$$ [10.2]

in which e is an empirical constant correction value, and $dis = (f_i, r_i)$ is the
measured distance. The similarity value gets lower if the distance between f_i
and r_i gets longer. When $dis = (f_i, r_i)$ is equal to zero or f_i-related rule does
not exist, the similarity is 1.

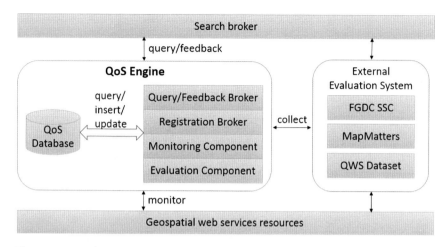

Figure 10.4 Architecture and workflow of QoS engine. *QoS*, quality of service; *FGDC*, Federal Geographic Data Committee; *SSC*, Service Status Checker; *QWS*, Quality of Web Service.

QoS Engine

To provide quality information for geospatial Web services, we developed a resource monitoring and evaluation system—the QoS engine (Xia et al., 2015) (Fig. 10.4). The QoS engine evaluates service quality by measuring its response time. This information can be used as a selectable sorting criterion to help end users distinguish services, which provide the same data/functionalities, and help service providers find a potential performance bottleneck (Wu et al., 2011). This system contains four components: (1) A Query/Feedback broker responds to query/feedback requests. (2) A Registration broker inserts new services to be monitored. (3) A Monitoring component monitors daily the performance of registered services. (4) An Overall evaluation component evaluates the overall quality of services. The major characteristics of this system are as follows:

- Quality information from multiple channels (active monitoring, end-users' feedback and other monitoring systems such as Service Status Checker provided by the FGDC);
- Quality query capabilities in different granularities (overall quality grade from long-time statistics and service/operations/dataset levels performance over a certain period).

Visualization Tool

To enhance visualization of search results, a series of interactive viewers and functions are employed in our system, including Polar Viewer, Paged List Viewer, Pivot Viewer, and Service Quality Viewer.

- Polar Viewer is utilized to reduce geometric distortion and display polar resources in a polar projection.
- Paged List Viewer provides a traditional, paged, list-based view to cluster records with categorized resource types (e.g., dataset, services, documents, Web sites, systems, and models) and multiple sorting rules (e.g., relevance, quality, provider, and search source).
- Pivot Viewer filters and sorts records through various rules such as quality and displays records with deep-zoomable column diagrams.
- Service Quality Viewer shows the response time of the *getCapabilities* operation, FGDC Scores and each layer's performance for a specified WMS in a certain period with diagrams. The quality information reveals the nonfunctional properties and is helpful to users for distinguishing the services with the same function and data.

Cloud-Computing Environments

To address challenges of computing and concurrent intensity and enable high-performance resource access for users, the entire system is deployed to the Amazon EC2 (Yang and Huang, 2014). The advantages of cloud-computing environments lie in three aspects. First, it automatically distributes end-user requests to multiple Polar CI Portal instances to reduce the response time, and handles a larger number of concurrent requests by utilizing multiple instances to achieve greater fault tolerance (Wang and Ng, 2010). Second, it enables the automatic provision of computing, storage, and network resources in two directions: scaling Polar CI Portal instance configurations (e.g., central processing unit (CPU) core number and memory size) on the fly to meet the computing demand and scaling to increase or decrease the number of Polar CI Portal instances for the cloud–workload balance (Walker, 2008; Vaquero et al., 2008). Last, the "pay-as-you-go" service model releases users from upfront cost for purchasing, hosting, configuring, and managing computing infrastructure, a perennial dream of distributed computing (Armbrust et al., 2010).

STATUS

Data Harvesting

Currently, the data-harvesting middleware is capable of harvesting data from five catalogs with customized CSW request (Table 10.2), including the GEOSS Clearinghouse, NODC, National Geophysical Data Center (NGDC), National Climatic Data Center (NCDC), and Data.gov; one

Table 10.2 Number of resources harvested with different techniques

Technique	Source	Number of resources (datasets and services)
CSW	GEOSS CLH, NODC, NGDC, NCDC, Data.gov	5606
OpenSearch	NSIDC	1891
REST API	NASA ECHO	433,189

CSW, Catalog Service for the Web; *GEOSS*, Global Earth Observation System of Systems; *CLH*, Clearing House; *NODC*, National Oceanographic Data Center; *NGDC*, National Geophysical Data Center; *NCDC*, National Climatic Data Center; *NSIDC*, National Snow and Ice Data Center; *REST API*, Representational State Transfer Application Programming Interface; *NASA*, National Aeronautics and Space Administration; *ECHO*, Earth Observing System Clearing House.

catalog (ECHO) with REST search; and one catalog (NSIDC) with Open-Search. A total of 440,686 metadata records are stored in our data warehouse, including datasets, series, static maps, services, and collection sessions. A total of 5606 are harvested with CSW; 1891 are collected with Open-Search; and the remainder are collected with the REST search API. As long as a data center exposes its resources through a search standard such as CSW, OpenSearch, or REST API, it can be used as one of our sources. A few more of them have been found, and we plan to add them into our source list in the near future.

Semantic Engine

An experiment was performed to test the performance of the semantic engine. A simple text-based search input (Fig. 10.5) "water wms" has been used to demonstrate search accuracy improvement with semantic support. Because "ice," "ocean," "lake," "river" and other water-related terminologies will be included as search keywords, a total of 42 results were found by a search against Polar CI Portal and all of them were Web map services. In contrast, without semantic assistance, the search returns 985 results, only 22 of which are the desired Web map services. There are 944 datasets and 19 other records (e.g., Web Coverage Service and Web Feature Service) that are incorrectly included. Therefore, by collaborating with the semantic supporting system, users can get more of what they really need and eliminate those irrelevant results.

Quality of Service Engine

A Service Quality Viewer (Fig. 10.6) has been developed to show the response time of the *getCapabilities* operation, FGDC Scores, and each layer's

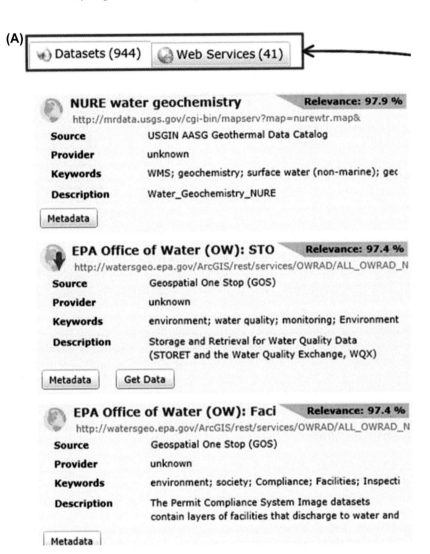

(A)

Datasets (944)   Web Services (41)

NURE water geochemistry   Relevance: 97.9 %
http://mrdata.usgs.gov/cgi-bin/mapserv?map=nurewtr.map&

Source	USGIN AASG Geothermal Data Catalog
Provider	unknown
Keywords	WMS; geochemistry; surface water (non-marine); gec
Description	Water_Geochemistry_NURE

Metadata

EPA Office of Water (OW): STO   Relevance: 97.4 %
http://watersgeo.epa.gov/ArcGIS/rest/services/OWRAD/ALL_OWRAD_N

Source	Geospatial One Stop (GOS)
Provider	unknown
Keywords	environment; water quality; monitoring; Environment
Description	Storage and Retrieval for Water Quality Data (STORET and the Water Quality Exchange, WQX)

Metadata   Get Data

EPA Office of Water (OW): Faci   Relevance: 97.4 %
http://watersgeo.epa.gov/ArcGIS/rest/services/OWRAD/ALL_OWRAD_N

Source	Geospatial One Stop (GOS)
Provider	unknown
Keywords	environment; society; Compliance; Facilities; Inspecti
Description	The Permit Compliance System Image datasets contain layers of facilities that discharge to water and

Metadata

Figure 10.5 Keyword and semantic-aided search comparison. (A) Results without semantic assitance. (B) Results with semantic assitance.

(B)

Figure 10.5 cont'd

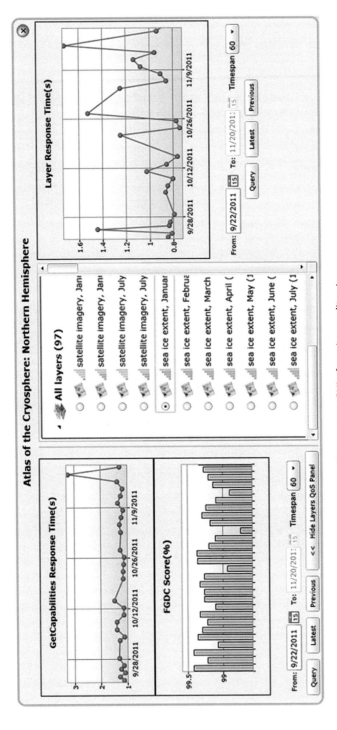

Figure 10.6 GUI of service quality viewer.

performance for a specified WMS in a certain period. All the information is plotted in intuitive diagrams to show changes within a time period. The service quality information reveals the nonfunctional properties for distinguishing the services with the same function and data.

Visualization Tools

Interactivity and intuitive information visualization methods are critical factors of any framework that needs widespread and voluntary adoption. Following this rule, we integrated four interactive views, including Polar Viewer, Bing Maps Viewer, Pivot Viewer, and Service Quality Viewer, as mentioned in section QoS Engine, into a single client–user interface (Fig. 10.7). Users can easily browse and interact with search results in any of the previous viewers or their combinations. Additionally, the relevance index, which measures the similarities between the user's requirement and the retrieved records, can be utilized as sorting and filtering rules for the resulting set. With the integration of these additional information tools, the user's experience and decision-making capacity would be improved.

CONCLUSIONS AND DISCUSSION

This chapter introduces the architecture and methodologies of the Polar CI Portal, a cloud-based polar resource discovery engine, a cyberinfrastructure building block for conglomerate patches of existing polar-related resources.

- To hide cross-domain barriers and heterogeneities from users and providers, a loosely coupled plug-in search framework is proposed based on brokering middleware mode and configurable search work flow. By adopting the framework, communication and maintenance costs between federated catalogs can also be reduced.
- To improve recall level and precision, a semantic engine is developed by extending the ESIP Semantic Testbed with a new methodology. However, further research is needed to improve the search accuracy and recall level.
- To assist service selection decision-making, a quality-of-service engine is developed to provide value for users in determining known limitations for services offered through our portal. Subsequently, users can make increasingly informed decisions regarding the quality and confidence in their Web services of choice.

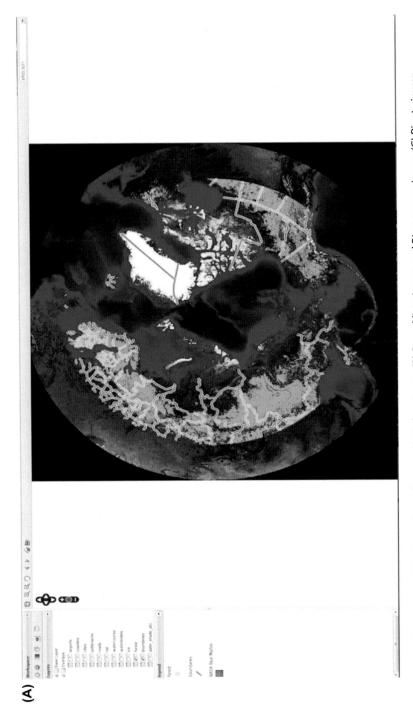

Figure 10.7 GUI of search result page. (A) Polar viewer. (B) Paged list viewer and Bing maps viewer. (C) Pivot viewer.

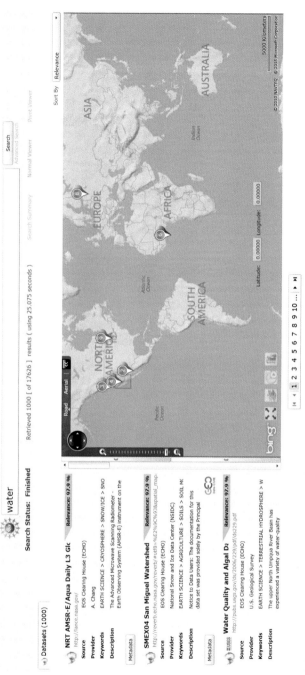

Figure 10.7 cont'd

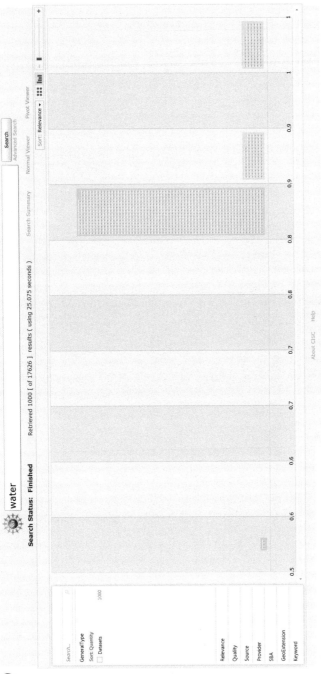

Figure 10.7 cont'd

- To better visualize search results, a series of viewers, external visualization tools, and functions are integrated through a sophisticated GUI. All those would enrich the user's experience and help users make selection more intuitively.
- To overcome the challenges of computing complexity and concurrent intensity that comes with massive data query requests from end users, the system is deployed to Amazon EC2 by leveraging elastic resource pooling and dynamic workload balancing.

This Polar CI Portal is expected to advance discovery, innovation, and education across disciplines in the Arctic and Antarctic. For additional improvements in geospatial resource discovery, access, and usage, future research would include:

- Integrating more useful information and functionalities to assist in resource selection. We intend to add a data quality module into our system. Although currently Polar CI Portal provides users with access to metadata, inspecting metadata content can be a difficult and time-consuming task, especially when information is not presented in a way that makes it easy to read and understand (Zabala et al., 2013; Goodchild, 2007). The data quality module aims to provide users with data quality information in an understandable and comparable way by parsing metadata and collecting user feedback. Additionally, we plan to extend the statistics and analytical functions on resource popularity and user feedbacks to help users identify resources through other users' selections and evaluation by collecting access frequency of metadata, records/data, data downloads, data visualization, and quality information (Agichtein et al., 2006). Meanwhile, we will improve the semantic search capabilities, such as multilingual and advanced word segmentation and matching methods (Lutz, 2007).
- Enhancing value-added search capabilities. Associating data with services can help users discover tools to work with the retrieved data and vice versa. This information can be interworkable both horizontally (data with data, result with result) and vertically (data with tool, with workflow, with result, or with experiment) (Duerr et al., 2011). We will provide the glue function to integrate the components of data, processing, applications, and infrastructure (Yang et al., 2010). Furthermore, we plan to develop a user registration and subscription service to help users find potential resources from profile and search context and history and also establish a professional networking system to help researchers and scientists find potential collaborators.

REFERENCES

Agichtein, E., Brill, E., Dumais, S., 2006. Improving web search ranking by incorporating user behavior information. In: Proceedings of the 29th Annual International ACM SIGIR Conference on Research and Development in Information Retrieval. ACM, pp. 19–26.

Armbrust, M., Fox, A., Griffith, R., Joseph, A.D., Katz, R., Konwinski, A., Lee, G., Patterson, D., Rabkin, A., Stoica, I., 2010. A view of cloud computing. Communications of the ACM 53, 50–58.

Bergamaschi, S., Domnori, E., Guerra, F., Orsini, M., Lado, R.T., Velegrakis, Y., 2010. Keymantic: semantic keyword-based searching in data integration systems. Proceedings of the VLDB Endowment 3, 1637–1640.

Buyya, R., Yeo, C.S., Venugopal, S., Broberg, J., Brandic, I., 2009. Cloud computing and emerging IT platforms: vision, hype, and reality for delivering computing as the 5th utility. Future Generation Computer Systems 25, 599–616.

Craglia, M., Nativi, S., Santoro, M., Vaccari, L., Fugazza, C., 2011. Inter-disciplinary interoperability for global sustainability research. In: GeoSpatial Semantics. Springer, pp. 1–15.

Duerr, R., et al., 2011. Discovery white paper short [Online], ESIP Federation Wiki. Available from: http://wiki.esipfed.org/index.php/Discovery_White_Paper_Short [Accessed 28 December 2015].

Foster, I., Zhao, Y., Raicu, I., Lu, S., 2008. Cloud computing and grid computing 360-degree compared. In: Grid Computing Environments Workshop, 2008. GCE'08. IEEE, pp. 1–10.

Goodchild, M.F., 2007. Beyond metadata: towards user-centric description of data quality. In: Proceedings, Spatial Data Quality 2007 International Symposium on Spatial Data Quality, June, 13–15. CiteSeer.

Gui, Z., Yang, C., Xia, J., Liu, K., Xu, C., Li, J., Lostritto, P., 2013. A performance, semantic and service quality-enhanced distributed search engine for improving geospatial resource discovery. International Journal of Geographical Information Science 27, 1109–1132.

Jiang, J.J., Conrath, D.W., 1997. Semantic Similarity Based on Corpus Statistics and Lexical Taxonomy. arXiv preprint cmp-lg/9709008.

Klein, S., 2007. Professional WCF Programming: .NET Development with the Windows Communication Foundation. John Wiley & Sons.

Lei, Y., Uren, V., Motta, E., 2006. Semsearch: a search engine for the semantic web. In: Managing Knowledge in a World of Networks. Springer, pp. 238–245.

Liu, K., Yang, C., Li, W., Li, Z., Wu, H., Rezgui, A., Xia, J., 2011. The GEOSS clearinghouse high performance search engine. In: Geoinformatics, 2011 19th International Conference on. IEEE, pp. 1–4.

Lutz, M., 2007. Ontology-based descriptions for semantic discovery and composition of geoprocessing services. GeoInformatica 11, 1–36.

Lutz, M., Klien, E., 2006. Ontology–based retrieval of geographic information. International Journal of Geographical Information Science 20, 233–260.

Mazzetti, P., Nativi, S., 2011. Multi-disciplinary interoperability for Earth observation sensor web. In: Geoscience and Remote Sensing Symposium (IGARSS), 2011 IEEE International. IEEE International.

NSF, 2011. Earth Cube Guidance for the Community. Available from: http://www.nsf.gov/geo/earthcube/ (accessed 13.03.15.).

Nasraoui, O., Zhuhadar, L., 2010. Improving recall and precision of a personalized semantic search engine for e-learning. In: Digital Society, 2010. ICDS'10. Fourth International Conference on. IEEE, pp. 216–221.

Schwarz, J., Morris, M., 2011. Augmenting web pages and search results to support credibility assessment. In: Proceedings of the SIGCHI Conference on Human Factors in Computing Systems. ACM, pp. 1245–1254.

Tedesco, M., 2013. NSF's Polar Cyberinfrastructure Program Initiative. Available from: http://www.arcus.org/witness-the-arctic/2013/1/article/19613 (accessed 14.03.15.).

Tran, T., Cimiano, P., Rudolph, S., Studer, R., 2007. Ontology-Based Interpretation of Keywords for Semantic Search. Springer.

Vaquero, L.M., Rodero-Merino, L., Caceres, J., Lindner, M., 2008. A break in the clouds: towards a cloud definition. ACM SIGCOMM Computer Communication Review 39, 50–55.

Walker, E., 2008. Benchmarking Amazon EC2 for high-performance scientific computing. USENIX Login 33, 18–23.

Wang, G., Ng, T.E., 2010. The impact of virtualization on network performance of amazon EC2 data center. In: INFOCOM, 2010 Proceedings IEEE. IEEE, pp. 1–9.

World Meteorological Organization, 2012. International Polar Initiative. Available from: http://internationalpolarinitiative.org/IPIhomepage.html (accessed 05.05.15.).

Wu, H., Li, Z., Zhang, H., Yang, C., Shen, S., 2011. Monitoring and evaluating the quality of Web Map Service resources for optimizing map composition over the internet to support decision making. Computers & Geosciences 37, 485–494.

Xia, J., Yang, C., Liu, K., Gui, Z., Li, Z., Huang, Q., Li, R., 2014. Adopting cloud computing to optimize spatial web portals for better performance to support Digital Earth and other global geospatial initiatives. International Journal of Digital Earth 1–25.

Xia, J., Yang, C., Liu, K., Li, Z., Sun, M., Yu, M., 2015. Forming a global monitoring mechanism and a spatiotemporal performance model for geospatial services. International Journal of Geographical Information Science 1–22.

Yang, C., Huang, Q., 2014. Spatial Cloud Computing: A Practical Approach. CRC Press. 348 p.

Yang, C., Nebert, D., Taylor, D.F., 2011. Establishing a sustainable and cross-boundary geospatial cyberinfrastructure to enable polar research. Computers & Geosciences 37, 1721–1726.

Yang, C., Raskin, R., Goodchild, M., Gahegan, M., 2010. Geospatial cyberinfrastructure: past, present and future. Computers, Environment and Urban Systems 34, 264–277.

Zabala, A., Riverola, A., Serral, I., Díaz, P., Lush, V., Maso, J., Pons, X., Habermann, T., 2013. Rubric-Q: adding quality-related elements to the GEOSS clearinghouse datasets. Selected Topics in Applied Earth Observations and Remote Sensing, IEEE Journal of 6, 1676–1687.

CHAPTER 11

Climate Analytics as a Service

J.L. Schnase
NASA Goddard Space Flight Center, Greenbelt, MD, USA

INTRODUCTION

Cloud technologies provide an unprecedented opportunity to expand the power and influence of computing in daily life. So far, those opportunities have unfolded in largely ad hoc ways, resulting in a creative chaos that at times can be confusing. Over the years, we have become comfortable with a classic von Neumann perspective on what constitutes a computer. We share mental models and patterns of thinking about how computers are built and how they behave—what in broad terms computing technologies can do, how they do it, and what they cannot do. But when it comes to cloud computing, those patterns have not been established—save one: the concept of *service*. We have developed a shared notion that cloud technologies in an essential way provide the basis for services. By definition, cloud capabilities reside there, not here—the action of helping is conveyed to the user: the user is served. Hence, terms such as Software-as-a-Service and Platform-as-a-Service have become common parlance in the world of cloud computing.

In our efforts to deal with the big data problems of climate science, we are trying to take a deeper dive into our understanding of cloud-computing services. To begin, we ask the fundamental question: What is it that needs to be served? Our answer is *analytics*. But analytics served in a particular way. For now at least, we believe it would be productive to focus on the basics—do simple things well and very fast. We need to garner the agile high-performance computing and storage resources of cloud computing to address climate science's big data problems in a new way—in a way that melds knowledge creation with data creation, curation, discovery, and workflow orchestration. This chapter is an effort to advance the cause.

Our story begins with the observation that big data challenges are generally approached in one of two ways. Sometimes they are viewed as a problem of large-scale data management, in which solutions are offered through an array of traditional storage and database theories and

Cloud Computing in Ocean and Atmospheric Sciences
ISBN 978-0-12-803192-6
http://dx.doi.org/10.1016/B978-0-12-803192-6.00011-6
Copyright © 2016 Elsevier Inc.
All rights reserved.

technologies. These approaches tend to view big data as an issue of storing and managing large amounts of structured data for the purpose of finding subsets of interest. In contrast, big data challenges sometimes are viewed as a knowledge-management problem, in which solutions are offered through an array of analytic techniques and technologies. These approaches tend to view big data as an issue of extracting meaningful patterns from large amounts of unstructured data for the purpose of finding insights of interest. It is the latter that we believe will dominate in the coming years: as climate datasets grow in size, the focus will increasingly fall on climate analytics— but not at the expense or exclusion of long-term digital preservation and not at the expense of delivering the simple answers that form the basis of much that we seek to understand about the climate.

From this grows our notion of a harmonized view of the problem—a converged approach to analytics and data management, in which high-performance compute–storage implemented as an analytic system is part of a dynamic archive comprising both static and computationally realized objects—a system the capabilities of which are framed as behaviors over a static data collection, but in which queries cause results to be created, rather than found and retrieved. Those results can be the product of a complex analysis, but, importantly, they also can be tailored responses to the simplest of requests.

We believe there are basic concepts that can guide the way we build these systems. This chapter lays out an approach that delivers climate analytics as a service (CAaaS) within this harmonized view—what we refer to as *CAaaS*. It is an architecture designed to exploit the scalable resources enabled by cloud computing and Web services. It also lays out a pattern for thinking more generally about cloud-based, domain-specific software for large-scale scientific data analytics. We begin in section An Architectural Framework for Climate Analytics as a Service by describing a basic architectural framework for climate analytics. In section Climate Analytics as a Service Reduced to Practice: The MERRA Analytic Service and the MERRA Persistence Service, we describe a system built along these lines to support a climate reanalysis dataset. Section The Climate Data Services Application Programming Interface describes a domain-specific application programming interface that makes it easier to use the capabilities of a climate data analytics system. In section Implications and Vision for the Future, we describe what we believe are the important implications of this work and where it might lead. Finally, in section Conclusions, we provide some closing thoughts.

AN ARCHITECTURAL FRAMEWORK FOR CLIMATE ANALYTICS AS A SERVICE

Our approach to climate analytics represents a convergence of analytics and data management, in which compute–storage implemented as an analytic system is treated as a dynamic archive comprising both static and computed objects. Fig. 11.1 shows the basic architecture for a *climate data analytics system* that implements this integrated view. On the *server side*, the system includes a high-performance data analytics platform and, at a minimum, two required services: an *analytic service* and a *persistence service*. The service also needs a *system interface* to expose its capabilities to the outside world.

High-Performance Data Analytics Platform

These systems acquire their parallelism from computing clusters—large collections of conventional processors that form "compute nodes" connected by network cables and fast switches. Associated with the cluster is a high-performance, distributed file system that can be integrated in various ways to support the activities of the system and provide the option of alternative

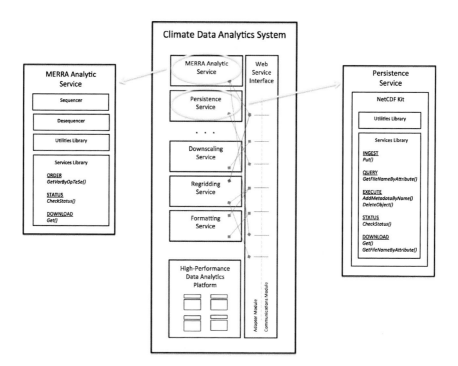

Figure 11.1 Architecture of the MERRA data analytics system.

storage configurations. These file systems feature much larger disc block sizes and replicate data to protect against media failures that can occur when data are distributed over many low-cost compute nodes (Leskovic et al., 2014). The overall architecture can vary, but it provides the high-performance compute–storage "fabric" upon which analytic services run.

Analytic Service

At the heart of the analytic service lies a collection of programs that implement its core functionality. These programs typically are designed to perform parallel operations that exploit the high-performance computing capabilities of the data analytics platform. In general, climate data comprise climate variables the values for which in the aggregate characterize the state of Earth's atmosphere at a given time and place—variables like temperature, precipitation, wind direction and speed, relative humidity, atmospheric pressure, etc. Operations over climate data, as a result, typically require at a minimum inputs that specify the name of the *climate variable* of interest, a *spatial extent* specifying the area of interest, and a *temporal extent* specifying a time span of interest. Space and time are *invariant attributes* of this class of data— every measurement, observation, or computed value of every variable related to the climate must be associated with the attributes of space and time to have meaning. This becomes important in the way we optimize analytic operations on climate data.

The relationship between data and workflows also contributes to the way we think about data analytics. Data-intensive analysis workflows, in general, bridge between a largely unstructured mass of archived scientific data and the highly structured, tailored, reduced, and refined analytic products that are used by individual scientists and form the basis of intellectual work in the domain. In general, the initial steps of an analysis, those operations that first interact with a data repository, tend to be the most general, whereas data manipulations closer to the client tend to be the most tailored—specialized to the individual, to the domain, or to the science question under study. The amount of data being operated on also tends to be larger on the repository side of the workflow, smaller toward the client-side products.

This is important, because this stratification can be used to optimize data-intensive workflows. The first job of an analytics system is to implement a set of near-archive, early-stage operations that are a common starting point in many of these analysis workflows. For example, it is important that a system be able to compute maximum, minimum, sum, count, average, variance, and difference operations of the general form:

$$\text{result} <= \text{average} \left(\text{variable}, (t_0, t_1), ((x_0, y_0, z_0), (x_1, y_1, z_1))\right)$$

that return, as in this example, the average value of a variable when given its name, a temporal extent, and a spatial extent. Because of their widespread use, these simple operations—*microservices*, if you will—function as "canonical operations" with which more complex expressions can be built. By virtue of implementing these basic descriptive statistics for a dataset in a high-performance compute–storage environment using sophisticated analytical software, the system also is able to support more complex analyses, such as the predictive modeling, machine learning, and neural networking approaches often associated with advanced analytics (Schnase et al., 2014a).

Persistence Service

A climate data analytics system also includes a persistence service, the basis of which is a software component we refer to as a *climate data server*. A climate data server stores and manages data objects created by the analytic service, along with virtualization and provisioning software that allow multiple climate data server instances to be deployed as virtual climate data servers.

The climate data server performs the traditional functions of collections-building, managing, querying, and accessing data, as well as applying and enforcing the policy-based controls and metadata management required for long-term digital preservation. It functions as a full-featured archive management system, implementing the classic "CRUD" operations of an archive: create, read, update, and delete data objects and the metadata associated with those objects. The capacity to create multiple instances of a climate data server and federate them into collections of servers conveys to the system a high level of tailorability and accommodates the particular needs of users or applications. This includes the construction of personalized data collections and even the delivery of full-featured, stand-alone personalized data management systems as virtual cloud images (Schnase et al., 2011).

System Interface

Finally, a climate data analytics system must include an interface that exposes the capabilities of the system to users and applications. One way of thinking about this interface is that it includes two components: an *adapter module* that triggers internal workflows based on external requests received from the outside via a *communications module* that translates communications-specific syntax into system-specific syntax. We distinguish two modules in the abstract system interface to highlight key places that support integration and extensibility. System implementers can modify the adapter module to accommodate new services, and the communications module can be adapted to accommodate different Web service protocols.

We also believe that there is a big advantage to basing the communications module's service request protocol on the data-flow categories of a long-term preservation digital archive reference model. By using a communications protocol based on an archive reference model, the climate data analytics system as a whole takes on the appearance of a *dynamic archival information system* capable of performing full information life-cycle management in an analytics context. Existing archive systems should therefore find it easier to integrate the system, because the interfaces and interactions with the system will be familiar to archive authorities and existing archive systems, and the behaviors implemented by the climate data analytics system can be organized around traditional archive operations workflows.

CLIMATE ANALYTICS AS A SERVICE REDUCED TO PRACTICE: THE MERRA ANALYTIC SERVICE AND THE MERRA PERSISTENCE SERVICE

Having considered things from an abstract perspective, we now turn to a real example. The Modern-Era Retrospective Analysis for Research and Applications (MERRA) Analytic Service (MERRA/AS) along with the MERRA Persistence Service (MERRA/PS) and a client-side, domain-specific Climate Data Services Application Programming Interface (CDS API) provide an end-to-end demonstration of CAaaS capabilities.

MERRA/AS enables MapReduce analytics over NASA's MERRA data. The MERRA reanalysis is produced by NASA's Goddard Earth Observing System Data Assimilation System Version 5 (GEOS-5). It integrates observational data with numerical models to produce a temporally and spatially consistent global synthesis of key climate variables (MERRA, 2015; Reineker et al., 2011). Spatial resolution is $1/2°$ latitude $× 2/3°$ longitude $× 42$ vertical levels extending through the stratosphere. Temporal resolution is 6-h for three-dimensional, full spatial resolution, extending from 1979 to the present, nearly the entire satellite era. MERRA data are typically made available to the public through NASA's Earth Observing System Distributed

Information System (EOS DIS) (EOSDIS, 2015). A subset of the data is made available to the climate research community through the Earth System Grid Federation (ESGF), the research community's data publication infrastructure (ESGF, 2015).

We are focusing on the MERRA collection because there is an increasing demand for reanalysis data products by a growing community of consumers, including local governments, federal agencies, and private-sector customers. Reanalysis data are used in models and decision support systems relating to disasters, ecological forecasting, health, air quality, water resources, agriculture, climate, energy, oceans, and weather (Edwards, 2010). Fig. 11.1 illustrates the overall organization of an example system providing support for MERRA data analytic services. The system includes a high-performance data analytics platform, the MERRA Analytic Service, the MERRA Persistence Service, and a system interface.

High-Performance Data Analytics Platform

Our MERRA/AS platform supports MapReduce analytics; however, alternative configurations are possible so long as they enable high-performance, parallel computing over a high-performance, parallel file system. The configuration we have used in our development work comprises a 36-node Dell cluster of 576 Intel 2.6 gigahertz (GHz) Sandy Bridge cores, 2304 gigabytes (GB) of random access memory, 1296 terabytes (TB) of raw storage, and has a 11.7 teraflops (TF) theoretical peak compute capacity. The compute nodes communicate through a Fourteen Data Rate Infiniband network having peak Transmission Control Protocol/Internet Protocol (TCP/IP) speeds in excess of 20 gigabits (Gb) per second, and an open-source Cloudera enterprise-ready distribution of the Apache Hadoop file system (HDFS) and MapReduce engine (Cloudera, 2015; HDFS, 2015).

MERRA Analytic Service (MERRA/AS)

MapReduce is of particular interest to us, because it provides an approach to analytics that is proving useful to many data-intensive problems (Dean and Ghemawat, 2008; Reed and Dongarra, 2015). It is an analysis paradigm that combines distributed storage and retrieval with distributed, parallel computation, allocating to the data repository analytical operations that yield reduced outputs to applications and interfaces that may reside elsewhere. Because MapReduce implements repositories as storage clusters, dataset size and system scalability are limited only by the number of nodes in the clusters. Hadoop and MapReduce are the core of an open-source ecosystem that is allowing many research groups to apply analytic techniques to large-scale scientific data without deep knowledge of parallel computing (Reed and Dongarra, 2015).

MapReduce distributes computations across large datasets using many computers, usually referred to as nodes. In a "map" operation, a head node takes the input, partitions it into smaller subproblems, and distributes them to data nodes. A data node may do this again in turn, leading to a multilevel tree structure. The data node processes the smaller problem and passes the answer back to a reducer node to perform the reduction operation. In a "reduce" step, the reducer node then collects the answers to all the subproblems and combines them to form the output—the answer to the problem it was originally trying to solve.

To execute MapReduce operations on the MERRA data, the data first need to be ingested into the Hadoop file system of the high-performance data analytics platform. MERRA GEOS-5 output files are in Hierarchical Data Format–Earth Observing System (HDF-EOS) format, which is an extension of the Hierarchical Data Format Version 4 (HDF-4) (HDF-5, 2015). The system converts these source files to Network Common Data Format (NetCDF) files, because many applications use the NetCDF file format. NetCDF is a self-describing format that contains both data and metadata (NetCDF, 2015). The system can support new versions of the MERRA dataset as they are produced, and a refined version, MERRA-2, will soon be released and incorporated into the system (Reineker et al., 2011).

In an initial load step, the MERRA/AS *sequencer* transforms MERRA NetCDF source files into the flat, serialized, block-compressed sequence files generally required by MapReduce programs and loads the sequence files into the Hadoop file system. The system creates a single custom sequence file for each NetCDF file, wherein the source file's data are logically stored as <key, value> pairs within the resulting sequence file. As a result, each sequence file has a one-to-one mapping to the original NetCDF file. One benefit of this approach is that NetCDF metadata are preserved within the sequence file (Schnase et al., 2014a).

The sequencer partitions native MERRA data files by time such that each record in the sequence file contains a composite key comprising a *timestamp* and *climate variable name* that is associated with a value that is the value of the named climate variable. Note that, in our case, the value of the climate variable is itself a complex data structure containing all the two- or three-dimensional data associated with that variable at that point in time over its entire spatial extent. In other words, in our implementation, the sequencer initially organizes the entire MERRA corpus in the HDFS for convenient indexed subsetting by time and climate variable name; spatial subsetting requires further computing over these indexed subsets.

The results computed by MERRA/AS's MapReduce operations are in the same format as those produced by the sequencer. The system therefore also includes a complementary program, the *desequencer*, that transforms the sequence files back into NetCDF files and moves them out of the HDFS for consumption by a calling program.

MERRA/AS's services library contains the third major component in the MapReduce ecosystem: the MapReduce codeset proper, which consists of a collection of methods that implement the core capabilities of the service (Fig. 11.1). The services library organizes these methods in a manner that contribute to CAaaS's integrated analytics/archive perspective. In our implementation, the functional capabilities of the service correspond to the *Open Archival Information System* (OAIS) Reference Model data-flow categories of an operational archive (OAIS, 2012).

An OAIS is an archive comprising people and systems that have accepted the responsibility to preserve information and make it available for a designated community. The term OAIS also refers to the International Organization for Standardization (ISO) OAIS Reference Model for an OAIS. This reference model is defined by recommendation Consultative Committee for Space Data Systems (CCSDS) 650.0-M-2 of the CCSDS, the text of which is identical to ISO 14721:2012. The CCSDS's purview is space agencies, but the OAIS model has proved useful to a wide variety of other organizations and institutions with digital archiving needs. OAIS provides a framework for the understanding and increased awareness of archival concepts needed for long-term digital information preservation and access, and it provides the concepts needed by nonarchival organizations to be effective participants in the preservation process (OAIS, 2012).

The OAIS-based categories used to organize a service's methods include ingest, query, order, and download (Fig. 11.2). *Ingest* methods input objects into the system. *Query* methods retrieve metadata relating to data objects in the service. *Order* methods dynamically create data objects. *Download* methods retrieve objects from the service. We assert, along with OAIS, that the behaviors of any and all services can be categorized this way—which gives us a powerful, unifying framework for system design. To these four we have added execute and status categories to accommodate the dynamic nature of

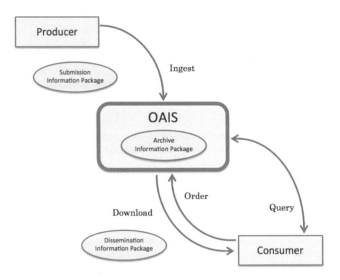

Figure 11.2 Data flows and interactions in an OAIS archive.

the archive. *Execute* methods initiate service-definable operations that can extend the system's functionality, and *status* methods check on the progress of running operations.

MERRA/AS's services library contains three OAIS categories of capability: order, status, and download. A single *GetVariableBy_Operation_ TemporalExtent_SpatialExtent* order method (abbreviated *GetVarByOpTeSe* from here on) implements the seven commonly used canonical operations described previously: maximum, minimum, sum, count, average, variance, and difference operations. Inputs to the method include the name of a climate variable in the MERRA collection, the type of operation to be performed, the specification of a temporal extent and a two- or three-dimensional spatial extent (depending on the variable), and any additional parameters needed by the operation. The method's output includes a unique session identifier for the order session that can be used to retrieve session status information about progress of the operation and download results when the operation is finished.

As described previously, MERRA/AS's *GetVarByOpTeSe* method implements its operations using MapReduce, wherein mapper and reducer programs direct the parallel subsetting and execution of the operations over MERRA data stored in the analytic platform's Hadoop file system. MERRA/AS's status and download capabilities are implemented by the service library's *CheckStatus* and *Get* methods. These methods can be used to check on the progress of a running order operation and download the computed results when the operation finishes. The functioning of MERRA/AS's methods is supported by a collection of additional programs composing a utilities library. We have implemented these libraries as Java classes, but a variety of alternative approaches could be used.

MERRA Persistence Service (MERRA/PS)

The MERRA Persistence Service is an important complementary service of the climate data analytics system. A *climate data server* implements the core functionality of the persistence service. The data server is a software appliance specialized to the needs of a managed collection of climate-related scientific data. The climate data server is designed to take advantage of the flexible resource-allocation capabilities afforded by cloud computing.

The climate data server in our persistence service uses the *Integrated Rule-Oriented Data System* (iRODS) data grid software running in an SUSE Linux Enterprise Server (SLES) 11 SP3 operating system environment (iRODS, 2015). Provisioning software encapsulates the operating system and iRODS in a virtual machine image. Various application-specific kits are then used to specialize the data server's functionality for particular uses. Our data server has a NetCDF and Geostationary Earth Orbit Tagged Image File Format (GeoTIFF) file management kit and can include other kits supporting different formats and capabilities.

As shown in Fig. 11.1, the *NetCDF kit* contains the methods that implement the core functionality of the service. Like MERRA/AS, MERRA/PS's methods contribute to the integrated analytics/archive perspective and correspond to the OAIS Reference Model data-flow categories. The kit has five OAIS categories of capability: ingest, query, execute, status, and download, which collectively implement the classic "CRUD" operations of an archive. In our implementation, the service includes a *Put* order method that stores a user-specified payload and a *GetFileNameByAttribute* query method that performs a metadata search operation on stored data objects. We also have a *Get* method to download a previously ingested data object. An *AddMetaDataByName* execute method can add metadata to a stored data object, and a *DeleteObject* method removes a stored data object from the service. Finally, a *CheckStatus* status method can check on the progress of a service request.

The data objects managed by MERRA/PS represent OAIS's Submission Information Package (SIP), Archive Information Package (AIP), and Dissemination Information Package (DIP) abstractions. In further compliance with the OAIS Reference Model, MERRA/PS also manages metadata in accordance with the OAIS Reference Model's metadata taxonomy, which recognizes four categories: Representation Information, Preservation Description Information, Policy Information, and Discovered Metadata. Specifically, it includes utilities that extract the embedded metadata in self-describing NetCDF files and stores that information as a set of iRODS database tables managed by method libraries. This externalized metadata supports classic static object discovery, and enables clients to perform searches over the NetCDF data objects stored in the persistence service without opening files to access embedded metadata (Schnase et al., 2011).

System Interface

Finally, MERRA/AS and MERRA/PS have a system interface that exposes the capabilities of the services to external users and applications. An "adapter" module maps an incoming service request to the appropriate method, and a representational state transfer (REST) "communications" module implements a Web server that external applications can access over a network. We implement the RESTful interface as a PHP: Hypertext Preprocessor (PHP) program.

A detailed description of the organization and operation of the system interface is beyond the scope of this chapter. However, it is important to note the value of this two-tier approach. The adapter module contains a table that maps the names of services, their various and idiosyncratically named methods, and the particulars of their method invocations to the standard OAIS categories of ingest, query, order, etc. Introducing the map table abstraction at this level enables convenient service extensibility: new services can be added and existing services can be modified at any time without having to change the communications protocol.

The RESTful communications module translates HTTP-specific syntax into system-specific service requests using the adapter module's map table. As described in the following, our REST module implements a communications protocol that also is based on the OAIS Reference Model, which provides a consistent and standard way of mapping OAIS data flow behaviors from the system's external point of entry to specific internal behaviors in a suite of arbitrarily named services and methods. By marshaling their unique protocols into our OAIS-based protocol, it also gives us a way of conveniently extending the communications tier to include other Web services.

MERRA/AS and MERRA/PS in Use

In an example use of the system, a user may want to know the historic average summer temperatures in North America over the time span of 1990 to 1999. The user or the user's client application would submit a RESTful order request to MERRA/AS via the system interface indicating the operation to perform and parameters that further specify the request. The service interface maps the incoming service request to the appropriate order method, in this case the *GetVarByOpTeSe* method, which launches the operation as a MapReduce computation on the data analytics platform and returns a session identifier through the interface to the calling application. In this example, the service request would specify the variable of interest (temperature), the spatial and temporal extent of interest (North America and the summer months from 1990 to 1999), the operation to be performed (average), and the service that will perform the operation (MERRA/AS).

Once the order request is launched, the calling application can issue status service requests wherein the *CheckStatus* method uses the session ID to monitor the progress of the order. When the order request is finished, the computed data object is desequenced and prepared for retrieval by the calling application as a NetCDF file. In a final step, the calling application could then submit a download service request via the system interface.

At this point, the application could issue a service request to the persistence service to store and manage the newly created object. In an example storage operation, a user or application would submit an ingest request to the persistence service along with a pointer to the data object to be stored. The service interface would map the incoming service request to the appropriate method, in this case, the NetCDF kit's *Put* method, which would store the object in the iRODS storage system and return a session ID. Subsequent query, status, and download requests would operate similarly using the methods provided in MERRA/PS's service library.

THE CLIMATE DATA SERVICES APPLICATION PROGRAMMING INTERFACE

Our consideration of CAaaS could stop here: all the basic server-side behaviors of such a system are accounted for in the preceding example. But to maximize the benefit of these capabilities, we think it is important that the system also include a *client-side application programming interface*. To that end, we are developing the Climate Data Services Application Programming

Interface (CDS API), a full-featured, domain-specific Python library to facilitate application development.

In building the CDS API, we hope to provide for climate science a uniform semantic treatment of the combined functionalities of large-scale data management and server-side analytics. As we have described, in our view, the best way to bring coherence to this archive/analytic dichotomy is to organize the basic elements of a CAaaS system around the widely accepted OAIS reference model, which asserts that a scientific archive comprises four data interactions—ingest, query, order, and download. Because the archive is dynamic, we have added status and execute to accommodate process control and run-time extensibility. These high-level OAIS abstractions provide a vocabulary that we have adopted throughout our service functions, Web service protocol, and the CDS API itself.

To see how this approach influences the design of the CDS API, it is helpful first to understand that in the Web services world there are two types of interfaces. As shown in Fig. 11.3, on the service side, a system interface maps the methods, functions, and programs that implement the service's capabilities to Hypertext Transfer Protocol (HTTP) messages that expose the service's capabilities to the outside world. Client applications can consume these REST endpoints directly to access services. The World Wide Web Consortium (W3C) views Web services as a way to insure predictable *machine-to-machine interoperability* (W3C, 2015). The messaging format can vary from community to community, often reflecting the specialized functions or audiences they serve. Additionally, significant standards activities have grown up around the design and implementation of such Web services.

There also are the classic client-side APIs familiar to application developers. Generally, these comprise local libraries that reside on the developer's host computer and can be statically or dynamically referenced by client applications. They speed development, reduce error, and often implement abstractions that are specialized to the needs of the audiences they serve. They can be used to build applications, workflows, and domain-specific toolkits, workbenches, and integrated development environments (IDEs).

The CDS API is a slightly different beast. It is a Python library that resides on the client machine, but consumes Web service endpoints and abstracts them into higher-order methods that software developers generally find easier to use than the raw endpoints. It functions primarily as a client-side orchestrator of service-side capabilities and is itself a client of the system's REST interface.

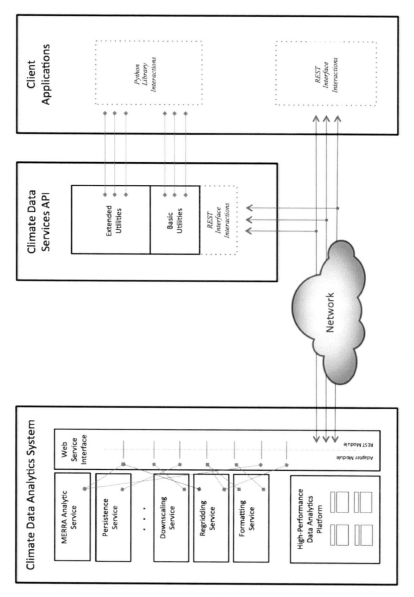

Figure 11.3 Web service architecture of the CDS API.

Our outfacing Web service protocol is based on a RESTful implementation of OAIS's data-flow interaction categories. Although the details of the services are beyond the scope of this chapter, the basic idea is that the service's client-side system interface communicatively links through Universal Resource Locators (URLs) of the general form:

```
http://<base_URL>/(ingest|query|order|download|execute|status).
php?service=<service_name>&request=<operation>&parameters=<param
eters>.
```

The Python-based, client-side CDS API encapsulates this low-level inbound and outbound traffic, abstracting it into higher-order methods that we organize into basic and extended utilities. The API's *basic utilities* provide a one-to-one mapping of OAIS-classified operations on the client side to corresponding OAIS-classified operations on the server side, and its *extended utilities* build on basic methods, placing them under programmatic control to create more specialized convenience methods and workflows. As shown in the following, these can be folded back into the API's libraries allowing it to evolve naturally in response to community needs.

It is important to note that Python applications can bind to the CDS API's library as well as server-side endpoints (Fig. 11.3). However, upward compatibility of client applications is easier to maintain by abstracting the idiosyncrasies of various Web service requests away from individual applications into a single, sharable client-side API, a traditional advantage of distributable programming libraries.

IMPLICATIONS AND VISION FOR THE FUTURE

In the preceding sections, we laid out many low-level details about climate analytics as a service. But is there really anything new here? Is there anything about this particular assemblage of technologies or way of thinking that fundamentally improves the way we interact with climate data? We think there is, and the larger story emerging from this work begins with a simple observation about how people use data about the climate.

The Intergovernmental Panel on Climate Change (IPCC) is the leading international body for the assessment of climate change (IPCC, 2015). It was established by the United Nations Environment Program (UNEP) and the World Meteorological Organization (WMO) in 1988 to provide the world with a clear scientific view on the current state of scientific knowledge about climate change and its potential environmental and socioeconomic impacts. The IPCC is open to all member countries of the UN and WMO.

Currently, 195 countries are members of the IPCC. The collective wisdom of scientists contributing to the IPCC is conveyed to the world through periodic assessment reports, the latest of which is the Fifth Assessment Report (AR5), which was released in 2014. The report contains many findings distributed across five printed volumes and is the basis for environmental policy-making throughout the world.

If we set aside for a moment the specific details of these findings, we see that there are actually very few *classes* of findings. IPCC findings comprise statements about the past, present, and future *values* of climate variables; statements about *climatology*—the maximum, minimum, and average values of those variables over given periods of time; statements about *trends*—how those variables change over time; and statements about *anomalies*—how climate variables at a given time or place might depart from corresponding values at a different time or place. AR5's five volumes contain around 4 million words, about 50,000 of them unique. Of those unique words, "maximum," "minimum," "average," "variance," "difference," "climatology," "anomaly," and "trend" collectively account for approximately 12,000 direct references to the data upon which IPCC's findings are based; over half of these are the word "trend" alone.

This small collection of data attributes provides the basis for an astonishing amount of intellectual work in the discipline. They comprise far and away the preponderance of climate data analyses as performed today. When viewed through the lens of contemporary climate analytics, we believe that the notion of data-proximal operations capable of quickly computing basic descriptive statistics—the classic measures of central tendency, variability, maximum, minimum, etc.—on terabyte, petabyte, and larger data sets—affords important benefits.

Analytic Microservices and Client-Side Workflow Orchestration

Recall that in our CAaaS architecture, a data service is "under contract" to implement a small set of canonical operations—maximum, minimum, sum, count, average, variance, and difference—that can be invoked through RESTful OAIS-based service calls—ingest, query, order, download, execute, and status. At a minimum, the canonical operations require as input the name of a variable of interest and spatiotemporal bounds. These operations run asynchronously: once initiated, they run to completion without interruption, independent of other operations that might be running on the service. They essentially function as atomic microservices, doing one specialized task as fast as possible on the high-performance compute–storage platform of the analytic service. The OAIS-based *service requests* that invoke these operations comprise the CDS API's basic methods, which essentially provide a one-to-one mapping of OAIS-classified operations on the client side to corresponding OAIS-classified operations on the server side.

The CDS API's extended methods build on these basic methods, placing them under programmatic control to create more elaborate convenience methods and workflows. These extended methods *orchestrate asynchronous server-side operations into synchronous client-side operations*. Take the following pseudocode as an example of a CDS API extended utility that implements a client-side averaging operation called MY_AVERAGE:

```
MY_AVERAGE(MAS, Variable_Name, (t_0, t_1), ((x_0, y_0, z_0), (x_1, y_1, z_1)),
Result)
    ORDER(MAS, GetVarByOpTeSe(Variable_Name, Average, (t_0, t_1),
    ((x_0, y_0, z_0), (x_1, y_1, z_1)), ID)
    Repeat
      STATUS(MAS, ID)
      until finished
    DOWNLOAD(MAS, ID, Result)
```

The ORDER, STATUS, and DOWNLOAD service requests—which the API maps to actual REST endpoints—launch single, asynchronous operations through calls to the MERRA/AS (MAS) service. The ORDER request uses MERRA/AS's *GetVarByOpTeSe* method to launch an averaging MapReduce job on the service and returns the unique ID of that process. The MY_AVERAGE extended utility shown in the example controls the flow of asynchronous calls such that the client-side MY_AVERAGE method returns a completed answer as a single call. MY_AVERAGE is, in effect, a scripted workflow.

This general approach can be applied to the client-side orchestration of more complex and meaningful workflows. For example, the approach can be used to move the work of data assembly and preparation from the scientist's desktop to the remote data service. Consider for a moment the data-gathering work associated with climate impact research, a classic application of climate data. In a study by Wei et al. (2013), MERRA reanalysis data were used to study the effects of irrigation on precipitation in four agricultural settings over a period of 23 years. The study areas included the Nile valley, northern China, the central valley of California, and northern India and Pakistan. The MERRA variables of interest included the average values of humidity, wind speed, and temperature. The period of time of interest was 1979–2002 at a high temporal resolution of every 6 h. Eighteen vertical levels of the atmosphere were used in the researchers' models.

Taken together, isolating the data of interest for the four study regions required the assembly of over 7 million "layers" of MERRA data (4 study sites × 3 variables × 23 years × 365 days/year × four 6-h intervals/day × 18 levels of the atmosphere). Working in the traditional mode, it took days to assemble the necessary source files from NASA archives, clip the files to the study areas, and prepare them for use in the researchers' models (Dr. Wei, personal communication). This is because the archived source files were supersets of the data that were actually needed—they contained greater spatiotemporal coverage and more variables than required by the models. The work of tailored subsetting, averaging, and file format transformations were the responsibility of the scientist using the data. Given that the size of potential regions of interest remain the same for a particular study, the time requirement for data gathering and preparation only increases as the overall size of our collections grow.

By using a CDS API client-side workflow for data assembly, the data subsetting and averaging required by the Wei experiments can be performed in less than 2 min by MERRA/AS (average runtime on various system loads is about 90 s), resulting in a tailored, site-specific data bundle of approximately 500 MB, which can be moved to the client workstation over the Internet in a few minutes. This dramatic reduction in data assembly time is possible because MERRA/AS's analytic software is able to carry out its averaging operations quickly and *in parallel* on *selected variables*, and it is able to do so at the *native spatial and temporal resolution of the data*. Moreover, two types of parallelism are possible: MapReduce inherently decomposes each subsetting average operation into a parallel computation, and the system's atomic, asynchronous microservice architecture allows multiple subsetting averages to run concurrently. The result is a dynamically computed, highly tailored subset of the 200TB MERRA collection crafted by an external client workflow that has orchestrated high-performance microservices to carry out the work at the data's source rather than the scientist's desktop.

Before leaving this example, it is useful to point out that the data package computed in the Wei example is in fact a collection of climatologies, which over the time span of the experiment, captured the two-decade trends in average humidity, wind speed, and temperature. Note also that the Wei data assembly workflow is a simple Python script that can be folded back into the CDS API as an extended utility, a process that over time could allow the research community to contribute to the construction of the API.

Ensemble Analytics

We have shown the advantages of doing simple operations quickly over a large data collection using high-performance computing and analytic software and delivering reduced results as a service. What would it take to generalize this capability across different collections and different systems? And what of interest could be achieved by doing so?

In the abstract, this type of interoperability is straightforward: participating services would have to honor three contracts. First, each participating data service would have to implement the *canonical operations* at the data's native spatial and temporal resolution. We have demonstrated the value of this in the MERRA/AS example. Although we used MapReduce, any technology capable of high-performance subsetting at native spatiotemporal resolutions could be used.

The second contract is trickier, but not impossible. Because each system operates in a separate namespace, there would have to be a policy for *global namespace management*. This can be coordinated in one or more places in a Web services architecture. An agreed-upon naming convention could be adopted for each conformant data service. Regardless of how the services are accessed, each node would have known capabilities referred to in known ways. This is analogous to all UNIX operating systems having an "ls" command—if one has the means of accessing a system, they can expect "ls" to be implemented on that system in a familiar way.

Another approach would be to adopt a standard naming convention in the Web services communication protocol, perhaps by specializing an existing standard wherein the syntax and semantics of required and optional fields of requests and responses are tailored to the needs of the federation. With this approach, regardless of how services are implemented or named, their means of access is commonly understood within the federation.

Finally, a client-side API, along the lines of the CDS API, that consumes the Web service endpoints regardless of how they are implemented and presents them to client applications as a "standard" library of easy-to-use function calls tailored to the needs of the federation could be used. Here, regardless of implementation and communication details, namespace reconciliation would occur on the client side encapsulated in the shared API, and programmers could access node capabilities using a familiar programming library. Notice that integrity control becomes less distributed with each succeeding alternative.

The required third contract addresses an often contentious issue. There would have to be *conformal attributes* of the data to insure the integrity of the canonical operations themselves. Key among them are attributes such as file format, resolution, gridding, projection, etc. These could be statically insured at the time the collections are built by adopting a community standard for the data, such as the Coupled Model Intercomparison Project Phase 5 (CMIP5) Climate and Forecast (CF) Metadata Conventions (CF, 2015; Taylor et al., 2012). A preferred alternative would be to build ancillary services to do the transformations dynamically.

Establishing community standards for canonical operations, global namespaces, and file formats may seem burdensome, but huge advantages are to be gained by the effort. For example, detecting anomalies is one of climate scientists' most common and useful determinations. However, it is often the case that the data of interest come from different sources. Nadeau et al. (2014) have demonstrated the effectiveness of using the CDS API to compute simple anomalies across multiple reanalysis datasets when those collections conform to the basic conventions previously described.

A climatological anomaly refers to the positive or negative departure of a climate variable from a long-term average or reference value for the variable. Anomalies can be determined in various ways. Finding the arithmetic difference between a reference value and an observed or modeled value for the variable in question is perhaps the simplest way to compute an anomaly. Normalized anomalies divide this arithmetic anomaly by the climatological standard deviation to remove dispersive influences and reveal more information about the magnitude of the anomaly. In either case, a set of basic operations are applied to what can be large data-sets to make an anomaly calculation: for a given variable, region, and time span of interest compute sum, count, average, variance, and difference—our canonical operations.

In Fig. 11.4, the image on the left shows the departure of 2010 summer surface temperatures as determined by the European Centre for Medium-Range Weather Forecasts (ECMWF) Interim Reanalysis (ERA-Interim) from a 34-year ensemble average of summer temperatures obtained by combining data from ECMWF, MERRA, and the National Oceanic and Atmospheric Administration (NOAA) National Center for Environmental Prediction (NCEP) Climate Forecast System Reanalysis (CFSR). The image on the right shows how CFSR's summer temperatures of 2010 depart from the same 34-year ensemble average. There are obvious differences between the two that could reveal valuable information about the underlying modeling systems, inputs, or atmospheric phenomena.

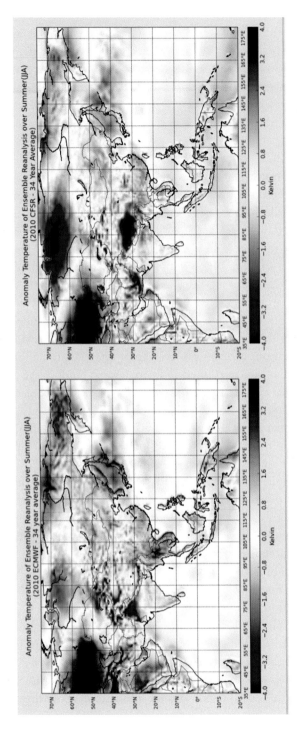

Figure 11.4 Temperature anomalies between 34-year summer average and summer 2010 using the ERA-Interim (left) and CFSR (right) reanalyses.

This experiment was done over a period of about 2 weeks using traditional manual processes for data gathering and preparation. In analogous runs using prototyped ERA-Interim and CFSR analytic services based on MERRA/AS, we have developed Python workflows that use a single CDS API library call to compute the same results in less than 3 min (our average runtime on various system loads is about 150 s) (Nadeau et al., 2014). The principles of interoperability described here, therefore, make possible a concept of *ensemble analytics*, turning an otherwise inert, dissociated collection of climate datasets into a coordinated, coherent, and dynamic platform for exploration.

Many of these concepts are being adopted in the work that we do at NASA and by the extended research community. Over the next 2 years, we will build a full-featured Cloudera Hadoop MapReduce-based *Reanalysis Ensemble Service* (RES). Using MERRA/AS as a model, the RES will deliver, in a uniform analytics environment, customized data products from six of the major reanalysis projects: NASA Modern Era Reanalysis for Research and Applications Version-2 (MERRA-2), ECMWF Interim Reanalysis (ERA-Interim), NOAA NCEP Climate Forecast System Reanalysis (CFSR), NOAA Earth System Research Laboratory (ESRL) 20th Century Reanalysis (20CR), Japanese 25-year Reanalysis (JRA-25), and the Japanese 55-year Reanalysis (JRA-55). In addition, we will develop a set of CDS API utilities to support ensemble analysis, uncertainty quantification, and reanalysis intercomparison.

Related larger-scale projects are underway as well, the most significant of which is an effort to improve interoperability within the Earth System Grid Federation (ESGF). The ESGF is the primary mechanism for publishing and sharing IPCC data as well as a wide range of ancillary observational and reanalysis products (ESGF, 2015). ESGF is an international collaboration that focuses on serving coupled-model intercomparison projects and supporting climate and environmental science in general (CMIP, 2015).

The ESGF Compute Working Team (ESGF-CWT)—the group having primary responsibility for architecture and capability development—is working to increase interoperability within the Federation, improve access to distributed resources, and enable scientists to build and share workflows of common interest. Similar objectives have been achieved, at least in part, by the geospatial community through a series of long-running standards-making activities, one of the most notable being the work of the Open Geospatial Consortium (OGC). OGC is an international industry consortium of over 500 companies, government agencies, and universities participating in a consensus process to develop publicly available interface standards (OGC, 2015).

The ESGF–CWT is adopting OGC's Web Processing Service (WPS) interface standard for its next generation architecture. WPS is essentially an Extensible Markup Language (XML)-based remote procedure call (RPC) protocol for invoking processing capabilities as Web services. It has been used in the geospatial community for delivering low-level geospatial processing capabilities. However, WPS can be generalized to other types of applications and data because of its simplicity: WPS uses a single operation (*Execute*) to invoke remote services; its two other operations (*GetCapabilities* and *DescribeProcess*) are used for discovery and to query services for the information necessary to build the signatures needed by *Execute* operations. ESGF is essentially adapting the WPS standard to its needs through ESGF-specific extensions to the WPS communications protocol. When combined with other elements of the interoperability contract described previously, the stage will be set to integrate compute and storage services across a heterogeneous collection of distributed resources.

These interoperability activities are contributing to the research community's ability to engage in entirely new types of scientific inquiry. One example is ESGF's *Collaborative REAnalysis Technical Environment—Intercomparison Project* (CREATE-IP). CREATE-IP is designed to facilitate the study of the similarities and differences among the major climate reanalysis efforts by providing access to output variables at all available temporal resolutions (monthly, 6-h, 3-h, and 1-h). Access will be provided through multiple distribution, visualization, analytic, and knowledge-based services (CREATE-IP, 2015). Data preparation for publication in the CREATE-IP project is led by NASA in collaboration with the world's major modeling centers (CDS, 2015; NCCS, 2015). Other participants include the European Copernicus Climate Change project (CCCS, 2015) and Reanalysis.org (2015).

Integrated Analytics and Archive Management

The abstract architecture and actual system implementation previously described bring together the capabilities of an analytic service and a persistence service. The resulting climate data analytics *system* provides the technology framework for dealing with the big data challenges of the climate sciences from both perspectives. The system treats interactions with the data analytics system as though they were the interactions a user or application might have with an archive system, in particular an archive that is specialized for the long-term preservation of digital scientific data. The system also treats the data objects that are generated by the analytic system as objects within the archive, specifically as dynamically created objects of the archive that have no real existence until they are computed.

It is important to note that the system can also include additional services that contribute to its overall usability. These services can augment capabilities by transforming data generated by the analytic service or persisted by the persistence service to yield data tailored to the specific requirements of the end user. Examples might include regridding services, downscaling services, reprojection services, and file formatting services.

An important class of additional capability are discovery services that allow users to find information about the data objects that either *can be computed* by the analytic service or *have been stored* in the persistence service. Searching for existing objects in the persistence service follows the traditional pattern of matching the metadata associated with objects with search criteria provided by the user. However, object discovery in the analytic service behaves differently.

Because discoverable objects in the analytic service do not come into existence until they are requested, discovery becomes a matter of knowing whether or not the analytic service can compute the desired object. If the object is computable, then *asking if an object exists actually means telling the service to create the object*: it is no longer a query, it is an assertion. It is in this way that the archive becomes dynamic, accommodating unanticipated applications of the data on an as-needed basis. Given sufficient computational resources—and in the coming world of exascale computing, it is reasonable to expect more compute resources—the system can create *virtual collections* of special interest that have no real existence or corresponding storage and management requirements. This stands in contrast to existing approaches that actually create multiple, real specialized collections to satisfy multiple varying needs, thereby contributing to the big data problem rather than solving it (Schnase et al., 2014b; Schnase et al., 2016).

Up to now, most of our direct experience in this area is through our work with the iPlant Collaborative. iPlant is a virtual organization created by a cooperative agreement funded by the US National Science Foundation (NSF) to create cyberinfrastructure for the plant sciences (iPlant, 2015). The project develops computing systems and software that combine computing resources, like those of TeraGrid, and bioinformatics and computational biology software. Its goal is easier collaboration among researchers with improved data access and processing efficiency. Primarily centered in the US, it collaborates internationally and includes a wide range of governmental and private-sector partners.

The CDS API has been used to integrate MERRA data and MERRA/AS functionality into the iPlant Discovery Environment. As shown in Fig. 11.5, the Discovery Environment (DE) is the primary Web interface and platform by which users access the high-performance, cloud-based computing, storage, and analytic resources of iPlant's cyberinfrastructure. The DE is designed to facilitate data exploration and scientific discovery by providing analysis tools that can be used individually or in workflows; seamless access to the iPlant Data Store, which uses iRODS in a manner similar to what we characterize as a virtual climate data system; flexibility to run tools on local or high-performance computing nodes, if needed; and collaboration tools for sharing data, workflows, analysis results, and data

Figure 11.5 iPlant Collaborative's Discover Environment interface.

visualizations with collaborators or with the community at large. Because the DE is integrated with iPlant's data management system and compute resources, researchers can access tools *and* data with an unprecedented degree of scalability, demonstrating the usefulness of the merged analytic/archiving concept we advocate.

In another application, MERRA/AS's Web service is providing data to the Rehabilitation Capability Convergence for Ecosystem Recovery (RECOVER) wildfire decision support system, which is being used for post-fire rehabilitation planning by Burned Area Emergency Response (BAER) teams within the US Department of Interior and the US Forest Service (BAER, 2015). This capability has led to the development of new data products based on climate reanalysis data that until now were not available to the wildfire management community (Schnase et al., 2014c). RECOVER's key feature is its ability to quickly aggregate site-specific data from heterogeneous, distributed data sources—including MERRA/AS reanalysis data—and deliver geographic information system (GIS) layers for use by fire managers.

Again, as shown in Fig. 11.6, we are building the RECOVER system around the idea of integrated analytics and archive management. Our first version of RECOVER is essentially a cloud-based software appliance, at the center of which lies an iRODS database that manages the dynamically assembled data for each individual wildfire. Automating this data preparation task has significantly reduced the amount of time GIS analysts spend gathering data—from hours and days, to only a few minutes—allowing them to engage more quickly in the important, time-critical wildfire management decisions needed for response and postfire site rehabilitation (Schnase et al., 2014c).

Toward a Climate Computer

MERRA/AS is currently in beta testing with about a dozen other partners across a variety of organizations and topic areas (Schnase et al., 2014a). As we gain experience, we have begun to look at ways of improving performance. Owing to the widespread use of MapReduce in the marketplace, industry is pushing hard to develop real-time MapReduce capabilities that are easy to use. We too have begun working with these technologies, and initial results have shown that optimizations can yield near real-time performance of MERRA/AS's canonical operations over native NetCDF files (Buck et al., 2011; Li et al., 2016). That has led us to think about some entirely new possibilities afforded by high-performance CAaaS.

What do we have if we implement real-time canonical operations on both the analytic service *and* persistence service and view the persistence

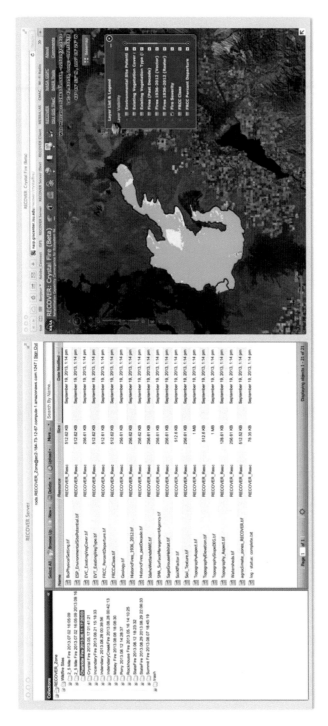

Figure 11.6 RECOVER server and client interfaces.

service as a way to store intermediate steps of a larger computation? With the distributed interoperability afforded by our CAaaS architecture, we begin to have the makings of a new abstract machine—a *climate computer* perhaps—in which small, atomic, microservices have a one-to-one correspondence to the computer's machine language instructions and the persistence service corresponds to registers and main memory. With such a computer, one can imagine, for example, performing distributed averages or more complex workflows across multiple, heterogeneous collections: find a partial sum and count on the MERRA service, find the same on the ERA-Interim service, store the partial results in a persistence service, order the persistence service to compute an average, then store and return the final answer—the client-side API's basic methods essentially acting like an assembly language for the climate computer.

Such a scenario may sound contrived, but the first demonstration challenge laid out by the ESGF-CWT has been to compute a "multi-model average" across various collections in the ESGF. So there are practical applications of the idea, and they go beyond the conventional notion of mashups—Web application hybrids that use open Web service APIs to knit services together to create new capabilities. The concept is more akin to OGC's WPS's approach to remote compute services, in which data are moved to a specialized service for processing. But it refines the WPS concept by building on the notion of distributed, lightweight atomic microservices—canonical operations that turn a disparate collection of data services into a coherent, fine-grained cooperating compute–storage ensemble, in which, like MapReduce itself, each node works in parallel on part of the problem, its overall behavior orchestrated by client-side scripts.

CONCLUSIONS

What we have described in this chapter may seem like an odd assemblage of topics and technologies, but they align with a larger conversation that is occurring around the issues of high-performance computing, cloud computing, and big data. Advanced computing and big data analytics are contributing to what some call the fourth paradigm of science (Tony et al., 2009). They are tools that are applicable to many areas. Yet the tools and cultures of advanced computing and big data analytics have diverged: each has grown up in an ecosystem of technologies and methods that differ greatly in their foci and technical approaches (Agrawal et al., 2011; Ernst and Young, 2011; Reed and Dongarra, 2015). Given science's increasing reliance on these tools, it is crucial that we find effective ways of bringing these ecosystems together.

Reed and Dongarra (2015) have examined this increasing interdependence and interplay of advanced computing and big data. One way to consider convergence is around the themes of locality and scale. In advanced computing regimes, the time and energy cost to move data will increasingly exceed the costs of computation itself. This leads directly to the overarching requirement for data-proximal analytics and everything that makes it possible: greater parallelism, finer-grained data exchanges among hundreds and thousands of processes, greater use of locality-aware software, and global namespace management. Efficiency in service-side analysis requires the restructuring of scientific workflows and applications, and maintaining load balance on all levels of a hierarchy of algorithms and platforms will be the key to this efficiency. For this, much can be learned from cloud computing and Web services, in which new abstraction layers and domain-specific toolkits, languages, and APIs allow developers to deploy custom environments and leverage services to potentially reinvent, redesign, and reformulate our solutions to science problems. Finally, mechanisms that allow the extended science community to contribute to the adaptive construction of these solutions will be key to the growth and sustainability of this new cyberinfrastucture (Baker and Bowker, 2007; Zittrain, 2008).

Our approach to CAaaS touches on many of these issues. In the end, however, our story is a simple one. We argue that tailored data products resulting from simple operations done quickly over large data collections and delivered as a Web service can have a significant impact on scientific data practices. Does this approach address the full range of requirements for big data analytics? No. Does it address the use of sophisticated, statistical approaches to analytics that can yield new knowledge and new insights? No. However, we set the stage for that by using high-performance analytic software to do our simple operations.

What our work does suggest is that fast and simple can make a big difference. It is estimated that in their modeling activities climate scientists spend 60% to 80% of their time gathering and preparing data for study (Lohr, 2014; NASA, 2012; Skytland, 2012). In our examples, we have consistently decreased that data wrangling time from days to minutes, a three orders-of-magnitude reduction. To put that in perspective, consider the work done by the IPCC. About 2000 scientists contributed to IPCC's Fifth Assessment Report. If these scientists spent only 10% of a 2000-h work year dealing directly with data, and half of that time was spent with data assembly, then the 6 years of research leading to AR5 could account for as much as 135 person-years of work on data preparation alone. That is a lot of time. Reducing that time by three orders of magnitude yields an aggregate effort of less than 2 months.

Is this an accurate estimate? Maybe not—but it is helpful to frame the issue this way, because it points to potential time and cost reductions that could substantively alter the way society gains benefit from climate data. At the very least, it suggests that the ideas described here should be given consideration.

ACKNOWLEDGMENTS

The work described here is the result of a long-running collaboration with a talented group of NASA scientists and engineers. In particular, my thanks go out to the core members of Goddard's climate informatics team: Glenn Tamkin, our lead engineer whose careful work has been instrumental in the development of these concepts and systems; Dan Duffy, who leads the NASA Center for Climate Simulation; and Mark McInerney, who leads NASA's Climate Model Data Services group. Denis Nadeau and Savannah Strong have provided much appreciated support in our beta testing activities, and Roger Gill, lead engineer on the RECOVER project, is an important source of expertise on real-time data gathering for real-world applications. I owe a particular debt of gratitude to the support and encouragement of Phil Webster, head of NASA Goddard Space Flight Center's Office of Computational and Information Sciences and Technology, as well as Tsengdar Lee, our program manager at NASA Headquarters. This work has been funded by the NASA Science Mission Directorate's High-End Computing Program and NASA's Applied Sciences Program.

REFERENCES

Agrawal, D., Das, S., Abbadi, A.E., 2011. Big data and cloud computing: current state and future opportunities. In: Electronic Proceedings of the 14th International Conference on Extending Database Technology (EBDT 2011) 4 pp. http://www.cs.ucsb.edu/~sudipto/papers/edbt-tutorial.pdf.

Burned Area Emergency Response (BAER), 2015. http://www.nifc.gov/BAER.

Baker, K.S., Bowker, G.C., 2007. Information ecology: open system environment for data, memories, and knowing. Journal of Intelligent Information Systems: Integrating Artificial Intelligence and Database Technologies 29 (1), 127–144.

Buck, J.B., Watkins, N., LeFevre, J., Ioannidou, K., Maltzahn, C., Polyzotis, N., Brandt, S., 2011. SciHadoop: array-based query processing in Hadoop. In: Proceedings of 2011 International Conference for High Performance Computing, Networking, Storage and Analysis (SC '11), pp. 1–11. Also available online at: https://users.soe.ucsc.edu/~carlosm/Papers/buck-sc11.pdf.

Copernicus Climate Change Service (CCCS), 2015. http://www.ecmwf.int/en/about/what-we-do/copernicus/copernicus-climate-change-service.

Climate Model Data Services (CDS), 2015. https://cds.nccs.nasa.gov.

Climate and Forecast Conventions and Metadata (CF), 2015. http://cfconventions.org.

Cloudera, 2015. http://cloudera.com.

Climate Model Intercomparison Project (CMIP), 2015. http://cmip-pcmdi.llnl.gov.

Collaborative REAnalysis Technical Environment – Intercomparison Project (CREATE-IP), 2015. https://www.earthsystemcog.org/projects/create-ip/aboutus/.

Dean, J., Ghemawat, S., 2008. MapReduce: simplified data processing on large clusters. Communications of the ACM 51 (1), 107–113.

Edwards, P.N., 2010. A Vast Machine: Computer Models, Climate Data, and the Politics of Global Warming. MIT Press, Cambridge, Mass. 518 pp.

Earth Observing System Distributed Information System (EOSDIS), 2015. https://earthdata.nasa.gov.

Ernst, Young, 2011. Cloud Computing Issues and Impacts. Publication in the Ernst & Young Global Technology Industry Discussion Series. https://www.hashdoc.com/documents/10977/cloud-computing-issues-and-impacts.

Earth System Grid Federation (ESGF), 2015. http://esgf.llnl.gov.

Hierarchical Data Format Version 5 (HDF-5), 2015. https://www.hdfgroup.org/HDF5/.

Hadoop Distributed File System (HDFS), 2015. http://hadoop.apache.org.

Intergovernmental Panel on Climate Change (IPCC), 2015. http://www.ipcc.ch.

iPlant Collaborative (iPlant), 2015. http://www.iplantcollaborative.org.

Integrated Rule-Oriented Data System (iRODS), 2015. http://www.irods.org.

Leskovec, J., Rajaraman, A., Ullman, J.D., 2014. Mining of Massive Datasets. Cambridge University Press. 513 pp. Available online at: http://www.mmds.org.

Li, Z., Schnase, J.L., Duffy, D.Q., Lee, T., Bowen, M., Yang, C., 2016. A spatiotemporal indexing approach for efficient processing of big array-based climate data with MapReduce. International Journal of Geographical Information Science, (in press).

Lohr, S., August 17, 2014. For Big-Data Scientists, "Janitor Work" Is Key Hurdle to Insights. The New York Times. Technology Section http://www.nytimes.com/2014/08/18/technology/for-big-data-scientists-hurdle-to-insights-is-janitor-work.html?_r=2.

Modern-Era Retrospective Analysis for Research and Applications (MERRA), 2015. http://gmao.gsfc.nasa.gov/merra/.

Nadeau, D., Schnase, J.L., Duffy, D.Q., McInerney, M.A., Tamkin, G.S., Thompson, J.H., Strong, S.L., 2014. Cloud-enabled climate analytics-as-a-service using reanalysis data (IN53A–3787). In: American Geophysical Union (AGU) Fall Meeting, December 15–19, 2014, San Francisco, CA.

NASA, 2012. A.40 Computational Modeling Algorithms and Cyberinfrastructure. NASA Research Announcement Solicitation NNH11ZDA001N-CMAC. http://nspires.nasaprs.com/external/viewrepositorydocument/cmdocumentid=257069/solicitationId=%7B074C12AB-FE57-8247-AC16D620E429359F%7D/viewSolicitationDocument=1/A%2040%20CMAC%20Amend%2034.pdf.

NASA Center for Climate Simulation (NCCS), 2015. http://www.nccs.nasa.gov.

Network Common Data Form (NetCDF), 2015. http://www.unidata.ucar.edu/software/netcdf/.

Open Archive Information System (OAIS) Reference Model, 2012. http://public.ccsds.org/publications/archive/650x0m2.pdf.

Open Geospatical Consortium (OGC), 2015. http://www.opengeospatial.org.

Reanalysis.org, 2015. http://reanalysis.org.

Reed, D.A., Dongarra, J., 2015. Exascale computing and big data. Communications of the ACM 58 (7), 56–68.

Rienecker, M.M., et al., 2011. MERRA: NASA's Modern-era retrospective analysis for research and applications. Journal of Climate 24 (14), 3624–3648. Available online at: http://dx.doi.org/10.1175/JCLI-D-11-00015.1.

Schnase, J.L., Webster, W.P., Parnell, L.A., Duffy, D.Q., 2011. The NASA Center for Climate Simulation Data Management System: toward an iRODS-based approach to scientific data services. In: IEEE Xplore Digital Library Proceedings of the 27th IEEE Symposium on Massive Storage Systems and Technologies (MSST 2011), (May 26–27, 2011, Denver, CO) 6 pp.

Schnase, J.L., Duffy, D.Q., Tamkin, G.S., Nadeau, D., Thompson, J.H., Grieg, C.M., McInerney, M.A., Webster, W.P., 2014a. MERRA analytic services: meeting the big data challenges of climate science through cloud-enabled climate analytics-as-a-service. Computers, Environment and Urban Systems. http://dx.doi.org/10.1016/j.compenvurbsys.2013.12.003.

Schnase, J.L., Duffy, D.Q., McInerney, M.A., Webster, W.P., Lee, T.J., 2014b. Climate analytics as a service. In: Proceedings of the 2014 Conference on Big Data from Space (BiDS '14), (European Space Agency (ESA)—ESRIN, November 12–14, 2014, Frascati, Italy), pp. 90–94. http://dx.doi.org/10.2788/1823.

Schnase, J.L., Carroll, M.L., Weber, K.T., Brown, M.E., Gill, R.L., Wooten, M., May, J., Serr, K., Smith, E., Goldsby, R., Newtoff, K., Bradford, K., Doyle, C., Volker, E., Weber, S., 2014c. RECOVER: an automated cloud-based decision support system for post-fire rehabilitation planning. International Archives of the Photogrammetry, Remote Sensing and Spatial Information Sciences, XL-1, 363–370. http://dx.doi.org/10.5194/isprsar-chives-XL-1-363-2014.

Schnase, J.L., et al., 2016. Big data challenges in climate science. IEEE Geoscience and Remote Sensing Magazine, (in press).

Skytland, N., 2012. Big Data: What Is NASA Doing with Big Data Today? Open.Gov open access article. http://open.nasa.gov/blog/2012/10/04/what-is-nasa-doing-with-big-data-today/.

Taylor, K.E., Balaji, V., Hankin, S., Juckes, M., Lawrence, B., Pascoe, S., 2012. CMIP5 Data Reference Syntax (DRS) and Controlled Vocabularies. 16 pp. http://cmip-pcmdi.llnl.gov/cmip5/docs/cmip5_data_reference_syntax.pdf.

Tony, H., Tansley, S., Tolle, K. (Eds.), 2009. The Fourth Paradigm: Data-Intensive Scientific Discovery. Microsoft Research, Redmond. 252 pp. Also available online at: http://research.microsoft.com/en-us/collaboration/fourthparadigm/4th_paradigm_book_complete_lr.pdf.

World Wide web Consortium (W3C), 2015. http://www.w3.org.

Wei, J., Dirmeyer, P.A., Wisser, D., Bosilovich, M.G., Mocko, D.M., 2013. Where does irrigation water go? An estimate of the contribution of irrigation to precipitation using MERRA. Journal of Hydrometeorology 14 (2), 271–289.

Zittrain, J., 2008. The Future of the Internet—And How to Stop it. Yale University Press, New Haven. 342 pp.

CHAPTER 12

Using Cloud-Based Analytics to Save Lives

P. Dhingra
Microsoft Corporation, Seattle, WA, USA

K. Tolle
Microsoft Research, Seattle, WA, USA

D. Gannon
School of Informatics and Computing, Indiana University, Bloomington, IN, USA

INTRODUCTION

A common problem with today's early warning systems for flood and tsunamis is that they are highly inaccurate or too coarse grained to be useful to first responders and the public. When information is overstated, people tend to become complacent and fail to follow directives provided in future warnings when there is real danger. If the information is too coarse, people are unsure of how they can protect themselves in a disaster situation and first responders are unsure of how best to deploy their resources.

Ed Clark from the National Oceanic and Atmospheric Administration's (NOAA) National Water Center in Tuscaloosa, Alabama, states the potential future flood damage this way, "In the next 30 years or so, 2538 people in the United States will lose their life[sic] to flooding and there will be approximately $300 billion in losses in infrastructure and property damage." (Keynote Presentation, "Water Resources and the Cyber-Infrastructure Revolution," 3rd Consortium of Universities for the Advancement of Hydrologic Science, Inc. (CUAHSI) Conference on Hydroinformatics, Tuscaloosa, AL, July 15th, 2015.) Worldwide, the Center for Disease Control reports that flood-related deaths are higher than any other type of natural disaster representing 40% of the total deaths (Centers for Disease Control and Prevention, 2014). With climate change, the severe weather may be more likely (http://news.nationalgeographic.com/news/2013/13/130215-severe-storm-climate-change-weather-science/), and there is real a possibility that death toll numbers could be even higher.

Cloud Computing in Ocean and Atmospheric Sciences
ISBN 978-0-12-803192-6
http://dx.doi.org/10.1016/B978-0-12-803192-6.00012-8

Copyright © 2016 Elsevier Inc.
All rights reserved.

The time to save people from a flood caused by a tsunami, superstorm, or extreme precipitation event is *before* it happens. Timely, precise, and granular predictions about natural disasters enabled by cloud computing and cyberinfrastructure are uniquely positioned to reduce the potential devastating impacts. Incorporating terrestrial information such as water levels, water table concentrations (ground water), and even the amount of permeable surfaces could help predict floods, mudslides, or sinkholes but would not necessarily improve the precision of the weather forecast itself. We can make prediction more precise by using high-powered computation and running different ensemble models in parallel. Extreme weather impact estimates can now be improved by using machine learning. And with advanced analytics and mobile personalization, individualized weather warnings can be sent via cell phones.

All of these are possible if you have access to the appropriate instrument data, high storage capacity, large processing power, easy-to-use statistical machine-learning software, visualization and simulation tools, and the power of the Internet, WiFi, and cellular networks. This chapter discusses an end-to end cyberinfrastructure that can make improved early warning systems and near real-time disaster prediction possible.

BACKGROUND

Previously, an effort of this scale could only be undertaken with government institutional infrastructure and financial support. And certainly in this case, government infrastructure and government research laboratories working in conjunction with government-funded academic researchers have laid the groundwork for what is now possible (Plale et al., 2006).

Open data initiatives supporting the US Government (https://www.data.gov) are providing oceans of data to the public, researchers, and industry. And it is data, along with on-demand, large-scale, networked computing resources, that will provide the means to create a new generation of applications to help tackle some of the toughest problems we face today—in particular, those that threaten our lives and our livelihoods.

But instead of a system hosted on government infrastructure, the system we propose can be set up in the cloud by *anyone*. With cloud computing, a startup company or a small research team can rent all the resources needed to collect data from millions of sensors, store petabytes of data, process data through pipelines, apply statistical modeling and machine learning to do forecasting and create relevant customizable notification services. It may be the case that direct access to the sensors may not be possible because they are owned and operated by academic or government entities. In this case, those entities could make the streams available to individual research teams through a high-throughput data hub.

CLOUD COMPUTING: ENABLING PUBLIC, PRIVATE, AND ACADEMIC PARTNERSHIPS

Governments provide the funding and resources for the vast majority of oceanographic, terrestrial, and atmospheric data. The analysis of this data takes place largely in government research laboratories or academia. Moreover, typically, the processing, storage, and computation take place on premises. However, to support the next generation of data gathering and analysis each institution must secure the funding to increase the local computing power and the additional physical infrastructure required to house, power, and cool these systems.

The increasing amount of openly shared data which fosters collaboration and discovery between academia, research institutions, and government agencies is a vicious circle. The vicious circle is that during these collaborations each institution conducting research requires more and more compute capacity to conduct, often redundant, experimentation on publicly available data sources.

As data increases in size and complexity, marshalling the needed compute resources becomes a barrier to research if it is undertaken by each institution independently. Cloud computing makes it possible for a large collaboration to share resources that can be scaled up to meet the demand and scaled back when not needed. By accessing a single data store, multiple experiments can be conducted by geographically dispersed teams with no replication of compute and data resources. Data can be stored once and accessed from separate virtual machines (VMs) anywhere in the cloud.

Simply put; cloud computing is a much more efficient and cost-effective solution over that of expanding the capacity of aging data centers. So if this is the case, why have institutions not generally adopted cloud infrastructure to increase research capacity? There are several reasons:
- Inexperience with cloud computing
- Fault/disaster tolerance concerns
- Ensuring capacity
- Inability to use existing tools/applications/languages
- Security and privacy issues

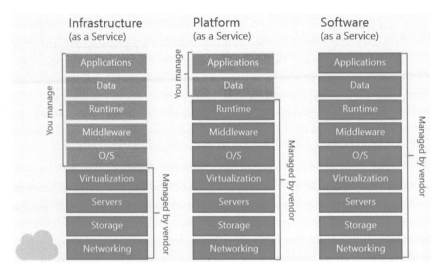

Figure 12.1 IaaS, SaaS, and PaaS diagrams. *Source: Microsoft Corporation.*

To adequately address these issues it is necessary for us to review some basics about cloud computing and also describe some more recent advances.

A Cloud Overview

Commercial cloud solutions are available from a number of providers with the most well-known being Amazon Web Services, Microsoft Azure, Google Cloud, and IBM Cloud Services. The examples we use in this paper are based on Microsoft's Azure. Although many of the features we describe are generic, we will also make use of some Azure-specific capabilities.

Cloud offerings typically fall into three categories: Infrastructure as a Service (IaaS), Platform as a Service (PaaS), and Software as a Service (SaaS) as shown in Fig. 12.1 below.

Infrastructure as Service (IaaS)

Public cloud systems provide users with the ability to "rent" virtual machines (VMs) running a variety of operating systems. For example, from the Azure Gallery, one can instantly create Windows, Linux, Structured Query Language (SQL) Server, Oracle, SharePoint, IBM, SAP, and BizTalk VMs.

In addition to the VMs that come standard with most cloud providers, users can also leverage the open-source community. On Azure, the VM Depot (https://vmdepot.msopentech.com) is a community-driven catalog of open-source VM images which contain a variety of Linux and FreeBSD VM images.

Platform as a Service (PaaS)

In PaaS, the cloud is not only provides the "iron" or computational hardware, it also provides the operating system and some high-level services. This solution still requires the user to provide the software, but removes the hassle of managing operating systems (O/S), such as applying O/S patches. This leaves applications developers free to focus on building solutions rather than the underlying software infrastructure.

Software as Service (SaaS)

SaaS goes one step beyond PaaS. In this instance, all of the software is built by a cloud provider. Users benefit from their favorite applications running in the cloud without having to install or update them. An example of such a system is Office 365 in which users log into and use Office applications like Excel. Though the experience is like using Excel on their own computer, the reality is that the application interface, the storage, and processing are all taking place in the cloud.

Deployment and Security Model

Commercial clouds have the ability to be very secure platforms. How secure is configurable by the user. There are three options in Azure: Public, Private, and Hybrid cloud. Each is described below.

Public Cloud

The public cloud is the least secure because all of the hardware, platform, and software services are visible via Internet by the cloud-provider data center. Public cloud is the most cost-effective and perfect solution for data that do not require extensive security.

Private Cloud

When an organization selects the private cloud option, they are provided with a dedicated system for computation, storage, software, and networking. Private cloud is suitable for any entity that wants to share data internally but protect it from external access. With the private cloud, an organization can ensure that organizational data and the hardware it sits on cannot be accessed outside the network. Private clouds do have a premium cost associated with these services.

Hybrid Cloud

Selecting the hybrid cloud can provide the best of both worlds. Workloads that need less security can be processed in the public cloud, but data that are more sensitive and processing can be run on premises or in the private cloud as needed.

Most large organizations have a mix of requirements and will therefore benefit from features of hybrid cloud. For example, a company trying to determine the best reservoir height to maximize profit while protecting communities from flood risk can leverage the public flood models in NFIE in the public cloud while maintaining their intellectual property workload in their private cloud.

Another example is support for government agencies. Much of the data they own and collect is made available to researchers to facilitate projects like the project described here. However, there are also related data that absolutely must remain secure to protect national security. This is a perfect application for a hybrid cloud solution. In the case of Azure, this is supported by an ability to easily provide hybrid connectivity and support services between public and private cloud deployments via ExpressRoute (http://azure.microsoft.com/en-us/documentation/services/expressroute/) and other services described as follows.

ExpressRoute

Azure ExpressRoute creates private connections between Azure data centers and on-premises infrastructure. These connections do not go over the public Internet and offer more reliability, faster speeds, lower latencies, and higher security than typical Internet connections. This service can also lower cloud-computing costs if users frequently transfer large data between public cloud and on-premises servers.

Service Bus

Azure Service Bus (http://azure.microsoft.com/en-us/documentation/services/service-bus/) provides the ability of message passing between public and private cloud applications. This allows the owner of private data services to filter out and provide as public only those streams that they desire to make public.

Azure Backup

Azure Backup Service (http://azure.microsoft.com/en-us/documentation/services/backup/) (ABS) provides a single solution for backup for private, public, and hybrid cloud solutions. In hybrid systems, it is possible to configure ABS to back up on-premises VMs and well as public cloud VMs.

Many, though not all, cloud-computing systems offer certified privacy and security compliance. If an extremely high level of security is needed, researchers are encouraged to ensure that the cloud provider they use has a suitable level of privacy compliance, the researchers select privacy and security options when configuring their cloud solution accordingly, and that the cloud provider has been accredited with meeting compliance audits by agencies relevant to their data.

For instance, Microsoft Azure has successfully been audited for Content Delivery and Security Association (CDSA), Criminal Justice Information Services (CJIS), Health Insurance Portability and Accountability Act (HIPAA), as well as many other security and privacy certifications (http://azure. microsoft.com/en-us/support/trust-center/compliance/).

Azure Site Recovery

Azure Site Recovery (http://azure.microsoft.com/en-us/documentation/services/site-recovery/) provides business continuity and disaster recovery (BCDR) strategy by orchestrating replication, failover, and recovery of virtual machines and physical servers. It enables one to quickly copy VMs in an Azure Data Center or on-premises data center.

Learning About the Cloud

Using cloud computing is often transparent to users, and management of these systems is similar and, in some cases, much easier than managing a local resource. There are training courses offered by cloud providers and third-party trainers that can bring knowledgeable data center personnel quickly up to speed as well as training that is more targeted for researchers themselves. For example, Microsoft Research offers an Azure for Research online course which is specifically targeted at onboarding researchers and students on how to use the cloud for scientific endeavors.

Fault Tolerance

Cloud providers automatically replicate data so access can be much more reliable than most on-premises data centers. For instance, Azure not only mirrors data and services in one location, it also replicates them in a different geographic location to ensure that if one cloud-computing center is compromised, the data and services associated with it can be preserved—even if the primary data center housing that system is destroyed by a disaster.

Compute Capacity

As previously mentioned, compute capacity in the cloud is scalable and "on-demand". There are several aspects to this capability. First, most cloud providers have data centers that are based on virtualized servers. Applications are configured as services that run in virtual machines (VMs) on the servers. Typically, an application service receives input commands as web requests, produces output in the form of data stored in the cloud, and replies to the requester as web responses. The application is code typically deployed as a VM image that the cloud O/S can load onto an appropriate server.

There are several different ways to create the VM image. The "traditional" approach is to configure a complete copy of an O/S (Linux or Windows) and install the application there. A more "modern" approach is based on the concept of a container which is a package containing the application service and its needed resources but not the entire O/S. The container is then deployed on a running VM that provides the core O/S resources. Many containers can be hosted on a single VM. Deploying a containerized application can be 10 to 100 times faster than deploying a full VM. Furthermore, a single containerized application can be run without change on data centers from many different cloud providers. In both the VM and the container approach to packaging, it is standard practice to store the image in a repository so that it can be easily accessed, downloaded, and deployed on a server in any of the cloud–provider data centers.

Depending on the requirements of the application, the user can select the appropriate server type. For example, if an application requires one central processing unit (CPU) core and 4 gigabytes (GB) of memory, a corresponding server can be configured. If the application requires 32 cores and 128 GB of memory, that too can be made available as a VM host. A second dimension of scalability is the number and location of the servers hosting copies of the application VM or container. If many users of a particular application request access at the same time, it may be necessary to deploy multiple instances of the application across the data center or different geographic locations. Cloud providers have standard tools to "auto-scale" an application if the load is heavy. This same mechanism can be used to reduce the number of running instances if the load drops.

Finally, many applications profit from being factored into smaller components which can each be containerized and deployed separately. For example, the application may have a Web-server component that deals with user interaction. The Web server may hand off "work" to separate computational components or it may invoke the services of other applications such as machine learning systems, databases, mapping services, etc. The application then becomes a web of communicating subsystems. Although this sounds complex, most large-scale cloud services are designed in this manner because it more efficiently scales the application and reduces maintenance.

Also, "data fabrics" enable the sharing of data across VMs in such a way that one need not replicate data—saving on storage costs as well as compute resources. And as mentioned before, these data sources are protected from loss. They can also be replicated intentionally, so that different datasets can be located closest to users who require the most immediate access.

Azure Storage

Azure provides a range of scalable storage for big data applications: blobs, tables, queues, and file systems. In addition, Azure storage stack includes support for relational databases, non relational databases, and caching solutions. We will discuss these in more detail in the following.

Azure Blob for Storing Unstructured Data

Blob storage is used to store unstructured data that can be accessed via Hypertext Transfer Protocol/HTTP Secure (HTTP/HTTPS) protocols. Data can have controlled access and be made available either publicly or privately. Blobs are stored in "containers," and a single Azure's blob storage container can store up to 500 terabytes of data. There is no limit to the number of containers you can create.

Azure Table Storage for Structured, Nonrelational Data

Azure table storage is a NoSQL (http://nosql-database.org/) data store in Azure for storing large amount of structured, nonrelational data. It can store terabytes of data as well as quickly serve the data to a Web application. This type of storage is suitable for those datasets that are already denormalized, do not need complicated joins, foreign key relation, or stored procedures.

Tools

Because cloud systems allow users to set up VMs, the tools and applications they use can be set to run on the cloud. In fact, with the container approach to application deployment it is easy to build and test a container running on a laptop. Virtually any tool or software library that the application needs can be installed in the container. The most popular container repository is on Docker Hub (https://www.docker.com/), and there are many prebuilt containers there with many of the standard programming tools (Python, C, Fortran, MySQL, etc.).

From such a hub, one can download a container and add an application with very little effort. It is also important to realize that the cloud providers have additional services pre-installed in the cloud that are easy to integrate into applications. For example, Microsoft Azure provides Azure Machine Learning, a complete machine learning tool kit, special services for managing event streams as well as services for building mobile applications.

CLOUD COMPUTING-ENABLED PARTNERSHIPS EXAMPLE: THE NATIONAL FLOOD INTEROPERABILITY EXPERIMENT

To illustrate the concepts in this chapter, we will use the National Flood Interoperability Experiment (NFIE). NFIE is a research initiative among government, academia, and industry to help demonstrate the next generation of national flood hydrology modeling to enable early warning systems and emergency response, as illustrated in Fig. 12.2.

The goal of NFIE is to answer the questions—What if it were easier to predict more accurately where floods will occur? What if more flood

	Emergency Responders	Public
Action	**Flood Response** Current and forecast storm rainfall Current and forecast flood conditions Impact on homes, critical infrastructure, and transportation routes	**Flood Warning** Public notification and warnings Current and anticipated road closures Real-time and forecast flood maps
Planning	**Flood Pre-Planning** Static flood risk maps Number of flooded homes, critical infrastructure and transportation routes for different risk conditions Pre-planning of flood response mobilization	**Flood Education** Local topography and relation to stream Putting together a "go bag" for evacuation Flood hazard prevention methods

Figure 12.2 Flood emergency response. *Source: David Maidment and Harry Evans, University of Texas at Austin.*

information could be shared in real time to aid in more effective response planning and prevent deaths and property damage? Those are the questions David Maidment, Professor of Civil Engineering at the University of Texas at Austin, wanted to answer in conjunction with his collaborators at University of Illinois Urbana–Champaign, Brigham Young University, Tufts University, and Renaissance Computing Institute (RENCI), among others.

It has been our privilege to work with Dr. Maidment and his collaborators on NFIE to enable a cloud-based system designed to protect lives and livelihoods from water-related disasters by using the cloud to receive and disseminate digital data to meet big data challenges.

The various collaborators involved with NFIE are working to interoperate data from multiple federal agencies that deal with physical aspects of flooding data sources as well as models which are provided by various government organizations and research facilities to do flood prediction. Another challenge was that each research institution was attempting to process data on *their* individual technological infrastructure—which, in general, was not available to the entire consortium. What was needed was a technology infrastructure to allow information to flow in from various agencies and academia and then flow out to citizens and first responders.

CLOUD COMPUTING AND BIG DATA: MADE FOR EACH OTHER

The term "big data" is often overused and ill-defined (http://www.extremetech.com/extreme/207309-forget-big-data-its-already-obsolete). In this paper, we use the term to mean the data generated by the "Internet of Things" or IOT, in which the focus is on "huge data, smart data"—effectively, data *in context*. What makes these data so interesting and valuable is that they do not come from a single source, but many sources which provide a much richer picture of our world. The challenge is interoperating these data in such a way that we can make decisions that are more meaningful.

The cloud provides an opportunity to receive, collect, and process a wide variety of data at a central location. Users can ingest large amounts of data from heterogeneous sources generated by multiple sensors transmitting periodic, near-continuous, or constant signals.

The cloud was invented by the Internet industry to respond to the continuous avalanche of requests from users for services (e-mail, Web search, media delivery, etc.). Within these companies the cloud data centers were used to do the data analytics needed to improve and support these services. By 2007, it was clear that the cloud could be provided to anybody with an internet connection.

Before the cloud, high-performance computing and supercomputing centers were already grappling with larger and larger data streams. The cloud brings two very important elements to the game—the ability to scale to thousands of data sources 24×7 and availability to everyone. Instead of being available on an institutional basis, in "batch computing mode" behind firewalls or open to only those that are granted supercomputer time, today, anyone with an Internet connection has access to significant compute power—and can pay for it as needed.

And now cloud computing has opened the doors to better research and business decision making, this combined with better security. This is why, increasingly, the cloud is becoming the new computing infrastructure for big data. And with advanced analytics being simplified and made available as a cloud service, the door is open for new scientific discovery.

Not long ago, corporations struggled with questions like, "What data do we need to store, archive or delete?" The expense of storing older data was the primary reason to archive or delete data. Log files, in particular, were generally deleted after some designated period unless there were legal requirements to maintain them. And in general, these companies relied on internally generated data—if they were doing data analytics at all.

Government organizations are expected to store and share the data they collect, however, combining this data or storing and serving up data products from other agencies is outside their mandates. Now, both public and private sector organizations realize the advantages of not only saving and mining their internal data, they realize that combining data with other sources can provide deeper insights into internal sources. And, in the case of government-owned data, academia or government research labs can act as bridges to enable deeper studies.

Example: NFIE—An Exercise in Using Big Data

What makes NFIE a big data challenge is the multiple data source integration from the US Geological Survey (USGS), the US Army Corps of Engineers (USACE), and the National Weather Service (NWS), as well as water hydrology models from the National Center for Atmospheric Research (NCAR). These data are, respectively, stream-flow levels, reservoir levels, precipitation, and terrestrial hydrological models. From there, academics calculate river-level models via Routing Application for Parallel Computation of Discharge (RAPID) (David, Cédric H. (2013), RAPID v1.4.0, Zenodo, DOI: 10.5281/zenodo.24756) (David et al., 2011). This can then be plotted over Federal Emergency Management Agency (FEMA) flood inundation maps days in advance of potential flooding—providing a foundation for accurate early warning systems.

Each of these different data sources has a cadence which must be considered. Some of the information comes in 2 h intervals, some in 15 min intervals, and others near real-time streaming. The needed forecast rate is every hour for a sliding window of 7–14 days into the future. The complexity of data and time intervals in addition to a need for high accuracy and near real-time forecasting make this a very interesting and challenging big data problem.

CLOUD COMPUTING, BIG DATA, AND HIGH PROCESSING: MEANINGFUL INSIGHT

Cloud platforms are well suited to big data processing—particularly when collaborators are geographically dispersed. The cloud is set up to enable easy ingesting, storing, processing, and sharing of high-volume, frequently changing, and heterogeneous data. In addition, there is the ability to convert the data into meaningful formats and identifiable patterns.

With the volume of data and the processing needs growing exponentially (Reed and Dongarra, 2015), it is increasingly hard for individual institutions to keep up with demand in part due to the resource costs and space limitations. Today, you can use data management and processing tools in the cloud without up-front purchasing costs, setup, physical infrastructure (e.g., buildings, wiring, and cooling), or ongoing maintenance costs.

Example: NFIE—Collect and Analyze Soil Moisture

Fig. 12.3 below shows an example of a research model deployed in the cloud that requires consolidating data from multiple data sources; soil moisture sensors from various parts of the country. Data are brought into centralized storage (Azure Table), the analytics program is run in a VM, and the result is deployed as a simple Azure Web site (http://statsnldas.azurewebsites.net/). The system not only estimates moisture in soil across the country, a user can also click and drill down at a specific location to see the calculated value of moisture (with uncertainty) along with a chart showing how that relates to historical measurements and make comparisons with previous months or years.

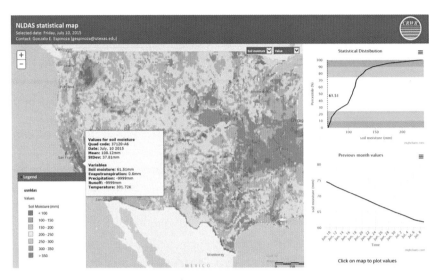

Figure 12.3 The California Drought detailed in the North American Land Data Assimilation System (NLDAS) statistical map. *Source: Gonzalo Espinoza Davalos, Center for River and Water Research, University of Texas at Austin.*

CLOUD COMPUTING, BIG DATA, AND MACHINE LEARNING

Accompanying big data has been the reemergence of machine learning. And although there is a long history of doing predictive analytics with statistical methods, increasing amounts of data have enabled advanced machine learning methods such as neural nets and decision trees to be combined with traditional statistics or, in some cases, to replace existing approaches with methods having similar or better accuracy. What makes machine learning so powerful is that by leveraging different algorithms it can perform a much broader range of tasks than a pure statistical approach alone.

In the past, machine learning was limited to the realm of computer scientists. However, as software companies have realized the demand for higher-level analytics services on their cloud-computing platforms, they have created machine learning and statistics tool kits that are more usable for nonexperts. Azure Machine Learning is an example of such a service.

Using machine learning, data scientists can directly and easily embed predictive analytics directly into cloud applications allowing them to draw insight from large amounts of data. In effect, big data and the cloud are driving the recent burst of innovation in the use of advanced statistical and machine learning in many domains. The cloud is becoming an "intelligent" as well as an efficient means of drawing insight from data—finding the signal in the noise.

Example: NFIE—Interpolation of Water Flow Levels Using Machine Learning

The USGS reports on nearly 10,000 stream gauges across the nation to measure water flow. From their Web site it is possible to retrieve near real-time water levels (http://waterdata.usgs.gov/nwis/rt). However, due to extreme conditions in some locations, these gauges are often damaged and/or fail to report accurately in disaster situations—when they are needed the most.

Later in this chapter we will describe how students involved with the National Flood Interoperability Experiment at the National Water Center in Tuscaloosa, AL, were able to use Azure Machine Learning to interpolate data using other gauges in the region to fill in missing data values.

NFIE ANALYTICS WITH MICROSOFT AZURE

Researchers from the University of Texas partnered with other researchers, federal agencies, and first responders to develop NFIE. They used Microsoft Azure to build a new national flood-data modeling and mapping system to save costs and protect citizens. The goals of the NFIE include standardizing data, creating a scalable solution, and closing the gap between national flood forecasting and local emergency response. So for this project these collaborators leveraged the cloud for both their research and educating students at an NOAA Summer Institute in June and July, 2015 (https://www.cuahsi.org/NFIE).

In October 2013, the Onion Creek area near Austin, Texas, faced a deadly flood (http://www.austintexas.gov/department/onion-creek), and in 2015 two floods have hit the Austin area with the second, on May 24, resulting in even more devastation and loss of life—including a first responder, Police Deputy Jessica Hollis (http://www.wunderground.com/news/austin-texas-flash-flooding-20140918). The May 2015 flood in Claremore, Oklahoma, also killed a first responder, 20-year veteran firefighter, Capt. Jason Farley (http://www.nydailynews.com/news/national/firefighter-3-killed-okla-texas-flooding-article-1.2234196).

It was just after the Onion Creek flood that the NFIE initiative was conceived with the goal to develop the next generation of flood forecasting for the United States and connect the National Flood Forecasting System with local emergency response to create real-time flood information services.

To carry off this audacious plan, the NFIE team needed a technology infrastructure that would allow information to flow in from various agencies, enable collaboration between government agencies and academia, and ultimately flow out to citizens and first responders in times of potential flooding—before loss of life occurs.

The NFIE system models the flows of rivers and smaller streams across the United States leveraging stream gauges, reservoirs levels, as well as soil moisture and terrestrial hydrologic models. Each of these services communicates with the others and they are then coupled with precipitation data to calculate future flood risk.

As a proof of concept, a regional scale model was built for the Texas Gulf (one of 20 water regions in the US). Compute resources came from the Texas Advanced Computing Center (TACC) (https://www.tacc.utexas. edu/). The model was based on data from the USGS for the lower Colorado River and was processed on demand rather than periodically.

To scale this up to the national level and eventually globally, the NFIE team needed a scalable system for storage and computation that could be run periodically—potentially hourly. This is well within the capabilities of the Microsoft Azure platform.

Azure VMs for Running Models

Because the NFIE models, RAPID, and NCAR's terrestrial water models were built on Linux, the simplest solution was to take an IaaS approach. The data was ported to Azure storage, and VMs were deployed to handle the various processes enabling the workload to be shifted from TACC to the cloud in which they benefited from built-in networking, load balancing, reliability, and connectivity to several heterogeneous data sources. Because of the ability to quickly scale by adding or removing compute nodes as needed, estimating the amount of compute resources needed for a national scale model could be adjusted on the fly to maximize performance of the system.

For NFIE, it was determined that the most powerful cloud machine possible would be necessary to generate near real-time results. So we used the largest VM available on any cloud at this writing—Microsoft Azure's G5 machine (http://azure.microsoft.com/blog/2015/01/08/largest-vm-in-the-cloud/). The system supports 32 cores and has 448 GB of virtual memory. This enabled researchers to execute the NFIE models every hour with a 15 min wall-clock time.

The NFIE team relied mostly on files in Network Common Data Format (netCDF) and estimated that executing a model every hour would produce approximately 5 GB of data per hour. Because models can be rerun at any time, there is no need to archive each model run. This meant that Azure blob storage was more than adequate to meet their needs for the national-scale model.

However, some data used for modeling were structured and would benefit from query support. So, for individual models, other types of storage were used.

To understand how much absorption might take place in a flood situation and monitor drought conditions, the NFIE team also wanted to create a national soil moisture model. The soil moisture dataset was large, but did not have complex relational data making it perfect for an Azure Table.

The team gathered the last 30 years of data of soil moisture data and loaded it into an Azure Table. They then built an interactive visualization for users to interact with the system deployed as an Azure web application.

Azure Web Applications

Azure Web applications are PaaS solutions. The soil moisture model described above is an example of an Azure Web application that was built using the Django framework in Python. A screenshot of the user interface was shown earlier in the chapter in Fig. 12.4 (http://statsnldas. azurewebsites.net/).

Azure Data Factory for Orchestration and Monitoring the Model Pipeline

Though the details of each of these processes is outside the scope of this chapter, to generate a national-level flood forecast system, the following high-level steps were undertaken by the NFIE team:

1. Ingest data from multiple public sources (e.g., stream flow and reservoir levels)
2. Generate high-resolution interpolation hydrological models which are massaged for input for next phase of processing
3. Run the land-surface hydrology model based on interpolation and combine with precipitation data to estimate future runoff
4. Route output from runoff model to estimate water levels in individual river basins
5. Visualize the results

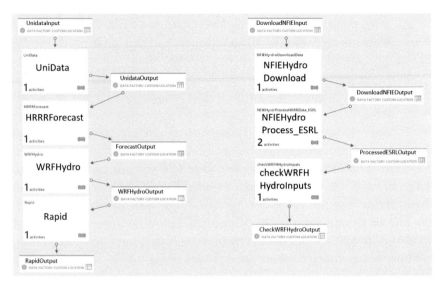

Figure 12.4 The Azure Data Factory pipeline for NFIE and WRF (Weather Research and Forecasting) Hydro Model.

What is important to note is that each of these is done by different research groups at different institutions.

The Azure Data Factory (ADF) enables users to build and orchestrate data pipelines. This is a very natural fit for the NFIE pipeline described previously. ADF allows users to visualize pipeline workflow, monitor dependencies, study, and compare across different processes executions and, at a glance, monitor each portion of the system's "health." ADF enabled the NFIE team to stitch multiple data processing models together, execute them in sequence, and monitor progress through a helpful user interface.

If at any time there is an issue with the execution, the team sees the problem immediately and can take corrective action immediately without stopping unaffected processes executing in parallel.

Fig. 12.4 shows how the NFIE pipeline is visually composed and configured—no code necessary.

Azure Machine Learning

Historical data provides a record of past occurrences of phenomena. Machine learning and advanced statistical analysis can identify patterns in these data to determine to potential causes. They can also be used, as in the case with NFIE, to generate future predictions or make up for missing data values.

NFIE Example: Interpolation of Missing Data Values Using Machine Learning

Being able to estimate stream-flow levels is critical for NFIE success.

Water-flow levels in streams is dependent on precipitation, snow melting/runoff, attrition by public consumption, industry, and farming, control by nearby dams and reservoirs along the water's path, as well as many other factors. There is a desire to have these data be more granular and precise. However, it is becoming increasingly cost prohibitive to set up and maintain these sensors—not including the yearly operational costs.

However, these gauges, though built to be as robust as possible, are often disabled in extreme or disaster situations. Occasionally, debris in the stream can cause the sensors to provide inaccurate or incomplete data. In some cases, debris has been known to wipe out a stream gauge station altogether, sometimes on an annual basis. During a storm or flood hazard event, the likelihood of a stream gauge getting broken is even higher. When a gauge is destroyed, budget limitations may result in months and even years before the gauge can once again become operational.

When gauges stop transmitting critical near real-time data, this hampers the ability of first responders to monitor floods and accurately deploy their resources.

Using Machine Learning, one can find patterns of water flow in river systems. A machine-learning algorithm can cluster a set of gauges that have a high correlation in related water flow. If one of those gauges is partially or fully disabled, other correlated gauges can be used to estimate water levels at the location by leveraging historical data.

Fig. 12.5 below illustrates a pattern which was identified by machine learning and visually verified by a scientist. What is shown is how stream site # 13185000 and site # 13186000 are highly correlated for water flow. This means that if either of these sites stops reporting, the other can act as a proxy and predict the water flow at the other location.

What is less apparent is that the negative correlation also has value. As can be seen in the same figure, when stream gauge # 13190500 is reporting high numbers, the other gauges' stream-flow levels approach zero. This negative relationship is a pattern that machine learning can use simultaneously with positive correlations to get the best prediction model.

A student team at the National Water Institute's Summer Institute took the NFIE data and did an experiment using Azure Machine

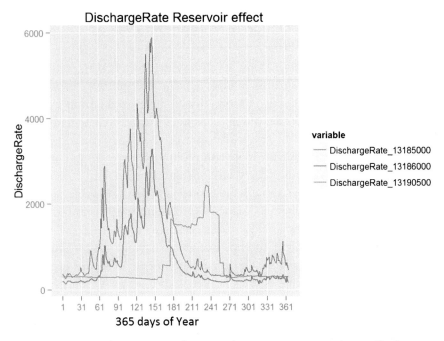

Figure 12.5 Discharge rate as a function of upstream reservoirs. *Source: Tim Petty.*

Learning. By testing different methods available in the Azure Gallery, they were able to cluster stream gauges and create a model that resulted in a predictive accuracy level of 80–95% for the Boise River system in the state of Idaho.

The steps to carry out this experiment and the Azure Machine Learning environment in which it was run are shown in the list and in Fig. 12.6 as follows:

- Fetch historical water-flow data from the USGS streamflow site for the gauges on the Boise River
- Use a boosted decision tree regression algorithm to train a model (selected from the Azure gallery)
- Use the sweep parameter module to find optimal parameters (selected from the Azure gallery)
- Score the model accuracy and cross-validate using held-out data
- Output the result to a comma-separated variable (CSV) file
- Take model output and feed it into an R script to visualize the predicted versus actual result

The result shown in Fig. 12.7 below is a comparison of actual versus predicted values for a stream gauge based on positive and negatively

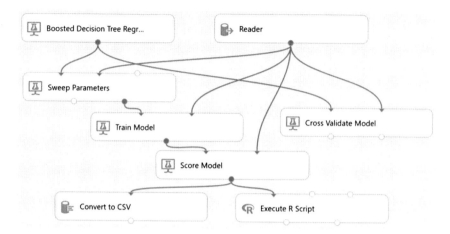

Figure 12.6 Azure machine learning workflow for identifying correlated stream gauges. *Source: Tim Petty.*

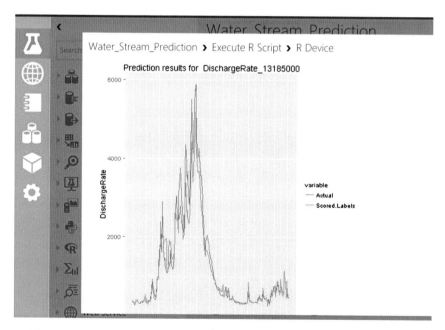

Figure 12.7 Estimated streamflow plotted against actual data. *Source: Tim Petty.*

correlated stream gauges. This shows that within the parameters of this simple experiment, Azure Machine Learning was able to make a reasonably accurate prediction about water flow in a stream based on other streams in similar area.

BENEFITS AND SUMMARY

Scalability

Cloud computing provides scalability because cloud architecture is by design elastic. This means that it is easy to add additional server instances on demand. And container technology creates the capacity to execute small independent tasks in parallel.

The original NFIE model was built for one water region and then later scaled to cover all of the water regions across the continental United States. Azure provided architecture that enabled NFIE the option to scale up, by increasing the computation size of the servers they used, or scale out by increasing the number of servers with one click without the need for re-architecture or redeployment. And if they had been unable to estimate their computational needs, they could have chosen the autoscale feature which would automatically scale up or out to best match the resources needed by their application.

Cost Saving

The cloud is a pay-as-you go solution. So you only pay for the resources you need as you need them. There is no up-front cost to set up machines in Azure as there would be if one had to set up or apply for the compute resources—as in the case with NFIE.

Using a cloud platform was the ideal option because when usage was low at the planning stage, little to no Azure costs were incurred. When the system was fully operational, they had the option to scale up to the largest possible configuration—and then release that system when it was not in use.

Performance

Azure has support for traditional VMs as well as very powerful ones. G5 machines have 32 core + 448 GB of memory. With a single G5 machine, NFIE model hourly runs were completed in less than 20 min.

Operational Quality

Azure VMs have high availability and reliability. In addition, by leveraging the Azure Data Factory on those VMs, NFIE researchers were able to compose a pipeline that provided monitoring, alerting, and issue resolving for NFIE components.

Disaster Management

In Azure, data and software are geographically distributed and replicated for fault tolerance. If one cloud center goes down during a disaster, an alternative cloud center in another region can take over using the replicated data and services.

CONCLUSIONS

Currently, the NWS makes forecasts at about 6500 locations on main rivers in the country. The NFIE expects to deliver specific and actionable data for 2.6 million locations nationally, including smaller streams. Ultimately, the greatest contribution of the NFIE is an increase in real-time responsiveness to improve public safety and save lives. Working closely with the FEMA, NFIE has had the chance to use the data modeling in their decision making, improving their responsiveness and ability to work early on with local first responders in a disaster situation.

The cloud enables improved scalability and lower maintenance costs; however, the biggest advantage of cloud computing was enabling collaboration. It removed organizational boundaries so that government, academics, and industry simultaneously worked together. It allowed researchers across multiple disciplines, people with expertise in hydrology, civil engineering, computer science, statistics, machine learning, and disaster management to work together toward providing a solution.

This is only one example of a project benefiting from a cloud approach. Ocean, atmospheric science, and earth scientists deal constantly with huge datasets and need multidisciplinary approaches to push the envelope on scientific discovery. They often need to interoperate their own data with that from multiple sources including federal and state agencies, and their collaborators in academia and industry. The cloud can provide the infrastructure to enable global collaborations on this data and facilitate experimentation without having to worry about cyberinfrastructure. And once an experiment is successful, it can easily be scaled and deployed as a service to be contributed to and validated by others in the field.

As with NFIE, anytime the research, development, and data of a project are outside the boundaries of a single institution, the cloud supports the ability to collaborate easily and build solutions that might not otherwise be possible.

REFERENCES

Centers for Disease Control and Prevention, 2014. Weather. Atlanta: JAMA 312 (10), 992.

David, C., Maidment, D., Niu, G., Yang, Z., Habets, F., Eijkhout, V., 2011. River network routing on the NHDPlus dataset. Journal of Hydrometeorology 12 (5), 913–934.

Plale, B., Gannon, D., Brotzge, J., Droegemeier, K., Kurose, J., McLaughlin, D., Wilhelmson, R., Graves, S., Ramamurthy, M., Clark, R., Yalda, S., Reed, D., Joseph, E., Chandrasekar, V., 2006. CASA and LEAD: adaptive cyberinfrastructure for real-time multiscale weather forecasting. IEEE Computer 39 (11), 56–64.

Reed, D., Dongarra, J., 2015. Exascale computing and big data: the next Frontier. Communications of the ACM 58 (7), 56–68.

CHAPTER 13

Hadoop in the Cloud to Analyze Climate Datasets

A. Sinha
Esri Inc., Redlands, CA, USA

INTRODUCTION

The climate of the Earth has continuously changed. Rising sea levels, global warming, ozone depletion, acid rain, biodiversity loss, and deforestation are evidence of such changes (National Aeronautics and Space Administration-Earth System Science Data Resources (NASA-ESSDR, 2012)). Variability and change in climate, both natural and anthropogenic, have profound impacts on humans and nature. It is critical that we understand how climate has been in the past and how it will be in future. Understanding these changes requires analyzing climate datasets efficiently. Knowledge gained from the discovery of patterns in climate datasets is of increasing interest not only to scientists, but also to sectors ranging from water, land, energy, health, to tourism that are impacted by climate variability, and need climate intelligence for better decision making. It is estimated that climate intelligence supports business decisions with revenues ranging in billions of dollars (Overpeck, 2011). As a result, there is a pressing need to harness and disseminate knowledge discovered from climate analytics to various stakeholders.

Climate datasets belong to the Big Data domain (Schnase, 2014). Due to their spatial dimensions, climate datasets are also classified as Spatial Big Data (Cugler, 2013). A report by Manyika (2011) described 3Vs—velocity, volume, and variety—as the main characteristics of Big Data. Climate datasets exhibits all these characteristics. Satellites monitoring the Earth are generating 25 gigabytes (GB) of observation data daily (velocity), the size of archive of satellite imagery about is 500 terabytes (TB) (volume), and their sensors continuously help measure more than 50 variables (variety) such as sea–surface temperature, air pressure, and vegetation (Moderate Resolution Imaging Spectroradiometer (MODIS, 2015; Sinha, 2015)). These massive datasets are growing rapidly, and the conventional approach of downloading the data to local workstations, and performing operations on the data locally is inefficient and

Cloud Computing in Ocean and Atmospheric Sciences
ISBN 978-0-12-803192-6
http://dx.doi.org/10.1016/B978-0-12-803192-6.00013-X

Copyright © 2016 Elsevier Inc.
All rights reserved.

impractical—it can take hours to days to complete analysis at modest scales (Dirmeyer, 2014). Clearly, a different approach is needed to analyze climate datasets. The Apache Hadoop framework (Apache-Hadoop, 2015) for distributed computing is of particular interest here because it is scalable, and can run Big Data analytics on climate datasets on a cluster of commodity servers.

As a practical implementation, the NASA Center for Climate Simulation (NCCS) is presently employing cluster computing in the cloud using the Apache Hadoop framework for high-performance climate data analytics (NCCS, 2015). Strategies for deploying clusters of computers range from physical machines on-premises to machines in the cloud that offer on-demand self-service, resource pooling, rapid elasticity, and measure-service provisioning (Mell, 2011). Studies have indicated that cluster computing with Hadoop in the cloud can offer better price performance and flexibility in altering virtual machines configuration with varying computational demands (Wendt, 2014). This Chapter presents the Apache Hadoop framework for analyzing climate datasets, the challenges inherent in analyzing climate datasets, and various approaches reported to rapidly access climate datasets for analysis in a cluster.

CHALLENGES

The urgent need for timely analysis of climate datasets is clear. Climate datasets are large and have spatial and temporal components that pose multiple challenges. From the data perspective, the binary data format of large climate datasets is not well suited for partitioning into smaller parts that can be used efficiently for distributed computing. The Hadoop framework offers ways to handle binary data of any size, but as the size increases, the time required to move the data across the network also increases thereby degrading the performance. From the data analysis perspective, the spatial and temporal nature of the data is a challenge. Traditional statistical methods used for data analysis assume that data are independent and stationary. These assumptions are not valid for climate datasets that are generally autocorrelated in space and time. For example, temperatures in nearby towns are usually in the same range. This is also known as Tobler's First Law of Geography (Miller, 2004), which states that nearer things are more related, and defines the inherent autocorrelation among proximal locations in space. The holy grail of Spatial Big Data analytics is to give users easy to use methods, such as hot-spot analysis, spatial regression, topological functions (union, intersect etc.), that can scale massively to analyze climate datasets (Cugler, 2013). Therefore, novel algorithms are required for climate analytics in a distributed environment. Many approaches have emerged to address these challenges, and will be presented in the following sections.

HADOOP FOR LARGE-SCALE DATASETS

Apache Hadoop is an open-source project that offers a reliable and resilient software framework for distributed storage and computing of large-scale datasets on a cluster of commodity servers (Apache-Hadoop, 2015). It is based on the idea of divide and conquer to analyze large datasets by creating subsets of data that can be processed in parallel on servers in a cluster. Hadoop is scalable, fault tolerant, and abstracts away the complexities of parallelization, so users can focus on data processing. Traditional approaches in parallelization require high-end computer hardware to achieve performance and reliability; Hadoop offers reliability by employing redundant storage and processing capacity on commodity servers that are expected to fail, and by building mechanisms in the software framework to address those hardware failures automatically in a resilient manner (Dean, 2008).

Hadoop has a master- and worker server-node architecture. The master server is called NameNode, and is responsible for managing resources across the cluster by worker servers known as DataNodes. The core of Apache Hadoop consists of two components: storage and processing. The storage part is called the Hadoop Distributed File System (HDFS), and the processing part is based on the MapReduce (MR) programming model.

Hadoop Distributed File System

The Hadoop Distributed File System (HDFS) is a file management system for cluster computing. DataNodes in the cluster provide storage that allows user data to be stored in files that are partitioned and saved as a sequence of storage blocks on disk on DataNodes as shown in Fig. 13.1. The storage blocks are shown in different colors to illustrate the replication of blocks across nodes in the cluster. Each storage block is replicated three times, which is the default replication factor, across DataNodes. HDFS allows high-speed access of data distributed across disks on DataNodes. Hadoop copies the source code to worker servers so they can be executed using local data, thus taking advantage of collocation of data and processing code. The NameNode server manages the file system and tracks the mapping of blocks to various DataNodes in the cluster. The NameNode periodically receives messages, known as a Heartbeat, about the health of DataNodes, and a report from each DataNode about the list of blocks known as a

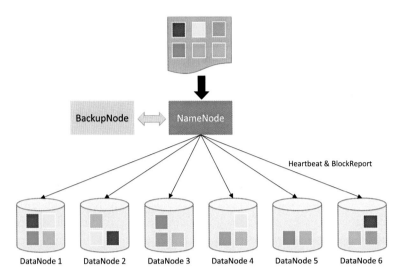

Figure 13.1 HDFS Architecture—data to be stored in files is partitioned and saved as a sequence of storage blocks on disk on DataNodes. The storage blocks are shown in different colors to illustrate the replication of blocks across nodes in the cluster.

BlockReport. If any of these messages indicate a machine failure, Hadoop has a built-in fault-tolerance mechanism to handle them that ensures reliability and resilience, and can reschedule jobs across servers to continue the data processing (Hadoop, 2015).

The Apache Hadoop stack includes applications for managing resources, jobs, and task scheduling such as Yarn and Oozie, and a nonrelational (NoSQL) database HBase, to make data processing efficient (Lam, 2010). A complete description of the Hadoop ecosystem is available at the official web site (Apache-Hadoop, 2015).

MapReduce

MapReduce (MR) is a programming model for processing and generating large datasets (Dean, 2008). Users specify a Map function that transforms a dataset to create intermediate results. A Reduce function processes the Map results to combine them to generate an output. Fig. 13.2 illustrates the Map and Reduce functions used to compute aggregated climate parameters, namely temperature (red), radiation (yellow), and humidity (blue) for points in an area from very large gridded climate datasets generated from satellite observations over a period of time. Bhardwaja (2015) provides another example of the MR process.

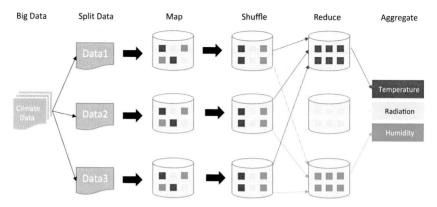

Figure 13.2 MapReduce (MR) Programming Model—using the Map and Reduce functions to compute aggregated climate parameters namely temperature (red), radiation (yellow), and humidity (blue) for points in an area from very large gridded climate datasets generated from satellite observations over time.

Let us consider point locations in a very large area with gridded climate parameter values that need to be analyzed to compute statistical mean, maximum, or minimum values of temperature, radiation, and humidity for a given time period. The example is simplistic but demonstrates the distributed workflow in the Hadoop framework. The input is a comma-separated variables (CSV) file containing records for latitude, longitude, time, and climate parameter names, and their values. The Hadoop framework takes the input climate CSV file and splits it into multiple parts that are sent to worker servers in the cluster. Each split file contains a set of CSV records for climate parameters in an unknown order. A Map function is applied to the dataset to sort CSV records for each climate parameter at each worker server. Sorting ensures that once the key value changes, there is no need to look for that variable any more. The output is shuffled to group temperature (red), radiation (yellow), and humidity (blue) records for each split file at worker servers. Finally, a Reduce function aggregates groups of climate records for each parameter across all worker servers to compute the average, minimum, or maximum values of temperature, radiation, and humidity for a given time period.

Hadoop Cluster Deployment in the Cloud

Hadoop clusters can be deployed in many ways that range from physical bare-metal clusters to virtual machine clusters hosted in the cloud (Wendt, 2014). A bare-metal cluster offers maximum control in handling spikes of load, but requires trained staff to manage the Hadoop cluster, and there are additional maintenance overheads and expenses incurred when it is necessary to replace outdated hardware. In a series of experiments with cluster configurations, Wendt (2014) demonstrated that a cloud-hosted Hadoop cluster could offer better price performance and greater flexibility in changing the central processing unit (CPU), random access memory (RAM), and disk storage of virtual machines that constitute the cluster to meet varying demands. Hadoop management is a sophisticated skill, so outsourcing maintenance of hardware and software to professional providers allows data scientists to better focus on analysis. There are many providers, such as Amazon Web Services (AWS), Google Compute Engine, International Business Machines (IBM), and others, that rent a Hadoop cluster in the cloud (Lam, 2010).

ANALYSIS OF CLIMATE DATASETS

Climate dataset analysis can range from computing simple statistics such as mean, minimum, or maximum values of climate parameters for a given area and time slice, to sophisticated statistical analyses that predict patterns of climate change over a long period of time. Climate data typically has following origins: in situ, remotely sensed imagery, model output and paleoclimate data (Faghmous and Kumar, 2013). These data are available in a variety of formats such as General Regularly-distributed Information in Binary form (GRIB)1, GRIB2, Network Common Data Format (NetCDF)3, NetCDF4, Hierarchical Data Format (HDF)4, HDF4-Earth Observing System 2 (EOS2), HDF5, HDF5-EOS5, Geographic Tagged Image File Format (GeoTIFF), and flat text files (National Center for Atmospheric Research (NCAR, The Climate Data Guide: Common Climate Data Formats: Overview, 2013)). To use them for analysis, this data can be either pre-processed using Extract-Transform-Load (ETL) models, or could be used unmodified as done by the NASA's Modern-Era Retrospective Analysis for Research and Applications (MERRA) project that uses the HDF dataset directly (Schnase, 2014).

Observations from sensors onboard Earth Observing System's (EOS) Terra and Aqua satellites are typically archived in the HDF-EOS format (NASA-ESSDR, 2012), whereas climate model-generated data are commonly stored in the NetCDF format (NCAR, 2015). GRIB is a format mostly used for weather analysis by meteorological departments, and is generally not used for climate datasets. Thus, a vast majority of climate datasets is available in some variation of HDF and NetCDF formats that store data in a gridded format.

Gridded data are point data, and can be transformed into matrices representing climate parameters for that location in multiple dimensions. These climate data matrices can be analyzed using numerical libraries, such as NumPy and SciPy (Stéfan der Walt, 2011), which can efficiently handle large matrices, and offer a variety of scientific computation capabilities applicable to climate datasets. These matrix algorithms are efficient, but do not scale as the size of climate datasets starts to exceed the computing capacity of the single computer running these processes sequentially. To scale, a distributed framework for matrix manipulation is needed. Apache Hama is a distributed computing framework built on top of HDFS for intensive scientific computations at scale to solve matrix, graph, and network problems (Hama, 2015). It uses the Bulk Parallel Synchronous (BPS) framework to communicate among peers in a cluster. The configuration of BPS jobs to perform tasks in parallel is similar to the MR interface, but is not the same (Hama, 2015).

Another way to harvest useful information from the climate datasets is to store them in a database, and query them to extract information using Structured Query Language (SQL) types of queries. However, Hadoop does not have natively built spatial and temporal search algorithms. New approaches have emerged for integrating spatial search techniques with the MR framework, and for storing indexed spatial datasets in the HDFS (Environmental Systems Research Institute (Esri), Geographic Information System (GIS) Tools for Hadoop, 2015; Eldawy and Mokbel, 2015a,b). Such spatial searches have direct applications to the domain of GIS, and are attractive because GIS offers many tools that can discover spatial relationships between climate data observed and predicted at sites with the environment and people around them. GIS also has cartographic tools that can be used to create maps showing these spatial relationships, and these maps can be published on the Internet using online mapping tools such as Esri's ArcGIS Online and Google Maps. Publishing climate knowledge and its impact, and making it accessible to the people can liberate the climate data locked in scientific formats. In addition, publishing climate impacts as maps on the internet is also likely to encourage discussions among stakeholders and policy makers to make better planning decisions that are guided by climate intelligence. These opportunities in GIS have created a lot of interest among researchers, and many alternative algorithms for querying large spatial datasets with applications to GIS have become available (Esri, GIS Tools for Hadoop, 2015; Eldawy and Mokbel, 2015a,b).

Gridded Datasets

Both HDF and NetCDF are binary data formats that typically store point observation data in a gridded format as shown in Fig. 13.3, in which each cell in the grid represents a value for a climate parameter for a given location and time. The time slice is shown for one climate parameter, but can be

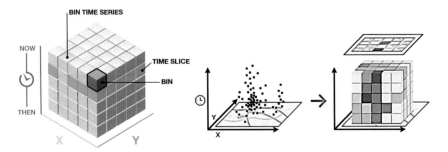

Figure 13.3 Climate parameter and time slice at a location over time as a unit of parallelization for the MR process (Esri, Space Time Cube, 2015).

easily extended to include altitude as an additional dimension. In addition, NetCDF and HDF datasets have metadata embedded in the data file: NetCDF has metadata in its header, whereas HDF has the metadata spread across the file. This encapsulation of data and metadata encourages machine independence, but is in conflict with transparency in data structure that Hadoop needs to partition the data. Hadoop can efficiently partition a text file such as CSV in which a valid partition can begin at any line. This is not readily accomplished for climate datasets in binary formats. For example, with binary climate data files, data slabs as shown in Fig. 13.3 can be used as a unit of parallelization for finding minimum, maximum, or averages, if they could be partitioned and stored as separate files on the HDFS. This mapping of logical data structure to physical storage structure on disk has been attempted using specialized tools, but with limited success (Buck, 2011). However, there are tools that can be used to transform HDF data slabs into CSV files, which can then be used for parallelization. However, this results in loss of HDF metadata, and it is an added overhead to analyze data.

The CSV representation of HDF dataset has been extensively used to migrate datasets to HDFS, and to build spatial indexes on them for efficient searches. In the following sections, selected Hadoop application frameworks are presented that are applicable to climate datasets.

Satellite Imagery

Traditionally, satellite sensing imagery has been a major source of climate data. The Landsat program has had satellites orbiting the Earth since 1972 to collect images at all locations every 16 days. These satellites have been an important source of information about vegetation and other land-use coverages. Landsat satellites are equipped with visible and thermal–infrared sensors to take images. An image covers an area of 12,000 square miles taken by sensors at an elevation of 700 km with a resolution of 100 to 120 m per pixel. The size of each image can be around 120 megabytes (MB) per band, and the data are not granular enough for the high throughput and network transfer required by the Hadoop framework.

Higher-resolution imagery has resulted in very large size of files that cannot be processed sequentially on a personal computer in a reasonable amount of time, hence parallel processing frameworks are needed (Almeer, 2012). Satellite images are binary files, and options to use them in a distributed framework range from using the file wholly as a unit of parallelization, or by splitting files by bands or into smaller tiles for parallelization. Example applications for both cases will be presented in the following sections.

DISTRIBUTED PROCESSING OF GRIDDED DATA

Processing large climate data using MR is not a novel approach (Buck, 2011). A multitude of frameworks have emerged to spatially enable MR to process such datasets. Many of these methods have some overlap and show various degrees of similarities, but a closer look reveals the differences. For example, some of these frameworks process the climate data unmodified or in situ, whereas others use an Extract, Transform, and Load (ETL) process to modify the climate data to ingest it in the HDFS for processing. ETL is a standard method for moving transformed data from one system to another. In situ processing of data has several advantages. Climate datasets are very large, and moving the data may not be feasible due to disk input/output (I/O) and network constraints, so algorithms that process the data unmodified at their repositories will scale better. However, given the binary nature of climate data, processing it in situ will also impose certain limitations in terms of partitioning it as units of parallelization. For cases in which users need to process a subset of climate data for a certain area and time, an ETL-based approach may be useful. NASA's Web site allows users to download subsets of climate data. ETL processes can be applied to convert these subsets of data to Geographic JavaScript Object Notation (GeoJSON), text, or CSV formats, and saved on HDFS. This is advantageous in that climate datasets are now available in desired formats, and custom algorithms can be developed. The disadvantage is that this may not be feasible for very large datasets. Further, using ETL sacrifices data portability as the metadata contained in NetCDF and HDF files are lost, and if this dataset is used to interface with other climate systems, the outputs will need to be converted back to NetCDF or HDF formats.

A review of various Hadoop frameworks in the literature show that, broadly speaking, most of the approaches for spatially enabling Hadoop for data analytics differ in the following criteria: input data, query language, spatial indexing, query, analytics, and visualization (Eldawy and Mokbel, 2015a,b). Based on the outlined criteria, a broad categorization of these ways of spatially enabling Hadoop is offered in the following. In the parentheses, the applicable methodologies described in the following sections are listed.

- Use ETL and add a spatial layer on top of MR (Esri GIS Tools for Hadoop, Hadoop-GIS)
- Use ETL and embed spatial capability into MR (SpatialHadoop, Esri Polygon-Map-Random (PMR) tree)
- Use ETL to move data into a distributed database like Apache HBase
- In situ optimization of unmodified binary climate data file for MR functions (SciHadoop)
- In situ processing of unmodified climate data as sequence files and delivering climate analytics as a web service (NASA MERRA/Analytic Service (AS))

GIS Tools for Hadoop

GIS Tools for Hadoop is an open-source project by Esri (2013) that allows users to spatially enable Hadoop by adding a layer of user-defined spatial functions on top of the MR layer. Fig. 13.4 shows a simplified illustration of stack for Esri's GIS Tools for Hadoop.

The stacks are color-coded for easier interpretation. The Big Data storage is shown in yellow, processing algorithms in blue, and query layer in light yellow colors. The application layer is shown at the top. The task managers, namely Ambari or Oozie, are shown in gray color. These task managers schedule the MR jobs and manage the workflow.

GIS Tools for Hadoop have the following additional components: Hive Spatial to be built using User Defined Functions (UDF) based on the Esri Geometry Application Programming Interface (API) (Esri-Geometry, Esri Geometry Java API, 2015) to add spatial capability, and ArcGIS Geoprocessing tools to move data in and out of HDFS. After the HDF data are transformed into GeoJSON files, these are copied to the HDFS using geoprocessing tools shown in the application layer at the top of the stack. In the Hive Spatial layer, the query is interpreted to create an MR job that performs the search.

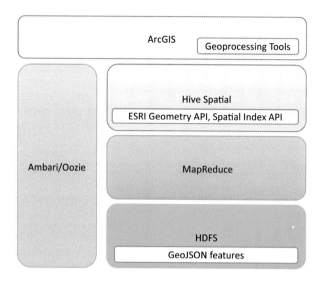

Figure 13.4 Esri's GIS Tools with PMR-Tree Spatial Index.

Apache Hive is a data warehouse system built on top of Hadoop for querying large datasets using an SQL-like language called Hive Query Language (HiveQL). The Hive Spatial component of this framework extends Hive by adding user-defined functions (UDFs) that are built using the Esri Geometry API and are modeled on the Geometry Supertype (ST_ Geometry) Open Geospatial Consortium (OGC)–compliant geometry type (Esri-Hive, 2015). Hive UDF allows adding custom Map or Reduce functions into a HiveQL query. This allows data scientists to avoid directly using MR and allows them to focus on analytics following a familiar workflow. Under the hood, Hive converts the query into an MR job, which runs on the cluster and generates a result.

Consider an example in which average temperature needs to be computed for counties in the United States given point features with temperature observations obtained from climate datasets such as NetCDF and HDF. The input data are county polygons and point features with temperature values in a geodatabase. ArcGIS Geoprocessing Tools can be used to convert ArcGIS features to CSV/JSON format in which geometries are expressed as text in Well Known Text (WKT) or Well Known Binary (WKB) format. The data are copied to the HDFS, and a HiveQL command is executed to find average temperature for each county by filtering points contained in each county, and then aggregating temperature values for those points to obtain the average temperature for each county. A sample HiveQL command is shown as follows:

```
SELECT counties.name, AVG (temperature) AvgTemp FROM counties
JOIN points
WHERE ST_Contains(counties.boundaryshape, ST_Point(points.
longitude, points.latitude))
GROUP BY counties.name
ORDER BY AvgTemp desc;
```

Notice that the query looks like a standard SQL command. This HiveQL query is converted into an MR job which partitions the input points (JSON/CSV) for worker nodes, copies all county polygons to each worker node, builds a spatial index in memory at each node for filtering, and executes a map job to find points within each polygon to compute average temperature for each county. A Reduce function is applied to aggregate results from each node to obtain the average temperature for county polygons in the United States. This result is converted back to JSON format using a geoprocessing tool for viewing in GIS software. Esri Hive Spatial supports an array of geometrical operations for constructing geometries, testing spatial relationships (ST_Contains, ST_Crosses, ST_Intersects, ST_Overlaps, ST_Relate, ST_Touches, ST_Within), supporting geometric operations (ST_Boundary, ST_Buffer, ST_Intersection, ST_Difference, ST_Union), and providing other operators (ST_Area, ST_Centriod, ST_Distance etc.) to gain valuable spatial insights from data.

In an experiment to find points in polygons, Esri reported a processing time of 4 min to search 175 million points in polygons using Hive ST_Geometry UDF on a cluster of 20 commodity servers (Murphy, 2015). Each server was running the Linux CentOS-6.5 operating system (OS), and was equipped with 1 TB hard drive, 16 GB of RAM, and an Intel Xeon 3.07 GHz quad-core CPU.

Spatial Polygon-Map-Random Tree Indexing and Analytics

A limitation of the GIS Tools for Hadoop framework is that all polygon features need to be copied to each node. Points are partitioned across each worker nodes. An MR job is invoked for each query that computes if a point is contained within a polygon. Thus, points at each node are examined against all county polygons resulting in inefficient search. To address these limitations, Whitman (2014) presents a method that combines spatial indexing, data load balancing, and data clustering to optimize performance across cluster. A key feature of this approach is that it builds a global spatial

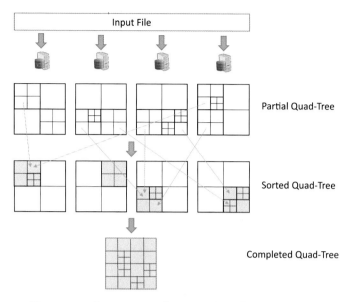

Figure 13.5 Construction of PMR quadtree for spatial search.

search index using the MR framework across the cluster, and stores it in the HDFS at the master node. The global index allows a random access query without requiring an MR job. This is useful for rapid visualization of features in ArcGIS software in which a pan and zoom operation requires a fast random spatial query to filter features for the current extent of display.

Authors use the PMR quad tree (Nelson, 1986) to build a spatial index that decomposes the global search space into disjointed quads in a regular pattern. This works well with point geometries, as they are most likely found in a disjoint quad, and offers strategies to handle points that are found at the edges of a quad.

The PMR tree is built across nodes in a cluster in a series of steps as illustrated in Fig. 13.5. The process begins with the generation of a tabular dataset from source point data in which each row describes a point geometry in WKT/WKB format along with its attributes. The algorithm can accept delimited text, CSV, JSON, or Hadoop sequence file. The Map function partitions the source CSV file into subsets of CSV records, and these are physically distributed to worker nodes in the cluster. On each worker node, a partial quadtree is built using subset of source features at that node. A custom partition function is written to assign each entry in a partial quadtree generated to a worker node. Finally, a Reducer function is invoked at worker nodes that combines these partial quads to build a locally consistent quadtree. These local quadtrees are combined to generate a global PMR quadtree using a merge and sort process. Finally, the global PMR tree structure is stored in the HDFS.

Whitman et al. (2014) further discusses strategies to optimize the construction of a PMR tree for point features. This is relevant to climate datasets as NetCDF and HDF can benefit from it. One of the strategies is to build the PMR tree by reorganizing the source data to take advantage of the ordering discovered within the index. For load balancing, the authors recommend merging quads to polygonal quads that are not necessarily rectangular. This allows faster indexing because the source data are spatially sorted and the load is better balanced across worker nodes. This technique is used to demonstrate an implementation of range and nearest neighbor (k-NN) spatial analytics. A range query finds data points within a certain range of values defined by user. Nearest neighbor analysis finds points that are proximal and most similar. A speedup of up to 20 times with range query was reported with ordered source datasets. The cluster configuration is same as described in the GIS Tools for Hadoop section.

Hadoop-GIS

Hadoop-GIS is a spatial data warehousing software for executing fast queries and handling complex spatial data on large volumes of spatial data using Hadoop (Aji, 2013). Hadoop-GIS differs in the way it generates the spatial index and queries data for analytics. The primary case studies reported are pathology image processing and an OpenStreetMap data querying (Aji, 2013). The authors do not address a climate use case, but this approach is relevant because it addresses spatial features and images that are large raster datasets.

An overview of architecture of Hadoop-GIS is shown in Fig. 13.6. Hadoop-GIS adds a data-splitting algorithm, query engine, query translation, and a query language stack on top of Hadoop. For a typical spatial operation, a spatial query is generated in a custom query language stored-procedure query language (QL^{SP}), which is optimized, translated, and submitted to a query engine called Real-Time Spatial Query Engine (RESQUE), and the query result is stored in HDFS. Features and spatial indexes are also stored in the HDFS.

Data partitioning is done by splitting the spatial data into subsets called tiles that can be processed in parallel across worker nodes. Two major considerations in generating these tiles is to ensure that the spatial density of

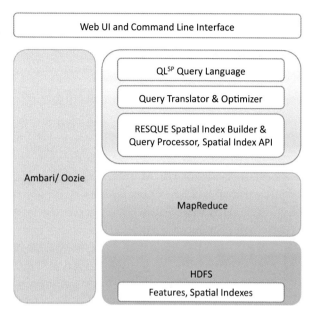

Figure 13.6 Hadoop-GIS architecture.

features for these tiles are in the same range for load balancing, and that appropriate checks are incorporated to handle boundary features that straddle tiles. Tiles are the unit of parallel processing across worker nodes, and an effort is made to keep them disjointed so a spatial query can be done within that tile only, but if there are overlapping features across tiles, RESQUE has remedying features in place to ensure accuracy in query results.

RESQUE implements a global- and local-level indexing scheme for fast spatial query. An ETL process is used to generate CSV data that is partitioned to create tiles that are assigned a tile identifier. The Minimum Bounding Rectangles (MBRs) of tiles are maintained in a global index and carried through in a binary format in HDFS. The global index is shared across nodes using Hadoop cache mechanism. For each tile, a local index of features is built on demand in memory. Thus, when a spatial query is submitted, the global index is used to search for the relevant tile, and a local index is built in memory for that tile to search for the feature. Hive^{SP} is the query component of Hadoop-GIS that extends Apache Hive with spatial query constructs, spatial query translation, and execution. It also offers an integrated query language called QL^{SP} which is an extension of Apache HiveQL. The spatial query engine of Hive^{SP} uses an open-source Goddard Earth Observing System (GEOS) software, and supports spatial relationships (intersects, touches, overlaps, contains, within, disjoint), spatial operations (intersection, union, distance, centroid, area), and spatial access methods (R^*-*Tree*, *Hilbert R-Tree*, *Voronoi Diagram*) for efficient query processing. Hadoop-GIS was run on a cluster with 8 nodes and 192 cores. Each node has 24 cores, 128 GB of memory, and 2.7 TB of disk storage. The OS was CentOS 5.6 (64 bit), and Cloudera distribution of Hadoop framework was used.

The current version of Hadoop-GIS also features a Spatial Data Partitioning Framework, named SATO, that can quickly analyze and partition spatial data with an optimal partitioning strategy for scalable query processing (Vo, 2014). SATO stands for the Sample, Analyze, Tear, and Optimize steps of partitioning spatial data. SATO begins with an input dataset in CSV format in which each record has features with geometry in WKT format and attributes. It samples the input dataset by applying a stratified sampling approach that uses a small fraction of the dataset (1–3%) to build a histogram that captures the spatial feature density distribution. This histogram is analyzed to spot areas of high density and examine the density distribution to generate a global partitioning scheme of tiles that seeks to minimize the number of intersecting features across tiles. SATO features six partitioning algorithms that can be configured to control the HDFS block size of tiles to avoid file fragmentation and improve I/O performance. This is followed by the Tear step in which these partitions are actually created, and tiles are generated for specific block sizes. Finally in the Optimize step, statistics on features in tiles and intersecting boundary features are computed. If a tile of very high density is discovered while rescanning the dataset, the partitioning algorithm is reapplied to divide high-density tiles to derive tiles the feature densities for which are close to the target density. Similarly, for very low-density partitions, neighboring tiles are merged to achieve the target density. Histograms and tile generation for data chunks of the input CSV file can be generated independently across nodes by a Mapper function, and a Reducer function can aggregate them. The computations to build an index were done on a 50-node Amazon Elastic MapReduce (EMR) cluster, in which each node is an extra-large EMR instance with 15 GB of memory, four virtual cores and four disks with 420 GB storage.

Spatial Hadoop

In Hadoop-GIS, the spatial index is added on top of the Hadoop stack; as a result, MR programs cannot access the spatial index to create new spatial operations. Spatial Hadoop is an MR framework the goal of which is to provide native support for spatial data consistent with PostGIS

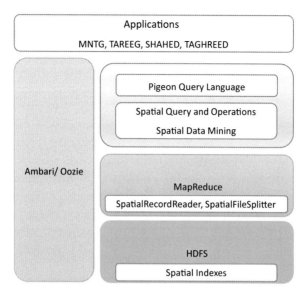

Figure 13.7 SpatialHadoop system architecture.

(Eldawy and Mokbel, 2015a,b). Spatial capabilities are enabled by embedding code to add spatial constructs and spatial awareness directly into the Apache MR framework. However, the native support is limited in terms of spatial predicates offered (Eldawy, 2015), but is in contrast with the approach taken by Hadoop-GIS that extends Apache Hive to add spatial capabilities to Hadoop, or other approaches like Multidimensional HBase (MD-HBase) (Nishimura et al., 2013) that extend HBase to support spatial indexes and queries. In addition, SpatialHadoop also supports spatial index types such as *R-Tree* that are stored in the HDFS. This kind of integration allows users to directly interact with Hadoop to develop additional spatial functions such as range query, k-NN, and spatial join.

The SpatialHadoop framework consists of a query language, operations, HDFS, and MR layers as shown in Fig. 13.7. Because not all spatial queries are available natively in the Spatial MR framework, SpatialHadoop provides a language layer called Pigeon which is an extension of the Apache Pig Latin Query language, and is based on spatial libraries that include Esri Geometry API (Esri-Geometry, Esri Geometry Java API, 2015) and Computational Geometry_Hadoop (CG_Hadoop) (Eldawy, 2013). Pigeon has an SQL-like syntax that supports OGC-compliant spatial data types, functions, and operations. This allows Pigeon to override the Hadoop native data type to support points, lines, and polygons. It also offers UDFs to provide spatial functions like Union, Buffer, k-NN, spatial filter, and spatial join. The query layer sits on top of the MR layer. Spatial indexes are stored in the HDFS, and the MR layer has spatial constructs directly embedded in it.

Similar to Hadoop-GIS, SpatialHadoop also employs a two-level indexing approach that builds global and local index to search features. But there are differences. Local indexes are stored in HDFS of worker nodes, and a global index is created in memory by aggregating local indexes. Spatial index building in SpatialHadoop is composed of three steps: partitioning, local indexing, and global indexing. Input data are partitioned in such a way that each partition occupies one HDFS block of 64 MB size, spatial features are nearby, and a roughly equal number of storage blocks is assigned to worker nodes to balance load. The Map function partitions the input data and assigns each record or feature to a partition, and the Reduce function builds a local index for each partition and stores it in the HDFS of the worker server. The Reduce function aggregates local indexes from all worker nodes, and generates a global index. The global index is loaded in the memory of the master node for querying. If the master node fails, the global index is regenerated by aggregating the local indexes. SpatialHadoop supports many indexing techniques including grid file, R-$Tree$, and R^+-$Tree$. The goal is to keep these partitions spatially disjointed as much as possible. If there are features that overlap partitions, they are addressed later by the query processor to ensure that correct query results are returned. The SpatialHadoop framework makes it possible to realize many spatial operations such as range query, k-NN, and spatial join. SpatialHadoop has been successfully implemented in many applications for querying and visualizing large scale satellite datasets (Eldawy and Mokbel, 2015a,b).

SpatialHadoop enhances the native MR functions of Hadoop to handle spatial features such as SpatialFileSplitter and SpatialRecordReader. The SpatialFileSplitter function uses the global index to split a file in a way that prunes data storage blocks that do not contribute to the answer. The SpatialRecordReader uses the local index to efficiently process the split file. For spatial queries, SpatialHadoop uses CG_Hadoop (Eldawy, 2013) for larger resolution, and for finer resolution, it uses Esri Geometry API (Esri–Geometry, Esri Java Geometry Library, 2015).

In an experiment, the SpatialHadoop framework was applied to analyze satellite imagery on vegetation indexes for the entire world for a period of 14 years that had 120 billion points with a total size of 4.6TB. The vegetation dataset was indexed using R-$Tree$ on an Amazon Elastic Compute Cloud (EC2) cluster of 100 large nodes, each with a quad core processor and 8 GB of memory running on Linux operating system. Spatial indexing took 15 h, and a query to select features took 2–3 min.

SciHadoop

SciHadoop is a plug-in to Apache Hadoop for querying unmodified or in situ array-based binary scientific datasets for common data analysis tasks that include mean, median, minimum, or maximum values in space and time (Buck, 2011). Scientific datasets such as NetCDF or HDF expose a logical view of data storage, and there is no information usually available about their physical layout and distribution on storage blocks of disk. SciHadoop presents algorithms to discover the physical layout of the storage. Using this knowledge, a Map function extracts an input data slab such that a partition of data is located completely on one node server. The goal is to extract a time slice of data as shown in Fig. 13.1 such that it is stored in HDFS blocks located on one worker node only. To discover the physical layout of data on disk, the storage structure pattern of data on disk such as column or row major order of array is exploited. For example, a NetCDF data array is stored in column major order on disk. The partitioning strategy uses this knowledge to split data.

Buck et al. (2011) note that the partitioning strategy achieved a locality of 71% for simple array structures with one variable (temperature), but dropped to 5% when multiple variables were present. Hence, the efficacy of the method to decompose logical space into local physical blocks depends upon the sophistication of the user and the degree of complexity of the data model. In general, there was a misalignment between the physical and logical data models that could not be discovered accurately in all conditions.

SciHadoop computations were performed on a Linux cluster of 31 nodes with 2×1.8 GHz Opterons, 8 GB RAM, and 4×250 GB disk drives. In an experiment with a 132 GB NetCDF dataset containing air pressure for multiple days and altitudes, the query for median air pressure with optimization strategies took about 25 min of computation time.

MERRA Analytics Service

MERRA is a reanalysis dataset that synthesizes data from Earth observing systems, such as satellites, and the results from numerical climate models to create an integrated dataset that is temporally and spatially consistent (Schnase, 2014). MERRA consists of 26 climate variables available, from 1979 to the present, at various spatial resolutions and altitudes, and is made available by NASA Center for Climate Simulation (Rienecker et al., 2011). MERRA datasets are available in NetCDF and HDF format, and are stored unmodified in HDFS as sequence files in storage blocks of 640 MB. This is significantly larger than the default size of 64 MB, but the advantage is that the larger size facilitates storage of large HDF and NetCDF files in a few storage blocks on a worker server, thereby ensuring locality. Sequence files are a special data structure in Hadoop that stores a key and value pair of file name (key) and file content (value). The sequence file has metadata about markers that define the beginning and end locations of a file in HDFS. This allows handling binary files that are larger than the default 64 MB block size of HDFS. There is no spatial index employed for a search, so a full scan of the dataset is needed to find the data of interest. Using a web interface, a user submits a request to get average, minimum, maximum, sum, variance, or count—known as canonical operations—on climate parameters for a selected spatial and temporal extent. A custom Map function processes sequence files to generate a list of parameter and time pairs, which is further aggregated using a specialized Reduce function that applies canonical operators to create a subset of sequence files with the results. Finally, these files are desequenced to generate NetCDF files and are available for download by the user. The MERRA AS runs on a 36-node Linux cluster that has collectively 576 cores rated at 2.6 GHz, 1300TB of storage, and 1250 GB of RAM.

Distributed Database for Climate Dataset Analytics

HDFS is limited in that is does not support a random-access read of the data. Apache HBase (HBase, 2015) is an open-source distributed, versioned, NoSQL, or nonrelational, database that natively allows random access and indexing of data. HBase typically stores data in HDFS in a cluster of computers, though it is not a requirement and other storage types are available. NoSQL databases are schema free and fault tolerant, and there are many choices available including Apache HBase, Apache Cassandra, and others (NoSQL, 2015). Li (2015) presents a cloud-based and service-oriented architecture built on MR and HBase to analyze climate and other geoscience datasets. HBase stores data as key and value pairs, so a naïve implementation would store climate datasets as (key, value) pairs in which key is file Uniform Resource Identifier (URI) or name, and the value is the binary NetCDF and HDF file. This approach can work for smaller-sized files, but will pose problems for files exceeding GBs (Li, 2015). To address this problem, the author presents a strategy that uses an ETL to decompose NetCDF and HDF datasets into slabs of data (reference Fig. 13.3 showing time slice) in which each slab contains data for a given climate parameter, and a selected time or spatial location at multiple altitudes. To identify the dataset, the dataset ID, time, and variable name are used to create a unique key that points to spatial data stored as functions of latitude, longitude, and altitude. The key and spatial data are stored in HBase for retrieval of the dataset later for other purposes. These atomic datasets are the units of parallelization used by the MR interface for analyzing the climate dataset.

DISTRIBUTED PROCESSING OF SATELLITE IMAGERY

Traditional methods of processing millions of high-resolution satellite images of on a single computer in a sequential manner can take an unreasonable amounts of time as they do not scale; therefore, a distributed framework such as Hadoop is needed for processing images in parallel (Almeer, 2012). Because the imagery data are binary, Hadoop sequence files can be used to ensure collocation of data and processing code. For Hadoop implementation, images can be split for parallelization, or the entire imagery file can be treated as a unit of parallelization. In the following section, example applications of Hadoop targeted toward image classification and image processing will be presented (Codella et al., 2011; Yadav and Padma, 2015).

Image Classification

Image classification has multiple uses. It can be used to identify different areas by the type of land use. Land-use data are used extensively for urban planning. High-resolution imagery is also used during to natural disasters such as floods, volcanoes, and severe droughts to look at impacts and damage. Codella et al. (2011) present a Hadoop-based distributed computing architecture for large-scale land-use identification from satellite imagery. To determine land use, semantic taxonomy categories such as vegetation, building, pavements, etc. are established. Training data are obtained from GeoEye public domain, and the imagery is divided into 128×128 pixel size tiles with $0.5\,m$ resolution. These tiles are units of parallelization for Hadoop implementation. These data are manually categorized for various land-use types to ensure that they are correctly identified in training data. IBM's Multimedia Analysis and Retrieval System (IMARS) is used to train the data. IMARS is a distributed Hadoop implementation of a Robust Subspace Bagging ensemble Support Vector Machine (SVM) prediction model for classification of imagery data. Using the SVM classifier, a collection, or bag, of features and training data for different semantics is generated. For each bag, an SVM model is generated. Imagery downloaded from Microsoft's BING Maps is used to test the accuracy of training. The crawled BING images are also processed to generate tiles of 128×128-pixel size. In the Reduce step, an SVM model validation score for each bag is evaluated, and the best SVM model parameters are used to test the efficacy of the training in correctly classifying the BING imagery data. An image classification workflow in Hadoop is shown in Fig. 13.8 that also shows different sets of images used for training, validation, and evaluation.

The system architecture consists of a dual-rack Apache Hadoop system with 224 CPUs, 448 GB of RAM, and 14 TB of disk space. The testing of 102,900 images in the San Diego area took 11.6 h to complete.

Image Processing

Sometimes images need to be enhanced for better resolution. For instance, oceans, seas, and large water bodies can appear dark in an image. To expose more details, these images need to be enhanced in intensity. Color sharpening, autocontrast, and Sobel filtering are common image-processing algorithms that can also be used in a distributed system using Hadoop to process large amounts of images (Almeer, 2012; Wang et al., 2015; Lin et al., 2013).

In a work by Almeer (2012), the original satellite images are processed to reduce their sizes so they could be read into the memory of a computer. The MR interface is enhanced to handle the imagery as one unit of data for processing. The sharpening algorithm uses a linear operator to compute the new pixel value from its original value and the surrounding pixel values. A Sobel filter is also used to enhance the edge using a different linear operator applied to update pixel values. Autocontrast is another method used for increasing the intensity in which pixel values are automatically adjusted based on the range of pixel values available. In an experiment by authors, 200 Landsat images at a spatial resolution of 30 m were used as a unit of parallelization, and no splitting was allowed. The images were processed on a cluster of nine computers with 18 GB memory and 1.2 TB of disk space, and time that varied with image size but was reported to be less than 5000 s.

Wang et al. (2015) partitioned the image files by band for image processing. An illustration of image splitting based on color band is shown in Fig. 13.9. In this example, the imagery is decomposed into three color bands, namely red, green, and blue bands. These bands are distributed across the worker nodes to process them. Once processed in the Reduce step, these bands are assembled to generate the actual color imagery. Notice that these images can have any number of bands representing different information for

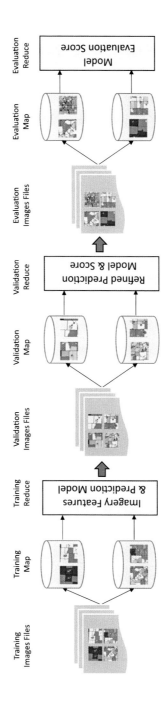

Figure 13.8 Image classification using predictive modeling in a Hadoop framework.

Images Files   Split Images   Map   Shuffle   Reduce
by Band

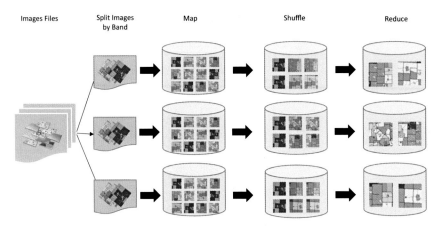

Figure 13.9 Image processing for edge detection in a Hadoop framework.

each band. The work was performed with Landsat imagery, using Apache Hadoop on a cluster of four nodes each with 16 GB memory and 256 GB Solid State Drive (SSD).

In another application, El-Kazzaz and El-Mahdy (2015) use Landsat imagery to detect land mines that have a different thermal signature than that of the ground. These spectral signatures are specially accentuated at dawn and dusk, and thus can be possibly used for the identification of land mines. For data management, HBase (NoSQL) database is used to store each image as one record, and no splitting of images is done. The resolution of the Landsat imagery is 100 m per pixel. The authors used image enhancement techniques to reconstruct high-resolution imagery to aid in the detection of land mines.

DISCUSSION

Observation satellite sensors and climate numerical models are rapidly generating vast amounts of data. In fact, growth of geoscience data has surpassed the capability of existing methodologies and infrastructure for data access, archiving, analysis, and mining (Li, 2015). Hadoop has emerged as a promising framework to analyze very large climate datasets. As an example, the Climate Corporation has been successfully using Hadoop for distributed computing to support farmers' decisions on precision agriculture and crop insurance (Khaliq, 2012). Their Hadoop system analyzes millions of weather observations and trillions of simulation data points to develop a weather forecast that is used to price insurance policies for farmers to protect them from losses. In the following sections, differences and applicability of various

Hadoop applications presented previously are discussed. In addition, some promising areas of future work toward efficient and scalable analytics on climate datasets are outlined. Finally, Apache Spark, which has recently emerged as another alternative for cluster computing in the cloud (Apache-Spark, 2015), is also discussed briefly.

Differences in Spatial Hadoop Frameworks

The multidimensional spatial structure of climate datasets makes spatial queries and operations complex. It is important to note that popular climate data formats such as NetCDF and HDF were designed as archival data formats, and they were are not designed for random and rapid spatial queries. Many approaches to handle spatial datasets in Hadoop have emerged to solve the problems of spatial queries and analytics with large climate datasets. In the previous sections, some selected approaches were presented. This field is still evolving. The methods presented previously differ in the way they accept input data, partition data, build spatial indexes, support spatial query and operations, use storage, and allow visualization of processed output climate datasets. Other differences include options present to extend the approach to build better frameworks either by adding new UDFs on top of the MR stack (Hadoop-GIS), or by embedding new spatial constructs and functionalities within the MR framework itself (SpatialHadoop, PMR Tree). Table 13.1 summarizes these differences. Spatial operations listed in Table 13.1 can be considered building blocks for more sophisticated queries in the future (Whitman, 2014).

As observed in Table 13.1, the first point of distinction among these approaches is the input data itself. The input data could be pipelined to HDFS either unmodified (in situ) or after processing it using an ETL procedure. The ETL data offers the most flexibility in that the data can be transformed into any format including text, JSON, etc. with geometry stored as WKB/WKT, and it can be readily partitioned by Hadoop. However, modifying climate datasets results in the loss of machine independence and portability of the data, making it difficult to use it in other systems that may require metadata. On the other hand, climate dataset repositories can be very large, and it may not be feasible to move data to process it using an ETL approach. The only option in such cases is to process data at its location by moving the processing to the data. To partition binary climate files, researches have tried to map logical data structures to physical data to identify partitions that can be treated as units of parallelization for MR functions. However, this approach has met with limited success for complex datasets with multiple variables. MERRA/AS used another approach in

Table 13.1 Differences between various Hadoop approaches

Case studies	Input data	Spatial indexing	Data partition	Spatial operations	Visualization
GIS tools for Hadoop	ETL	In-memory local *R-Tree* spatial indexing at data node	Text split	Spatial Hive extension, Esri geometry API	ArcGIS suite
Esri spatial indexing and analytics	ETL	*PMR-Tree*-based global spatial index stored in HDFS	Text split	Embedded spatial operations, range query, *k*-NN	ArcGIS suite
Hadoop-GIS and SATO	ETL	*R*★*-Tree, Hilbert R-Tree, Voronoi* structure carried through globally, and local index created in memory on demand	Text and image split	Spatial Hive extension, spatial join, aggregation, containment queries	Tools for viewing outputs
SpatialHadoop	ETL	*R-Tree, R+-Tree, Quad-Tree, K-d-Tree* carried through at local data node, and global index created in memory	Text split	Embedded extensible interface, range query, *k*-NN, spatial join, skyline, convex hull, polygon union	MapReduce-based output image generation at multiple resolutions
SciHadoop	In-situ	None	Binary file split	Canonical operations—*average, variance, max, min, sum, count*	Charts for outputs
MERRA/AS	In-situ	Index to search sequenced NetCDF file	Sequence file	Canonical operations—*average, variance, max, min, sum, count*	Web app to aggregate maps and data. NetCDF visualizer
Distributed database	ETL	NoSQL column indexing	NoSQL columns	Canonical Operations—*average, variance, max, min, sum, count*	GeoServer to publish results as Web Mapping Service

which it converted NetCDF files to Hadoop Sequence files and loaded it in HDFS with large default storage block size. Records in a sequence file represent a NetCDF/HDF file, and are valid units of parallelization.

Spatial indexing is required for efficiently searching large climate datasets. Consequently, many alternatives to index spatial data based on variations of *R-Tree* have emerged that are either kept in memory or carried through to HDFS. Researchers report that spatial indexes provide faster searches when the dataset queried is a small fraction of the total dataset, but when the percent of data queried is large, in the neighborhood of 40%, MR can perform competitively (Whitman, 2014). However, note that in practice, queries are performed to get a fraction of the data, so spatial indexing is a good strategy to consider. Table 13.1 describes some of the indexing strategies, storage options exercised, and spatial operations supported as key factors used as points of distinction among spatial Hadoop frameworks.

Visualization of results was either done by generating output images using MR, or results were exported out of the HDFS into JSON files that were imported into GIS software for viewing. For storing data, HDFS is the default storage available with Hadoop, but options that make use of HBase distributed database, built on top of HDFS or other NoSQL databases that also offer indexing facility, can also be used.

Apache Spark for Cluster Computing

Apache Spark is a cluster-computing framework that processes data in memory of the servers in a cluster (Zaharia, 2010). Spark is based on the concept of Resilient Distributed Datasets (RDD) (Zaharia, 2012). These datasets are partitions of source data that are loaded by a software library into the memory of worker nodes for computation. Thus, there is no need to write files to the disk as in Hadoop; consequently, there is no I/O operation overhead to write intermediate results to disk. As a result, Spark achieves performance gain of orders of magnitude over MR (Apache-Spark, 2015). Application of Spark to NetCDF and HDF files requires NetCDF (NCAR, 2015) and HDF (Heber, 2015) software libraries to read data slabs or time slices (Fig. 13.1) in memory without reading the entire dataset. These data slabs are the possible units of parallelization, and can be used to build Spark RDDs. Spark processes data by applying functions to an existing dataset to transform it to generate a new dataset. If data loss occurs, Spark has fault-tolerance mechanisms built in place that allow functions to be reapplied to regenerate the lost dataset. A sequence of transformations produces the final result. A prototype of Spark application to HDF datasets looks promising (Heber, 2015).

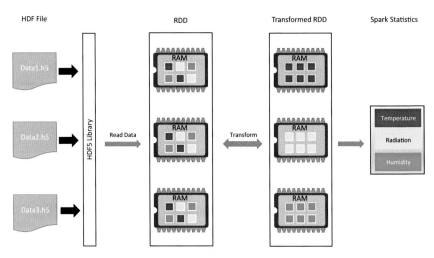

Figure 13.10 Spark Architecture—The HDF files contain temperature (red), radiation (yellow), and humidity (blue) data for locations and time period. Data are physically stored in blocks on disk. An HDF library is used to read slabs of climate data, and load them in memory of worker node server.

Fig. 13.10 shows the architecture of Apache Spark. The HDF files contain temperature (red), radiation (yellow), and humidity (blue) data for locations and time period. Data are physically stored in blocks on disk. An HDF library is used to read slabs of climate data, and load them in memory of worker node server. These data slabs are transformed by applying a function that groups data at worker node for each climate variable. These grouped data slabs at each worker node are transformed by applying a function that sorts them across worker nodes, and groups them at a worker node for a climate variable. Finally, the climate variable grouped at a worker node is transformed using Spark statistics functions that compute summary statistics that include minimum, maximum, average, variance, and histogram on aggregated data slabs. Note that the transform function is shown with a bidirectional arrow suggesting transform in either direction to obtain a backward or forward state of data.

Future Work

Whitman et al. (2014) has identified a few topics for further research work in Spatial Hadoop. In particular, new techniques are needed to enhance the two-dimensional (2D) spatial indexing algorithms to three-dimensional (3D) or higher dimensions to allow faster queries in time and space. This will require extending the 2D PMR quadtree index to a 3D PMR octree. In addition, it should be possible to incrementally update these indexes for data added in the future. Further, current data structure of climate datasets, such as NetCDF and HDF, were designed for archiving data, and there is a need to enhance it for indexing. Finally, though some progress has been made, adding advanced spatial constructs and operations inside the MR codebase would yield better performance and offer more customization opportunities.

CONCLUSION

Satellite observations and numerical climate models are rapidly generating large volumes of climate data. Climate data are needed by stakeholders in wildfire management, agriculture, tourism, water, energy, and other sectors. Timely analysis of climate data is critical to its utility. Apache Hadoop, based on the idea of dividing the problems into smaller parts and solving them on a cluster of commodity servers, has emerged as a potential solution to analyzing large climate datasets. There is no standard Hadoop framework that can be applied to all climate analysis problems. Selection of a method will depend on the nature of the problem being solved and factors related to data processing location, spatial indexing, spatial operations, and visualization. Hadoop is a widely used and mature framework, though it has limitations. Recently, Apache Spark has emerged as an alternative to the disk-based Hadoop. Because Spark is a memory-based cluster-computing platform, it is likely more expensive to use because disk is cheaper. Studies have found that Hadoop's management requires significant skill, so deploying Hadoop in cloud is an attractive option as it allows outsourcing its management to professional providers, allowing users to focus on data analytics. Cloud-based Hadoop also offers greater flexibility in rapidly altering the cluster configuration, and can offer a better price–performance ratio. Overall, cluster computing using commodity hardware running Hadoop is a promising option for the future of scalable and affordable climate data analytics needed in pursuit of climate intelligence for public use.

REFERENCES

Aji, A.W., 2013. Hadoop-GIS: a high performance spatial data warehousing system over MapReduce. Proceedings of the VLDB Endowment International Conference on Very Large Data Bases 6 (11), 1009.

Almeer, M.H., 2012. Hadoop map reduce for remote sensing image analysis. Journal of Emerging Trends in Computing and Information Sciences 637–644.

Apache-Hadoop, July 2015. Welcome to Apache™ Hadoop®!. Retrieved from Welcome to Apache™ Hadoop®!: https://hadoop.apache.org/.

Apache-Spark, July 2015. Apache Spark – Lightning-fast Cluster Computing. Retrieved from Apache Spark: https://spark.apache.org/.

Bhardwaja, N., July 2015. Hadoop Slides. Retrieved from Salesforce: http://www.slideshare.net/narayan26/data-sciencewebinar-061312-13372205.

Buck, J.B., Watkins, N., LeFevre, J., Ioannidou, K., Maltzahn, C., Polyzotis, N., and Brandt, S., 2011, November. Scihadoop: Array-based query processing in hadoop. In: Proceedings of 2011 International Conference for High Performance Computing, Networking, Storage and Analysis (p. 66). ACM.

Codella, N.C., Hua, G., Natsev, A., Smith, J., 2011. Towards large scale land-cover recognition of satellite images. In: 8th International Conference on. IEEE, pp. 11–15.

Cugler, D.C., 2013. Spatial big data: platforms, analytics, and science. GeoJournal http://www.spatial.cs.umn.edu/geojournal/2013/geojournal.pdf. https://scholar.google.com/citations?view_op=view_citation&hl=en&user=jokN2OcAAAAJ&citation_for_view=jokN2OcAAAAJ:5nxA0vEk-isC.

Dean, J.G., 2008. MapReduce: simplified data processing on large clusters. Communications of the ACM 51 (1), 107–113.

Dirmeyer, P.A., 2014. Comparing evaporative sources of terrestrial precipitation and their extremes in MERRA using relative entropy. Journal of Hydrometeorology 15 (1), 102–116.

Eldawy, A., Mokbel, M., 2015a. SpatialHadoop: a MapReduce framework for spatial data. In: Proceedings of the IEEE International Conference on Data Engineering (ICDE'15). IEEE.

Eldawy, A., Mokbel, M., 2015b. The ecosystem of SpatialHadoop. SIGSPATIAL Special 6 (3), 3–10.

Eldawy, A.L., 2013. CG_Hadoop: computational geometry in MapReduce. In: Proceedings of the 21st ACM SIGSPATIAL International Conference on Advances in Geographic Information Systems. ACM, pp. 294–303.

Eldawy, A., July 2015. OGC Esri Spatial. Retrieved from SpatialHadoop: http://spatialhadoop.cs.umn.edu/doc/edu/umn/cs/spatialHadoop/core/OGCESRIShape.html.

El-Kazzaz, S., El-Mahdy, A., 2015. A Hadoop-based framework for large-scale landmine detection using Ubiquitous Big Satellite Imaging Data. In: Parallel, Distributed and Network-based Processing. IEEE, Turku, pp. 274–278. http://dx.doi.org/10.1109/PDP.2015.121.

Esri, 2013. ESRI Tools for Hadoop. Retrieved July 2015, from: http://esri.github.io/gis-tools-for-hadoop.

Esri, July 2015. GIS Tools for Hadoop. Retrieved from GIS tools for Hadoop: http://esri.github.io/gis-tools-for-hadoop/.

Esri, July 2015. Space Time Cube. Retrieved from ArcGIS help: https://desktop.arcgis.com/en/desktop/latest/tools/space-time-pattern-mining-toolbox/space-time-cube.htm.

Esri-Geometry, July 2015. Esri Geometry Java API. Retrieved from: https://github.com/Esri/geometry-api-java.

Esri-Geometry, July 2015. Esri Java Geometry Library. Retrieved from Esri Java Geometry Library: https://github.com/Esri/geometry-api-java/wiki.

Esri-Hive, July 2015. Hive Spatial. Retrieved from Hive Spatial: https://github.com/Esri/spatial-framework-for-hadoop/wiki/Hive-Spatial.

Faghmous, J., Kumar, V., 2013. Spatio-temporal data mining for climate data: advances, challenges, and opportunities. In: Advances in Data Mining. Springer.

Hadoop, A., July 2015. HDFS Architecture. Retrieved from HDFS Architecture: http://hadoop.apache.org/docs/r2.7.0/hadoop-project-dist/hadoop-hdfs/HdfsDesign.html.

Hama, A., September 27, 2015. Apache Hama. Retrieved from Apache Software Foundation: https://hama.apache.org/.

HBase, July 2015. Apache HBase. Retrieved from Apache HBase: http://hbase.apache.org/.

Heber, G., July 2015. From HDF5 Datasets to Apache Spark RDDs. Retrieved from HDF Blog: https://hdfgroup.org/wp/2015/03/from-hdf5-datasets-to-apache-spark-rdds/.

Khaliq, S., 2012. Helping the World's Farmers Adapt to Climate Change. O'Reilly Publishers, New York. Retrieved from: http://strataconf.com/stratany2012/public/schedule/detail/25140.

Lam, C., 2010. Hadoop in Action. Manning Publications Co.

Li, Z.Y., 2015. Enabling big geoscience data analytics with a cloud-based, MapReduce-enabled and service-oriented workflow framework. PLoS One 10 (3), e0116781. http://dx.doi.org/10.1371/journal.pone.0116781.

Lin, F.-C., et al., 2013. Storage and processing of massive remote sensing images using a novel cloud computing platform. GIScience & Remote Sensing 322–336.

Manyika, J.C., 2011. Big Data: The Next Frontier for Innovation, Competition, and Productivity. Retrieved from: http://www.mckinsey.com/Insights/MGI/Research/Technology_and_Innovation/Big_data_The_next_frontier_for_innovation.

Mell, P., 2011. The NIST Definition of Cloud Computing.

Miller, H.J., 2004. Tobler's first law and spatial analysis. Annals of the Association of American Geographers 94 (2).

MODIS, July 2015. MODIS Web. Retrieved from MODIS Web: http://modis.gsfc.nasa.gov/data/.

Murphy, J., July 2015. Setting up a Small Budget Hadoop Cluster for Big Data Analysis. Retrieved from ArcGIS Blog: http://blogs.esri.com/esri/arcgis/2014/07/28/setting-up-a-small-budget-hadoop-cluster-for-big-data-analysis/.

NASA-ESSDR, 2012. Earth System Science Data Resources (ESSDR). Goddard Space Flight Center, NASA, Greenbelt, MD 20771.

NCAR, December 17, 2013. The Climate Data Guide: Common Climate Data Formats: Overview. Retrieved from National Center for Atmospheric Research: https://climate-dataguide.ucar.edu/climate-data-tools-and-analysis/common-climate-data-formats-overview#sthash.f0Obk27u.07GwPLaC.dpuf.

NCAR, July 2015. NetCDF Operators. Retrieved from NetCDF Operators: http://nco.sourceforge.net/.

NCCS, July 2015. Retrieved from NASA Center for Climate Simulation: http://www.nccs.nasa.gov/.

Nelson, C.S., 1986. A consistent hierarchical representation for vector data. ACM SIGGRAPH Computer Graphics 20 (4), 197–206 (ACM).

Nishimura, S., Das, S., Agrawal, D., Abbadi, E., 2013. MD-HBase: design and implementation of an elastic data infrastructure for cloudscale location services. DAPD 31 (2), 289–319.

NoSQL, July 2015. NoSQL Database. Retrieved from NoSQL Database: http://nosql-database.org/.

Overpeck, J.T., 2011. Climate data challenges in the 21st century. Science (Washington) 331 (6018), 700–702.

Rienecker, M.M-K., et al., 2011. MERRA: NASA's Modern-Era Retrospective Analysis for Research and Applications. Journal of Climate 3624–3648.

Schnase, J.L., Duffy, D.Q., Tamkin, G.S., Nadeau, D., Thompson, J.H., Grieg, C.M., McInerney, M.A., and Webster, W.P., 2014. MERRA analytic services: Meeting the big data challenges of climate science through cloud-enabled climate analytics-as-a-service. Computers, Environment and Urban Systems. http://dx.doi.org/10.1016/j.compenvurbsys.2013.12.003. http://ntrs.nasa.gov/archive/nasa/casi.ntrs.nasa.gov/20140013036.pdf.

Sinha, A., 2015. Atmospheric satellite observations and GIS. In: Armstrong, L. (Ed.), Mapping and Modeling Weather and Climate with GIS. Esri Press, pp. 51–64.

Stéfan van der Walt, S.C., 2011. The NumPy Array: a structure for efficient numerical computation. Computing in Science & Engineering 13, 22–30. http://dx.doi.org/10.1109/MCSE.2011.37.

Vo, H.A., 2014. SATO: a spatial data partitioning framework for scalable query processing. In: Proceedings of the 22nd ACM SIGSPATIAL International Conference on Advances in Geographic Information Systems. ACM, pp. 545–548.

Wang, C., Hu, F., Hu, X., Zhao, S., Wen, W., Yang, C., 2015. A Hadoop-based distributed framework for efficient managing and processing big remote sensing images. ISPRS Annals of the Photogrammetry, Remote Sensing and Spatial Information Sciences 63–66.

Wendt, M.E., 2014. Cloud-based Hadoop Deployments: Benefits and Considerations. Accenture Technology Labs.

Whitman, R.T., Park, M.B., Ambrose, S.M., and Hoel, E.G., 2014, November. Spatial indexing and analytics on Hadoop. In: Proceedings of the 22nd ACM SIGSPATIAL International Conference on Advances in Geographic Information Systems (pp. 73–82). ACM.

Yadav, R., Padma, M., 2015. Processing of large satellite images using Hadoop distributed technology and MapReduce: a case of edge detection. International Journal on Recent and Innovation Trends in Computing and Communication 3456–3460.

Zaharia, M.C., 2010. Spark: cluster computing with working sets. In: Proceedings of the 2nd USENIX Conference on Hot Topics in Cloud Computing, vol. 10. USENIX, p. 10.

Zaharia, M.C., 2012. Resilient distributed datasets: a fault-tolerant abstraction for in-memory cluster computing. In: Proceedings of the 9th USENIX Conference on Networked Systems Design and Implementation. USENIX Association.

CHAPTER 14

LiveOcean

R. Fatland, P. MacCready
University of Washington, Seattle, WA, USA

N. Oscar
Oregon State University, Corvallis, OR, USA

INTRODUCTION

LiveOcean (see Fig. 14.1) is a hybrid-technology solution for producing and conveying ocean state forecasts as actionable information to the marine industry. This chapter describes the LiveOcean solution in terms of technical components, project motivation from ocean acidification (Ocean Acidification), scenarios for data use, forecast validation, and extension of the project through community growth and emerging cloud technology.

LiveOcean (LO) represents a bridging project, one that originates strictly from the research domain of physical oceanography, that then bridges across to practical use: Initially intended for shellfish farming with applicability to many other marine uses.

LiveOcean originated from a study commissioned by the State of Washington, United States, under the guiding question: What are drivers of ocean acidification in the region, how do different species respond to it, and what are possible mitigation strategies for commercial fisheries? The study was conducted from February to November 2012 by the Washington State Blue Ribbon Council on Ocean Acidification. Results are available from and coordinated by the Washington State Department of Ecology (DOE). The study produced five recommendations; the resulting Request for Proposals for an ocean acidification forecasting system was successfully answered by Parker MacCready, Professor of Oceanography at the University of Washington. The consequent development of LiveOcean as this forecast system is the subject of this chapter.

LiveOcean has proceeded as three parallel efforts: Operationalizing the Regional Ocean Modeling System (ROMS) ocean circulation model, building the LiveOcean server cloud middleware, and producing client views of the forecast data. This third component proceeds from the LiveOcean design emphasis on decoupling the physical model from the use of the forecast data.

Cloud Computing in Ocean and Atmospheric Sciences
ISBN 978-0-12-803192-6
http://dx.doi.org/10.1016/B978-0-12-803192-6.00014-1

Copyright © 2016 Elsevier Inc.
All rights reserved.

Figure 14.1 The LiveOcean project logo enables quick visual identification of project materials.

LIVEOCEAN PROJECT MOTIVATION

The report motivating LiveOcean is entitled "Ocean acidification: From knowledge to action, Washington's strategic response" [1]. From this report:

> Today's ocean acidification is important not only for the amount of change that has occurred thus far but also for how quickly it is happening. The current rate of acidification is nearly ten times faster than any time in the past 50 million years, outpacing the ocean's capacity to restore oceanic pH and carbonate chemistry. The rapid pace of change also gives marine organisms, marine ecosystems, and humans less time to adapt, evolve, or otherwise adjust to the changing circumstances. At the current rate of global carbon dioxide emissions, the average acidity of the surface ocean is expected to increase by 100–150 percent over preindustrial levels by the end of this century…
>
> More than 30 percent of Puget Sound's marine species are (directly) vulnerable to ocean acidification by virtue of their dependency on the mineral calcium carbonate to make shells, skeletons, and other hard body parts. Puget Sound calcifiers include oysters, clams, scallops, mussels, abalone, crabs, geoducks, barnacles, sea urchins, sand dollars, sea stars, and sea cucumbers. Even some seaweeds produce calcium carbonate structures.

The implications of this for the marine industry are both direct and indirect. Directly, the acidification of the ocean has a uniformly deleterious impact on calcifiers. Indirectly, the impact on those calcifiers in turn affects the entire marine food web. LiveOcean allows us to model, forecast, and visualize the acidity of ocean waters off Washington and surrounding regions. Furthermore, the LiveOcean model described here forecasts other critical conditions in the ocean: hypoxia, harmful algal blooms, temperature, current extremes, and more.

PAST WORK: ROMS VALIDATION

The Regional Ocean Modeling System (ROMS) is a computational fluid dynamics ocean model widely used by the scientific community for a diverse range of applications. These include—as in our case here—the predictive modeling of water flow and water properties in and near estuaries given forcing input from wind, runoff, and tides. ROMS has been validated primarily using mooring data (locations where a vertical string of instruments to measure water properties is deployed for days to months) and Conductivity, Temperature, and Depth (CTD) station casts (locations where a research vessel will measure conductivity, temperature, and depth profiles from the surface to the sea floor). Both mooring locations and CTD stations are shown in the following figure (see Fig. 14.2) in relation to the LiveOcean forecast region (see also Fig. 14.3).

LIVEOCEAN TECHNICAL COMPONENTS

ROMS Model Forecast Generation

LiveOcean is based upon the modeling capabilities of ROMS adapted to the coastal waters off of Washington, Oregon, and British Columbia (see Fig. 14.4). The following figure shows the extent of the ROMS forecast superimposed on a BING map (Bing) with salinity color coded from orange-brown (most saline) to white (freshwater).

LiveOcean Data Structure

ROMS is a community computational fluid dynamics model widely used by oceanographic researchers around the world. It has been developed and validated over the course of two decades (MacCready et al., 2009; Liu et al., 2009; Sutherland et al., 2011; Giddings et al., 2014; Davis et al., 2014; Siedlecki et al., 2015). For LiveOcean it has been tailored, mainly by the grid and bathymetry specification, to the ocean waters of the northeast Pacific Ocean.

An ROMS model run proceeds from this set of initial conditions:

1. River forecasts for 16 rivers from National Oceanic and Atmospheric Administration (NOAA) (NOAA Rivers) or USGS with infill from climatological data. Discharge from other (ungauged) rivers is estimated from past studies (Mohamedali et al., 2011).
2. Openocean conditions from Hybrid Coordinate Ocean Model (HYCOM), sponsored by the National Oceanographic Partnership Program (NOPP) (HYCOM; NOPP).

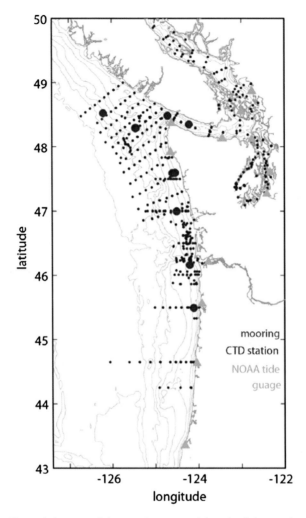

Figure 14.2 Physical domain of the LiveOcean model and validation data. The model covers both coastal waters of the northeast Pacific, and the inland waters of the Salish Sea (Puget Sound, Strait of Georgia, and Strait of Juan de Fuca). Model resolution is optimized to work efficiently in coastal waters, and a nested high-resolution Puget Sound sum model is planned for the coming year.

3. Current high-resolution atmospheric fields from WRF simulations by Cliff Mass' group at the University of Washington.
4. Tidal forcing for each day from maps of global tidal phase and amplitude generated at Oregon State University.
5. Initial ocean state is taken from the Hour 24 result from the previous day's ROMS model forecast.

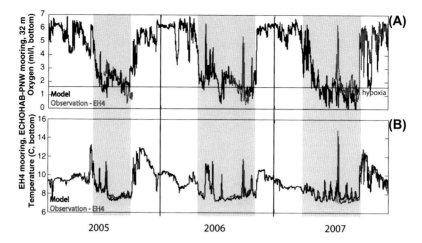

Figure 14.3 Model simulations are compared to observations at many locations, and time series from moorings have the highest value to this process owing to their stability and data volume. Here, bottom dissolved oxygen (top) and temperature (bottom) are compared over 3 years, with the summer upwelling season shown in gray. The station is in the middle of the Washington shelf. Model predictions (black) show good skill in this location at representing both the seasonal cycle and the storm events evident in the moorings (red). The equipment used to gather these observational records, collected from research vessels by Dr. Barbara Hickey at University of Washington, are difficult and expensive to maintain in this environment, as reflected in the patchy time series. *Figure adapted from Siedlecki, S.A., Banas, N.S., Davis, K.A., Giddings, S., Hickey, B.M., MacCready, P., Connolly, T., Geier, S., 2015. Seasonal and interannual oxygen variability on the Washington and Oregon continental shelves. Journal of Geophysical Research Oceans 120, doi:10.1002/2014JC010254.*

From these initial conditions the ROMS model produces an hourly forecast of ocean conditions for the ensuing 72 h. This forecast is analogous to a weather forecast, but it extends from the ocean surface down to the sea floor across the region of interest (ROI).

A ROMS forecast is a dataset representing hour-by-hour three-dimensional water state across approximately 20 scalar parameters including temperature, salinity, pH, aragonite saturation state, dissolved oxygen, phytoplankton, zooplankton, and detritus; as well as the ocean current vector field that advects and mixes these properties.

In operation, ROMS runs on a 72-core high-performance computing (HPC) cluster to produce 72 hourly Network Common Data Format (NetCDF) files representing water conditions at 40 stratified layers across ~200 × 400 map-plane cells covering a rectangular region from 43° to 50°N latitude and 127.5° to 122°W longitude. The resulting 344,000 square

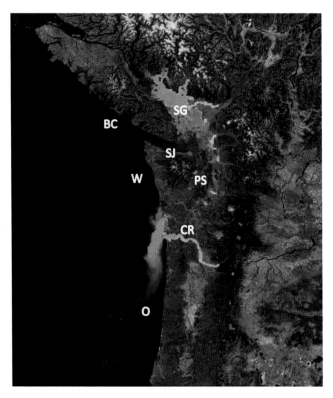

Figure 14.4 LiveOcean region of interest including coastal Washington, Oregon (United States), and British Columbia (Canada), the Strait of Juan de Fuca, the Strait of Georgia, Puget Sound, and the Columbia River estuary. Labels are, respectively, W, O, BC, SJ, SG, PS, and CR. Color indicates salinity forecast estimate at a single moment in time, white corresponds to freshwater.

kilometer region off the coast of Washington state (United States) results in individual grid cells approximately 2×2 km. Map-plane grid spacing is not even: Finer resolution is used in proximity to the coast to better support complex fluid flow in those areas.

The vertical stratification of each map-plane cell is fixed in proportion to mean water depth for that cell. Due to ocean bathymetry, therefore, the cells are not vertically aligned from one map-plane column to another. The cells are more finely spaced near the surface and near the seabed, and an algorithm is provided for converting between cell index, bathymetry, and physical depth. The ROMS data structure is thus driven by a compromise between precision and computing power.

The ROMS output files are in NetCDF format, a standard binary format with extensive metadata used by scientists working with gridded data.

These files are not readily consumed by the LiveOcean target audience. This suggests a translational layer that we describe in the following as cloud middleware built using a modular design approach.

Motivating the Applications Programming Interface (API)

Files as the traditional computer medium of information exchange present no trouble when shared "one-to-one" between like-minded software systems. Over time as information technology evolves, a profusion of file types emerges, and software is written to adapt to those file types. This creates a degree of fragility in the software ecosystem when a research program turns to a new data resource. The solution is often painful and time-consuming: Find, install, learn, and use a library or a tool to get information from new data resources of interest.

We now consider the case in which file transfer and interpretation is replaced by an Applications Programming Interface (API). An API defines functionality without defining how the functionality is implemented by defining standard inputs, outputs, and operations. Cloud computing is associated with services, and the API is the common implementation of the "how" of that service model. The API addresses the file Babylon problem by dispensing with files. That is, the API provides structured access to data within the file as a service. To deconstruct this further: "structured access" means well-defined specification of some part or subset of the larger file or data structure; and "service" means that the preparation and delivery of that subset is undertaken by the resource, by the Producer. In this manner, the file recedes into the background to be replaced by structures that are more fluid such as tables or objects.

All common programming languages now support API-based data access. In consequence, the programming skill and the necessary steps to obtain information using an API are reduced. Furthermore, in best practice the consumer can bootstrap from the API, i.e., learn what information is available via meta-information services that are part of the API. Finally, the data Producer may have information stores that span many files, types, and data structures. The API represents a contract for how data are accessed without stipulating how it is produced. This gives the Producer flexibility in operation while providing the Consumer with a consistent interface.

Finally, an API is not a simple panacea. Developing good APIs requires careful consideration and effort, as APIs can be found to be elegant and efficient or difficult and unwieldy. As noted, one argument for API implementation is the standardization of data access, in turn resulting in less time developing software. To carry this argument a step further, the API model in concert with the

```
# 'get-info' tells what data are available
apiReply = requests.get(urlparse.urljoin("http://liveocean.azurewebsites.net/api/", "get-info"))
print 'Call =', apiReply.url, '; reply status is', apiReply.ok, '; status code is', apiReply.status_code, '\n'

# convert this reply to json and print some diagnostics; unicode u'xxxx' converted to ascii (easier to read)
rjson = apiReply.json()
pk=[]
for i in range(0,len(rjson)):
        pk.append(rjson[i]['PartitionKey'].encode('ascii','ignore'))
print 'JSON reply as ascii, available data types from Live Ocean API:',pk,'\n'

Call = http://liveocean.azurewebsites.net/api/get-info ; reply status is True ; status code is 200

JSON reply as ascii, available data types from Live Ocean API: ['rho', 'salt', 'temp']
```

Figure 14.5 This iPython Notebook fragment demonstrates a prototyping environment for building an understanding of the LiveOcean API. Python modules support API interaction over the commonly used JavaScript Object Notation (JSON) format. Content is abstracted, however, so that "data format" considerations are replaced by a simple working knowledge of Python syntax. In this simplified case, LiveOcean tells the Client that it can provide three types of forecast data: Water density, salinity, and temperature.

burgeoning open-source community produces easily adopted tools for providing and consuming data and information. Consequently, modular design is encouraged over end-to-end solution building, meaning that it is easier for new consumers to use data for previously unanticipated purposes.

LiveOcean Cloud Middleware

LiveOcean cloud middleware acts to translate forecast data into consumable information by means of an API. This approach modularizes the task of disseminating information and can be contrasted with the more common "Web site/portal" approach. In the latter case, a research group will design, build, and maintain a Web site—often called a data portal—that is intended to anticipate and meet the Consumer's information needs (see Figs. 14.5 and 14.6). Although such data portals may solve the task of data dissemination, they face several common pitfalls. Data may become stale; the portal server or software may break down; operational costs can become burdensome; new data can be difficult to integrate, and so on. As described previously, the API approach separates data forecast contents from the Consumer's interest in a limited subset of the forecast information.

Once built, an API may need refinement to reduce query latency (delays in pulling data from a provider). LiveOcean uses cache generation (stockpiling typical results) to reduce query latency. This process is initiated as soon as a new forecast arrives in cloud storage and typically takes less than 1 h to complete, comparable to the time necessary to generate the forecast results.

Another important feature of the LiveOcean cloud middleware is to provide administrative oversight and information for Developers who would like to understand and use the API. These features are provided by

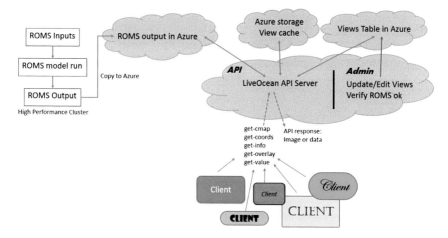

Figure 14.6 LiveOcean system architecture. ROMS model is configured and run (left) on an HPC cluster. Results are migrated to the Azure cloud in native NetCDF format. The LiveOcean cloud middleware (large cloud at center right) is responsible for translating that data into images or smaller data blocks in response to API requests. Support is also provided for color maps, grid coordinates, and a list of available parameters or views into the full dataset. The middleware also supports administrative tasks. A host of Clients at the bottom of the diagram are all shown interacting with LiveOcean via the same API.

means of a LiveOcean administration Web site (LiveOceanServer). This Web site includes API documentation, working API examples, reports on ROMS model run history (success/fail status on component tasks), and other state-of-health system information.

LiveOcean Clients

As described below LiveOcean is built on a modular design principle: Build system components as needed and minimize dependencies between those components.

In the context of cloud computing we generally see a connection from person to cloud by means of some device and attendant software. That software is referred to as a Client because of its relationship to the cloud as a results provider. In the case of LiveOcean, the shellfish grower would traditionally go to a Web site specifically designed to provide an ocean pH forecast from a LiveOcean "pH" service for the next 3 days and that would satisfy the purpose of the project. The Client is simply a Web browser loading information via a Universal Resource Locator (URL). However, suppose LiveOcean proves skillful at forecasting harmful algal blooms, or heavy zooplankton concentrations, or concentrations of salmon. Such skills might be discovered through unsupervised learning algorithms applied to the 20 LiveOcean prediction parameters combined with external data such as catch records and satellite imagery. Suppose further that LiveOcean can predict salinity gradients in coastal waters that imply mixing of freshwater from rivers with upwelled water from off the continental shelf. None of these practical or research applications are served by the original ocean pH forecast Web site built for the shellfish growers but the information could be provided by new services. Hence, we suggest here the idea of a profusion of *types* of data consumers connecting to a variety of services.

Because LiveOcean produces a very rich forecast in terms of types of information, LiveOcean Clients can consequently be seen as aftermarket applications that enable Consumers and Researchers to receive precisely the information they are interested in. As a technical example: An aftermarket LiveOcean Client onboard a fishing vessel could poll the LiveOcean cloud instance to notice the arrival of the day's forecast. It could then pull selected information—suppose these are ocean temperature, salinity, pH, and chlorophyll forecasts—to arrive at the forecast for an optimal fishing location within 80 km of Astoria, Oregon. These coordinates could be transmitted via very high frequency (VHF) to other company fishing vessels at sea with the forecast optimal location appearing as a symbol on a chart plotter. This assigns a certain degree of intelligence to the Client; but one of the strong points of the cloud paradigm is the option of placing that intelligence within the LiveOcean cloud resource as a derived service.

The current examples of LiveOcean Clients include a Web browser "data explorer" interface in which the forecast can be shown and adjusted as a BING map overlay with time-series animated playback (Yoyodyne). This overlay includes transparent pixels in which no data are present (land) and can be adjusted in opacity with respect to the BING map. Another Client might just as easily use a different map service.

A second LiveOcean Client connects LiveOcean forecast data to a visualization engine called Worldwide Telescope (WWT) with support for four to six dimensions of data including time. A third existing LiveOcean Client is an iPython Notebook (iPython) which serves as a demonstration platform for API access through the Python programming language.

LiveOcean Modular Design

In the research context the ROMS model is a tool for studying the physical ocean, an end in itself. LiveOcean will carry ROMS model results to a new endpoint: Maritime commercial operations. As an example: Young oysters ("spat") are distributed over oyster beds when ocean acidity will hopefully remain low (high pH) for several days to encourage shell growth. High confidence in an ROMS forecast of such conditions will enable shellfish growers to operate with less risk to their cultivation process.

In general, there is a natural break between a ROMS model run and the dissemination/consumption of the resulting forecast data. Commonly a forecast is expressed as a layered map with overlays loaded in a Web browser. As noted previously, in the LiveOcean project we modularize forecast consumption into two components: Middleware and Client. The Middleware provides a data service API; and the Client is software that acts on behalf of the information consumer. The Consumer is presumed interested in only a small subset of the ROMS model output, for example the aragonite saturation forecast at a set of oyster beds for the next 3 days. This Consumer represents one of potentially dozens of different forecast consumers. LiveOcean if successful, therefore, has a one-to-many relationship. By adopting a modular approach for LiveOcean, we address two aspects of this relationship: The information consumption is easily tailored, and the interpretation of the forecast information is not proscribed.

The LiveOcean middleware is the translational layer between the ROMS model output and the Client. It provides data according to three principles: The query–response mechanism is simple and well documented, the data are simple and readily understood, and the response latency is minimal. This middleware is implemented in the Python programming language using the Django Web Application Framework (Django). The software was developed in the Microsoft Visual Studio programming environment and deployed to the Microsoft Azure cloud. Hence, LiveOcean as a solution represents a synthesis of four technologies: Python, Django, Visual Studio, and Azure.

One facet of cloud implementation projects such as LiveOcean is the Developer experience. How should a system be designed, built, tested, maintained, and modified? As a subfacet: Integrated Development Environments (IDEs) such as Eclipse and Microsoft Visual Studio require upfront investment of time to learn but can reward that time by simplifying and reducing time to develop, deploy, and debug cloud solutions. These considerations sketch a meta-problem: Are there justifiable benefits to treating software development as an engineering process incorporating best practices to produce robust, reliable solutions at the cost of time? This question is increasingly impacted by the proliferation of open-source software.

The potential value of modularity in LiveOcean design extends further: Suppose that a need develops for LiveOcean forecast data as a Web

Mapping Service. Such a service could be built as a separate module that gets input using the standard API. The types of allowed LiveOcean queries could be extended forward and backward in time, to wit: "Where did the water at this location come from?" "Where will the water at this location go?"

Modular design also allows the development and rapid implementation of new prediction tools. Common phenomena such as Harmful Algal Blooms (HABs) and nutrient upwelling along the continental shelf have important consequences in the relationship between coastal ecosystems and the people who live there. The original ROMS model framework was created as part of a large National Science Foundation–NOAA-funded project to study the ecology and oceanography of HABs (Siedlecki et al., 2015; Davis et al., 2014; Giddings et al., 2014). If LiveOcean provides insight to better understanding these events, then we will be significantly closer to predicting them in advance.

These and other examples suggest that modular design principles are critical to keeping the system from becoming an overwhelming heap of source code.

LiveOcean API Details

The LiveOcean API is a computer-to-computer interface for Clients to get forecast information from the LiveOcean middleware. A query is in the form of a base URL with text appended, technically a Hypertext Transfer Protocol (HTTP) POST operation. There is no authentication required, and the interface is stateless and a service based on Representational State Transfer (RESTful). Therefore, two things are necessary for a Client to successfully interact with LiveOcean: The Client must correctly encode a query in the URL, and it must correctly parse the returned data.

The API vocabulary consists of five actions.

1. get-info provides information on what information is currently available from LiveOcean (JSON)
2. get-cmap provides a color map for rendering overlay images (JSON)
3. get-coords provides the coordinate grid of the LiveOcean map space (JSON)
4. get-value provides a map space sheet of floating point values (JSON)
5. get-overlay provides a grayscale image of parameter/time/depth (PNG)
   There are in some cases appended qualifiers.
1. A parameter choice: "salt" for salinity, etc.
2. A date-time.

3. A depth, either as model index or as a negative-valued depth, in meters. Due to the way the model output is created and stored, the extraction of output at a specified depth is surprisingly difficult and easy to do incorrectly, hence is hidden from the Client via the API.

The following URL is a get-value API call with three qualifying clauses:

```
http://liveocean.azurewebsites.net/api/get-value?date=2015-05-
28T22%3A00%3A00Z&depthMeters=-10.0&param=salt
```

The response is a block of text in JSON format, a dictionary with three entries: max, min, and data. "max" and "min" define the allowed range of the data. "data" is a two-dimensional array of floating point salinity values for the ocean cells at a depth of 10 m for May 28, 2015, at 2200 h. Nonocean cells in this array have salinity value 'null'.

```
{
    "max": 9.999999933815813e+36,
    "data":
        [[32.61193,
          32.61436,
          32.61267,
          --etcetera--
          null]],
    "min" : 0.0
}
```

LiveOcean Validation

The initial "use" of LiveOcean as a forecast system is the extension of its validation history using external observations to determine where and when it is most accurate. As noted previously, the ROMS model has been subjected to careful validation scrutiny from independent sources of data, primarily from moorings and sampling stations (Giddings et al., 2014; Davis et al., 2014; Siedlecki et al., 2015). Validation against moorings will continue as a matter of course. Extension to validation against other data can encompass CTD casts from research cruises, Array for Real-time Geostrophic Oceanography (ARGO) data, glider data, and expansion of the Ocean Observatories Initiative (OOI) cabled observatory (OOI). Independent verification of ROMS model output could also be aided by instrumentation placed about vessels of opportunity such as fishing boats and cargo ships.

One of the most important tests of ROMS/LiveOcean validity will be its track record in the study of HAB development and transport (such as the evolution of the large persistent HAB found in the northeast Pacific in

Figure 14.7 2015 NOAA Suomi National Polar-orbiting Partnership (NPP) Satellite-derived chlorophyll concentration off California, Oregon, Washington, British Columbia, and southeast Alaska (HAB).

summer 2015) (see Fig. 14.7). The LiveOcean ROMS model includes a biochemical component that predicts both phytoplankton and zooplankton concentrations (but not ecological parameters such as species distributions or toxicity). Demonstrating skill at predicting plankton distributions would be an invaluable first step in establishing the value of LiveOcean to HAB research.

The validation process addresses the question of whether and where LiveOcean can reliably provide higher-level or derived information to the marine industry. This is an opportunity for high-level cloud services—in cloud industry jargon, "data analytics"—to contribute to the validation story by searching for correlations and patterns in the large and growing body of forecast data accumulating from ROMS runs. As an example, suppose that commercial catch records compare favorably with the history of LiveOcean forecasts of zooplankton concentrations. This could facilitate a win–win scenario in which more efficient LiveOcean-guided fishing practices produce good economic results for the industry and enable choices that replenish fish stocks and reduce bycatch. There is also potential for misuse (overfishing for example) of hypothetical predictive skill.

FURTHER SCENARIOS FOR LIVEOCEAN USE

The following ideas are presented to indicate the future use and potential value of LiveOcean. To begin with, a phenomenal success would be a case in which commercial fishers—starting with shellfish growers—consult LiveOcean forecasts on a daily basis in the manner of a sailor consulting a weather forecast. As new use cases emerge, there is always the backstory of the opportunity to create new Clients and new capabilities within the LiveOcean cloud solution.

LiveOcean Data Use in Year-Over-Year Time Scales

Suppose the initial positive validation results for the ROMS model hold up in future studies as LiveOcean becomes operational, including hindcast reanalysis. The implication of this result will be that the forecast data represent the actual state of the ocean at some confidence level. After LiveOcean has produced such results for an extended period of time (months to years), a new opportunity will emerge for longitudinal analysis.

One example is the analysis of coastal upwelling: Does the LiveOcean forecast record suggest seasonal patterns of nutrient supply to the coastal/shelf waters of the region of interest? Does this upwelling signal correlate with trends in marine coastal ecology? Does it correlate with independent high trophic level data such as industrial fisheries catch records?

LiveOcean circulation results represent physical ocean transport and extend to plankton estimates. Plankton are defined to be living creatures that do not overcome prevailing currents with their own motility and can therefore be taken as "swept along in physical water mass." A fascinating potential consequence of an accurate LiveOcean forecast is the implication for higher trophic levels above those of plankton (see Fig. 14.8). For example, a carnivorous and highly mobile fish such as a salmon can sample and respond to gradients in prey density which in turn will have dependencies along the trophic chain down to plankton. LiveOcean forecasts do not currently predict distributions of high trophic level predators. However, we anticipate that LiveOcean could become a component of a predictive system in concert with other data streams as might be provided by in situ autonomous robots (gliders, ARGO profilers, etc.), moorings, and tethered sea-floor sensor networks.

The following figure was produced as a simple dynamic visualization of hypothetical whale paths grazing along maximum local surface zooplankton gradients in the LiveOcean forecast over 72 h (see Fig. 14.9).

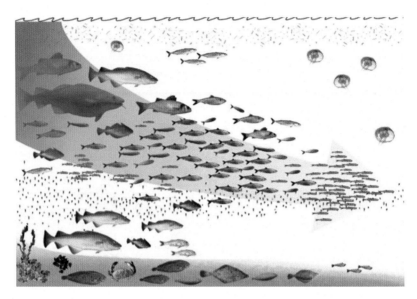

Figure 14.8 Trophic diagram implying spatial correlation between plankton and commercial fish. Such an hypothetical correlation could be explored by comparing LiveOcean forecasts with catch records.

Figure 14.9 Oblique view of coastal waters west of Washington State showing hypothetical cetacean trajectories based on following zooplankton gradients generated by the LiveOcean ROMS model.

The zooplankton concentration is rendered as a grayscale overlay on which the fuzziness in the near–field image over land reflects the model resolution. This represents another interesting possibility in the validation process supposing that live telemetry was available from tracking devices on feeding baleen whales.

Growth and Adoption of the LiveOcean Cloud Solution

LiveOcean has many possible ways to grow. From its forecast data it could provide current directions and turbulence estimates in regions with complex flow fields such as Puget Sound. By outsourcing wind forecasts, it could provide high-resolution sea-state forecasts, rip-current warnings, and more. LiveOcean's intrinsic maritime operations forecast data could be supplemented to pass along other marine information such as tide data available from NOAA. As a validated research tool it could provide biomass estimates and inform carbon cycle models. These "value adds" suggest further possibilities of unanticipated use given support for model publication.

Publication support is a growing practice that is familiar in the context of YouTube. In the case of LiveOcean, an external group might publish software to infer some new type of information from the LiveOcean forecast base, perhaps in combination with other resources. An example following from the foregoing might be a map of safe/unsafe paddleboard or kayak zones across Puget Sound extrapolated out over the course of the day (CalSurf). Where this software would reside depends on the publication architecture. It could be implemented in another cloud or server environment, or it could be integrated with the LiveOcean middleware.

An important consequence of the modular design of LiveOcean is perceptual: The project can be understood as a template for other forecast systems. For example, in 2014 LiveOcean middleware (i.e., excluding the ROMS model) was adapted by Anthony Arendt as Ice2Ocean (Ice2Ocean), a system for forecasting river discharge from glaciated watersheds along the southern coast of Alaska. The ongoing combined software development efforts will become an open-source solution that may reduce the required effort in other cloud-based forecast-to-public-benefit projects.

Integration Paths

LiveOcean is about the value of new types of information from the sea, considered as both an ecosystem and a vital resource. The LiveOcean project originated from the recognition of the fragility of that resource in view of ocean acidification. We assert that LiveOcean will fulfill its potential only if it successfully integrates with other programs working to understand, preserve, and recover ocean habitat. Organizations such as the Ocean Foundation (Ocean Foundation) and the World Ocean Council (World Ocean Council) are developing capabilities and strategies for helping to protect the world's oceans through research and through changing human behavior. One of the foremost challenges for such organizations is understanding ocean dynamics in detail. Although efforts at direct observation are scaling up the available data, the need for validated model results scales up commensurately to fill in the huge gaps in what is a very large study area.

To consider LiveOcean's integration path: LiveOcean forecasts are already integrated with the Northwest Association of Networked Ocean Observing Systems (NANOOS) Visualization System (NVS). As noted, LiveOcean forecasts could also be used by NOAA to improve its current forecast services. In fact, integration with NOAA is an extremely rich opportunity as NOAA has initiated a shift in its data holdings to the cloud in partnership with several cloud providers (NOAA Big Data). It would be ambitious but quite feasible to make LiveOcean implementations for other coastal regions and estuaries, whereupon each could be registered with a central organization such as NOAA to reproduce globally the value of the original LiveOcean.

Turning back to research, we note that LiveOcean could be registered with other data brokerage services such as Community Inventory of Earth-Cube Resources for Geosciences Interoperability (CINERGI) (CINERGI) (affiliated with Earthcube) to make LiveOcean services more easily discoverable by the ocean research community.

Client Paths

Previously, we described LiveOcean results arriving as symbols on the chart plotter of a fishing vessel. The modular {ROMS—Azure cloud middleware—API—Client} design is intended to *not* restrict the implementation of other creative or valuable uses of the forecast data. Other examples of customized LiveOcean Clients might include:

- Passing along health and safety conditions for a particular beach to a surfer's wrist device.
- On a tablet map application: Tap a location to determine "Where will this water go in the next 3 days?" to give a sense of the dispersion of an oil spill, or simply to educate.
- Aboard a ferry: A kiosk providing a rich four-dimensional immersive visualization to elevate public awareness of the dynamics of the waters they are passing over. Obvious extension to the classroom.

CONCLUSIONS

The cloud is evolving rapidly in both perception and the services it offers for adoption. In this chapter, the cloud story of LiveOcean has two elements. The first is the modularity of the solution architecture that decouples Consumer information and experience from the technical machinery used to produce the forecast data. The second element is a three-stage progression of cloud technology, from cloud storage to cloud services to cloud analytics (as a future development). The latter is the basis of the advanced argument for cloud adoption: Sophisticated analysis services on the Microsoft Azure cloud platform can scan forecast data for patterns (HABs, fish concentrations) with relatively small investment in software development because the core algorithms are already in place. In summary, the system addresses its charter to provide ocean pH values to shellfish growers but in fact goes far beyond that by making all results openly accessible and future growth comparatively easy.

In terms of practical utility, we note first that the use of scientific models such as ROMS in predictive forecasts is developing as a necessity. At the same time it is perilous to place blind faith in scientific modeling results, recognizing the current limitations in predicting biology and chemistry of ocean systems. Fluid dynamics is fraught with uncertainty and error when modeled using finite-extent grid cells in software. Scientists are comfortable with this uncertainty; but how should such uncertainty be translated and interpreted in fisheries science? On one extreme, the forecasts prove unreliable, and at the other extreme a reliable system could fail at a critical time resulting in financial loss. These refinements wait upon validation of LiveOcean, whereupon (we hope) both anticipated and unanticipated value of LiveOcean forecast information will emerge.

REFERENCES

ARGO: http://www.argo.net/.
Bing: https://www.bingmapsportal.com/.
CalSurf: http://www.surf-forecast.com/weather_maps/California?over=none&type=htsgw.
CINERGI: http://earthcube.org/group/cinergi.
Django: https://www.djangoproject.com/.
DOE: http://www.ecy.wa.gov/water/marine/oa/2012panel.html.
Davis, K.A., Banas, N.S., Giddings, S.N., Siedlecki, S.A., MacCready, P., Lessard, E.J., Kudela, R.M., Hickey, B.M., 2014. Estuary-enhanced upwelling of marine nutrients fuels coastal productivity in the U.S. Pacific Northwest. Journal of Geophysical Research Oceans 119, 8778–8799. http://dx.doi.org/10.1002/2014JC010248 (highlighted in Eos).
Geyer, W.R., MacCready, P., 2014. The estuarine circulation. Annual Review of Fluid Mechanics 46 (1), 175.
Giddings, S.N., MacCready, P., Hickey, B.M., Banas, N.S., Davis, K.A., Siedlecki, S.A., Trainer, V.L., Kudela, R.M., Pelland, N.A., Connolly, T.P., 2014. Hindcasts of potential harmful algal bloom transport pathways on the Pacific Northwest coast. Journal of Geophysical Research Oceans 119. http://dx.doi.org/10.1002/2013JC009622.
HAB: http://www.businessinsider.com/toxic-algae-bloom-is-killing-animals-on-the-pacific-coast-2015-8 (in lieu of a better link to NOAA which was the source of the figure; so this is temporary).

HYCOM http://hycom.org/data/glbu0pt08/expt-91pt1.

Ice2Ocean: http://ice2ocean.azurewebsites.net/.

IPython: http://ipython.org/– [github ipython client needed (Kilroy)].

LiveOceanServer: http://liveocean.azurewebsites.net.

Liu, Y., et al., 2009. Evaluation of a coastal ocean circulation model for the Columbia River plume in summer 2004. Journal of Geophysical Research Oceans (1978–2012) 114 (C2).

MacCready, P., et al., 2009. A model study of tide-and wind-induced mixing in the Columbia River Estuary and plume. Continental Shelf Research 29 (1), 278–291.

Mohamedali, T., Roberts, M., Sackmann, B., Kolosseus, A., 2011. Puget Sound Dissolved Oxygen Model Nutrient Load Summary for 1999–2008. Washington State Dept. of Ecology Report 11-03-057, 172 pp. Available at: https://fortress.wa.gov/ecy/publications/summarypages/1103057.html.

NANOOS: http://nvs.nanoos.org/.

NOAA Big Data: https://www.commerce.gov/news/press-releases/2015/04/us-secretary-commerce-penny-pritzker-announces-new-collaboration-unleash.

NOAA Rivers: http://www.nwrfc.noaa.gov/rfc/.

NOPP http://www.nopp.org/.

Ocean Foundation: http://oceanfdn.org/our-story/staff.

OOI: http://oceanobservatories.org/.

Ocean Acidification: From Knowledge to Action, Washington State's Strategic Response. https://fortress.wa.gov/ecy/publications/SummaryPages/1201015.html. Blue ribbon panel (authors) website: http://www.ecy.wa.gov/water/marine/oa/panel.html.

Siedlecki, S.A., Banas, N.S., Davis, K.A., Giddings, S., Hickey, B.M., MacCready, P., Connolly, T., Geier, S., 2015. Seasonal and interannual oxygen variability on the Washington and Oregon continental shelves. Journal of Geophysical Research Oceans 120. http://dx.doi.org/10.1002/2014JC010254.

Sutherland, D.A., et al., 2011. A model study of the Salish Sea estuarine circulation. Journal of Physical Oceanography 41 (6), 1125–1143.

World Ocean Council: http://www.oceancouncil.org.

WWT: http://worldwidetelescope.org.

Yoyodyne: http://yoyodyne.azurewebsites.net.

CHAPTER 15

Usage of Social Media and Cloud Computing During Natural Hazards

Q. Huang
University of Wisconsin–Madison, Madison, WI, USA

G. Cervone
Pennsylvania State University, University Park, PA, USA

INTRODUCTION

Natural hazards are severe events that pose a threat to the sustainment and survival of our society. Every year extreme weather and climate events, such as typhoons, floods, tornadoes, hurricanes, volcanic eruptions, earthquakes, heat waves, droughts, or landslides, claim thousands of lives, cause billions of dollars of damage to property (Smith and Matthews, 2015) and severely impact the environment worldwide (Velev and Zlateva, 2012). Natural hazards become disasters when they cause extensive damage, casualties, and disruption (Vasilescu et al., 2008). Disasters have been increasing in both frequency and severity in the 21st century because of climate change, increasing population, and reliance on aging infrastructure. Recently, major events have caused havoc around the world, such as the 2015 earthquakes in Nepal, the 2015 heat wave in India, the 2011 tsunami in Japan, the 2010 earthquake in Haiti, and the extremely cold winter of 2014/2015 in the United States and in Europe.

Most disasters occur rapidly with little or no warning, and therefore are often extremely difficult to predict. However, effective actions and management strategies can mitigate the potential effects. For several decades, emergency managers and disaster researchers have typically relied on a four-phase categorization (mitigation, preparedness, response, and recovery) to understand and manage disasters (Neal, 1997; Warfield, 2008):

- Mitigation: Concerns the long-term measures or activities to prevent future disasters or minimize their effects. Examples include any action that prevents a disaster, reduces the chance of a disaster happening, or

Cloud Computing in Ocean and Atmospheric Sciences
ISBN 978-0-12-803192-6
http://dx.doi.org/10.1016/B978-0-12-803192-6.00015-3

Copyright © 2016 Elsevier Inc.
All rights reserved.

reduces the damaging effects of a disaster, e.g., building levies, elevating a building for a potential hurricane, or public education.

• Preparedness: Plans how to respond a disaster. Examples include developing preparedness plans, providing emergency exercises/training, and deploying early warning systems.
• Response: Minimizes the hazards created by a disaster. Examples include search and rescue, and emergency relief.
• Recovery: Restores the community to normal. Typical activities during this phase include providing temporary housing, grants, and medical care.

The four disaster management phases do not always, or even generally, occur in isolation or in this precise order. Often phases of the cycle overlap and the length of each phase greatly depends on the severity of the disaster (Warfield, 2008). However, the four-phase categorization serves as a time reference for practitioners to predict challenges and damage, prioritize functions, and streamline activities during the course of disaster management (U.S. Department of Education, 2010; FEMA, 1998). It also provides a common framework for researchers to organize, compare, and share their research findings.

When natural hazards occur, disaster management and coordination rely on the availability of timely actionable information. Crucial information includes an assessment of damage and available resources that can be used for planning evacuation and rescue operations to minimize the losses and save lives. This information augments our understanding of the overall disaster situation, and facilitates the decision-making toward a better response strategy. Such information is referred as "situational awareness," i.e., an individually as well as socially cognitive state of understanding "the big picture" during critical situations (Vieweg et al., 2010). However, such information is difficult to obtain because of limitations in data acquisition and techniques in processing the data efficiently in near real time. Additionally, such information may not be effectively disseminated through traditional media channels.

Because of the massive popularity of social media networks and their real-time production of data, these new streams offer new opportunities during emergencies. Social media data are increasingly used during crises. A Red Cross survey in 2012 indicated that 18% of adults, if a call to 911 was unsuccessful, would next turn to social media, whereas 76% expected help to arrive within 3 h of posting their need to social sites (American Red Cross, 2012). Social media networks have even become widely used as an intelligent "geo-sensor" network to detect and monitor extreme events or disasters such as earthquakes (Sutton, 2010). The fundamental premise is that human actors in a connected environment, when augmented with ubiquitous mechanical sensory systems, can form the most intelligent sensor web (Sheth et al., 2008). Such intelligent sensor webs have the most realistic implications for operations such as disaster, in which information is the most valuable and hard to obtain asset (Verma et al., 2011; Vieweg et al., 2010). Additionally, it has been widely acknowledged that Humanitarian Aid and Disaster Relief (HA/DR) responders can gain valuable insights and situational awareness by monitoring social media-based feeds from which tactical, actionable data can be mined from text (Ashktorab et al., 2014; Gao et al., 2011; Huang and Xiao, 2015; Imran et al., 2013; Kumar et al., 2011).

Faced with real-time social media streams from a multitude of channels during emergencies, identifying authoritative sources and extracting critical, validated messages information for the public could be quite challenging in a time of crisis. The volume, velocity, and variety of accumulated social media data produce the most compelling demands for computing technologies from big data management to technology infrastructure (Huang and Xu, 2014). For big data management, many nontraditional methodologies such as non-relational and scalable Structure Query Language (NoSQL) are implemented (Nambiar et al., 2014). Meanwhile, to address big data challenges, various types of computational infrastructures are designed, from the traditional cluster and grid computing to the recent development of cloud computing and central processing unit/graphics processing unit (CPU/GPU) heterogeneous computing (Schadt et al., 2010). Specifically, cloud computing has been increasingly viewed as a viable solution to utilize multiple low-profile computing resources to parallelize the analysis of massive data into smaller processes (Huang et al., 2013b).

Similarly, the computational requirements for an operational system that can be deployed for event predictions and subsequent disaster management are very demanding. However, most natural hazards occur very quickly, have immediate impacts, but only last a relatively short period of time. Therefore, it is necessary to support operations by scaling up to enable high-resolution forecasting, big data processing, and massive public access during a crisis, and by scaling down when no events occur to save energy and costs. Cloud computing provides an ideal solution due to its intrinsic capability of providing a large, elastic, and virtualized pool of computational resources which can be scaled up and down according to the needs. The goal of this chapter is to discuss the opportunities and challenges associated with the usage of social media to gain situational awareness during natural disasters, and the feasibility of using cloud computing to build an elastic, resilient, and real-time disaster management system.

SOCIAL MEDIA FOR DISASTER MANAGEMENT

Disaster management aims to reduce or avoid the potential losses from hazards, assure prompt and appropriate assistance to victims of disaster, and achieve rapid and effective recovery (Warfield, 2008). Social media can be used as new sources to redefine situational awareness and assist in the management of various disaster stages.

Social Media Fundamentals

There are a variety of definitions of social media (Cohen, 2011). In general, social media are broadly defined as any online platform or channel for user-generated content. A large number of social media services or Web sites are developed to enable the public to distribute and share different types of content, such as videos, text messages, photos, etc. In this regard, Wiki, WordPress, Sharepoint, and Lithium qualify as social media as do YouTube, Facebook, and Twitter. Based on the content generated and functions provided, social media services are generally categorized into several classes (Nations, 2015):

- Social networking: Interacts by adding friends, commenting on profiles, joining groups and having discussions. Such social network sites can produce, spread, and share relatively short messages, photos, or videos over the Internet at a high speed. These messages can be immediately accessible by the linked groups or friends. Therefore, they are commonly used by the public to post relevant information via microblogs about the disaster, and share their own knowledge about the disaster situation with others, thus contributing the situational awareness. For example, users in the impacted communities can report what they are witnessing and experiencing. Twitter, Facebook, and Google+ belong to this category.
- Social bookmarking: Interacts by tagging Web sites and searching through Web sites bookmarked by other people. Typical examples include Del.icio.us, Blinklist, and Simpy.
- Social news: Interacts by voting for articles and commenting on them. Good examples are Reddit, Propeller, or Digg.
- Social photo and video sharing: Interacts by sharing photos or videos and commenting on user submissions. The most two popular examples are YouTube (videos), and Flickr (photos).
- Wikis: Interacts by adding articles and editing existing articles. In fact, any Web site that invites users to interact with the site and with other visitors falls into this category. Wikipedia is the one of the earliest and popular examples to enable users to create articles and web pages. OpenStreetMap is another example to enable the public to share geographic data and maps.

However, social media more narrowly defined include only channels for user-generated content, as distinguished from platforms, which are referred to as social technologies. According to this definition, for example, YouTube, Facebook, and Twitter are social media, and WordPress, Sharepoint, and Wikis are social technologies (Cohen, 2011). Although both channels for user-generated content and social technologies can be leveraged to support disaster management in many different ways, they have different functions and usages for disaster management. For this reason, this chapter adopts the narrow definition.

Opportunities

Social media are widely used during natural disasters as a news source and tool by both the public and emergency service agencies. For example, citizens normally use social media in four different ways (Velev and Zlateva, 2012): (1) To communicate with family and friends. Social network sites provide a bridge to connect with family members between affected and unaffected communities or areas (or within affected communities for situation updates and planning responses). This is the most popular use. (2) To update and share critical information between each other such as road closures, power outages, fires, accidents, and other related damage. (3) To gain situational awareness. In a number of cases citizens rely less and less on authority communication, especially through traditional channels (television, radio, or phone). Finally, (4) to assist in service access, citizens use social media channels to provide each other with ways and means to contact different services they may need after a crisis. On the other hand, emergency service agencies are utilizing social media to instantly alert emergency warnings to the public, and collect feedback and updates from the public users.

In fact, social media are useful in different disaster stages (Velev and Zlateva, 2012). Before a disaster, social media can help people better prepare and understand which organizations will help their communities. After the disaster, social media help bring the community together to discuss issues, share information, coordinate recovery efforts, and communicate information about aid. As we discussed earlier, social media can play a more significant role by helping users communicate and share information in real time directly to their families, reporters, volunteer organizations, and other residents during a disaster. In particular, research and reviews of different cases have identified the following benefits of social media and social technologies in emergency response (Prentice et al., 2008; Yates and Paquette, 2011):

- Near real-time: Social media (including social network sites and social technologies) are essentially real-time offering unique strengths as a data source, methods, and tools for the sharing of information in emergencies (Prentice et al., 2008). Previously, methodologies such as phone calls, direct observation, or personal interview are commonly practiced by disaster responders and damage evaluators to gain situational awareness and investigate impacted population during a disaster. However, these data collection methods are time-consuming and laborious in processing the data. Social media data, however, can provide near "real-time" information for the emergency managers to make effective decisions through multiple stages of the disaster management.
- Facilitates knowledge sharing: social media facilitate better knowledge sharing between communities and organizations. Connections can be made among individuals without limitations introduced by bureaucratic boundaries (Yates and Paquette, 2011).
- Provides broad access: Internet-connected devices allow sharing the information in real time (Yates and Paquette, 2011).
- Offers contextual cues for understanding a given perspective: Typically, when people engage in a conversation about the larger context of a disaster, they tend to be clearer about the situation (Yates and Paquette, 2011).
- Conversational, discussion-based style: Social media sites can provide a platform for discussion and feedback from those who care the most and have the most lasting impact on the story of the disaster (Prentice et al., 2008).
- Limits restrictions and maintains strengths of "old media": Social media also bypass the deadlines and restrictions placed on "old media" (Prentice et al., 2008).
- Two-way medium: Organizations can respond directly to comments and feedbacks posted on blogs, Twitter, or Facebook, or even leverage other social technologies such as YouTube to distribute timely and accurate information directly to those concerned (Prentice et al., 2008).

State-of-the-Art Work and Practice

Many recent studies have applied social media data to understand various aspects of human behavior, the physical environment, and social phenomena. Studies and applications of using social media for disaster related analysis include following major areas: (1) event detection and sentiment tracking; (2) disaster response and relief coordination; (3) damage assessment; (4) social media message coding during a disaster; and (5) user rank model.

Event Detection and Tracking

The network of social media users is considered a low-cost, effective "geo-sensor" network for contributed information. Twitter, for instance, has more than 190 million registered users, and 55 million Tweets are posted per day. As an example, Asiana Flight 214 from Seoul, Korea, crashed while landing at San Francisco International Airport on July 6, 2013. News of the crash spread quickly on the Internet through social media. With eyewitnesses sending Tweets of their stories, posting images of the plumes of smoke rising above the bay, and uploading video of passengers escaping the burning plane, the event was quickly acknowledged globally.

As a result of the rapid or even immediate availability of information in social networks, the data are widely applied for the detection of significant events. Sakaki et al. (2010), for instance, investigated the real-time interaction of events such as earthquakes and Twitter. Their research produced a probabilistic spatiotemporal model for the target event that can find the center and the trajectory of the event location. Kent and Capello (2013) collected and synthesized user-generated data extracted from multiple social networks during a wildfire. Regression analysis was used to identify relevant demographic characteristics that reflect the portion of the impacted community that will voluntarily contribute meaningful data about the fire. Using Hurricane Irene as example, Mandel et al. (2012) concluded that the number of Twitter messages correlate with the peaks of the event, the level of concern dependent on location and gender, with females being more likely to express concern than males during the crisis.

Disaster Response and Relief

Social media data are real-time in nature, and it has been widely acknowledged that HA/DR responders can gain valuable insights and situational awareness by monitoring and tracking social media streams (Vieweg, 2012). As a result, an active area of research focuses on mining social media data for disaster response and relief (Ashktorab et al., 2014; Gao et al., 2011; Huang and Xiao, 2015; Imran et al., 2013; Kumar et al., 2011; Purohit et al., 2013). Aiming to help HA/DR responders to track, analyze, and monitor Tweets, and to help first responders gain situational awareness immediately after a disaster or crisis, Kumar et al. (2011), for example, presented a tool with data analytical and visualization functionalities, such as near real-time trending, data reduction, and historical review. Similarly, a Twittermining tool, known as Tweedr, was developed to extract actionable information for disaster relief workers during natural disasters (Ashktorab et al., 2014). Gao et al. (2011)

described the advantages and disadvantages of social media applied to disaster relief coordination and discussed the challenges of making such crowdsourcing data a useful tool that can effectively facilitate the relief process in coordination, accuracy, and security. Recent findings also suggest that actionable data can be mined and extracted from social media to help emergency responders act quickly and efficiently. Purohit et al. (2013) presented machine learning methods to automatically identify and match needs and offers communicated via social media for items and services such as shelter, money, clothing, etc.

Damage Assessment

Damage assessment of people, property, and environment, and timely allocation of resources to communities of greatest need, and is paramount for evacuations and disaster relief. Remote sensing is capable of collecting massive amounts of dynamic and geographically distributed spatiotemporal data daily, and therefore often used for disaster assessment. However, despite the quantity of big data available, gaps are often present due to the specific limitations of the instruments or their carrier platforms. Several studies (Schnebele and Cervone, 2013; Schnebele et al., 2014, 2015), have shown how crowdsourced data can be used to augment traditional remote sensing data and methods to estimate flood extent and identify affected roads during a flood disaster. In these works, a variety of nonauthoritative, multisourced data, such as Tweets, geolocated photos from the Google search engine, traffic data from cameras, OpenStreetMap, videos from YouTube, and news, are collected in a transportation infrastructure assessment construct an estimate of the extent of the flood event.

Message Coding

As the messages broadcasted and shared through the social media network are extremely varied, a coding schema is needed to separate the messages into different themes before we can use them to produce a crisis map or extract "actionable data" as information that contributes to situational awareness. During Typhoon Bopha in the Philippines in 2012, volunteers using the PyBossa, a microtasking platform, manually annotated the Tweets into various themes, such as damaged vehicle, flooding, etc., and a crisis map was produced to be used by humanitarian organizations (Meier, 2012). A few attempts (Huang and Xiao, 2015; Vieweg, 2012) have been made to uncover and explain the information exchanged when Twitter users communicate during mass emergencies. Information about casualties and damage, donation efforts, and alerts are more likely to be used and extracted to improve situational awareness during a time-critical event. As a result, messages are typically categorized into these categories. Imran et al. (2013), for instance, extracted Tweets published during a natural disaster into several categories, including caution and advice, casualty and damage, donation and offer, and information source. The content categories (or topics) defined in those studies (Vieweg et al., 2010; Vieweg, 2012; Imran et al., 2013), are very useful to explore and extract the data involved in the disaster response and recovery phases. The content categories (or topics) defined in previous studies (Imran and Castillo, 2015; Vieweg et al., 2010; Vieweg, 2012), however, only consider the "actionable data" involved in the disaster response and recovery phases while ignoring useful information that could be posted before or after a disaster event. Huang and Xiao (2015) made an initial effort by coding social media messages into different categories within various stages of disaster management. Based on the coding schema, a supervised classifier was also trained and used to automatically mine and classify the social media messages posted by Twitter users into these predefined topic categories during various disaster stages.

User Rank Model

Research along this line focuses on identifying Twitter users who contribute to situational awareness. The topic of measuring "contribution" or "influence" within the online social network has been intensively investigated. Cha et al. (2010) examined three metrics of user influence on Twitter, including in degree (the number of people who follow a user), retweets (the number of times others "forward" a user's Tweet), and mentions (the number of times others mention a user's name). They also investigated the dynamics of an individual's influence by topic and over time. The results show that in degree alone reveals very little about the influence of a user. Bakshy et al. (2011) use the size of a diffusion tree, which represents the information diffusion by retweets, to quantify the influence of a Twitter user. Cheong and Cheong (2011) investigated popular and influential Twitter users in the digital social community during several Australian floods. The concept of centrality in social network analysis technique was adopted, and various centrality measures were used to identify the influential users, including local authorities, political personalities, social media volunteers, etc.

Challenges

As with any new technology, there remain many hurdles between current use and optimal exploitation of social media for disaster analysis and management.

Digital Divide

Although these media are used by people of both sexes and an expanding range of ages, it is important to recognize and explore the technology's limitations in reaching at-risk, vulnerable populations. The "digital divide" refers to the gap between those who have and those who do not have access to computers and the Internet (Van Dijk, 2006). An active area of research is focused on exploring the factors that lead to the social inequality in the usage of social media. Witte and Mannon (2010) claimed that marginalized populations often lack or have limited net access in their households, making access to Twitter a socially stratified practice. Livingstone and Helsper (2007) conducted a survey among children and young people, and found that inequalities by age, gender, and socioeconomic status relates to the quality of access to and use of the internet. For example, it was reported that the public under 35 are more likely to participate in social media every or nearly every day (63% vs. 37% for those 35 and over) (American Red Cross, 2012). It was also shown that racial digital divides continue to remain pervasive (Nakamura, 2008), and Twitter is no exception to this.

It has been recognized that certain groups (i.e., low income, low education, and elderly) may lack the tools, skills, and motivations to access social media, and therefore they may be less likely post disaster-relevant information through social media (Xiao et al., 2015). Additionally, certain areas may be severely damaged by the disaster, which result in extremely low participation in social media usage after the disaster. As a result, the situational awareness information extracted from social media data may be biased and can underestimate and mis-locate the needs of the significantly impacted communities. Therefore, the social and spatiotemporal inequality in the usage of social media data must be fully considered when using them to predict damage, investigate impacted populations, and prioritize activities during the course of disaster management by practitioners.

Data Quality

Despite many advantages of social media data, concerns have been raised about their quality (Goodchild and Li, 2012; Oh et al., 2010). The first concern originates from the user and information credibility. It is quite challenging to know whether social media users are who they claim to be or whether the information they share is accurate. For example, during the Haiti earthquake, rumors circulated in Twitter that United Parcel Service (UPS) would "ship any package under 50 pounds to Haiti" or "several airlines would take medical personnel to Haiti free of charge to help with earthquake relief" (Leberecht, 2010). These turned out to be hearsay rather than eyewitness accounts, and subsequently clarified by UPS and airline companies as false information. As a result, Twitter, has been long criticized as it may propagate misinformation, rumors, and, in extreme case, propaganda (Leberecht, 2010). In fact, based on the analysis of credibility of information in Tweets corresponding to 14 high-impact news events of 2011 around the globe, Gupta and Kumaraguru (2012) claim that only "30% of total tweets about an event contained situational information about the event while 14% was spam." In addition, about "17% of total tweets contained situational awareness information that was credible."

Another concern comes from the location reliability. Users with location services enabled on smart mobile devices can post content (e.g., text messages or photos) with locations, which typically are represented as a pair of coordinates (latitude and longitude). The locations along with the place names mentioned in the content text are then used to identify the areas of damaged infrastructure, affected people, evacuation zones, and the communities of great needs of resources. During this process, we rely on the assumption that users will report information about the events (e.g., flooded roads, closure of bridges, shelters, or donation sites) they witnessed and experienced at the exact locations where these events occurred. However, the locations in the time of posting content and locations of event occurred are not necessarily consistent and the supposed locations of greatest needs could be misleading. To address these concerns, other data sources (e.g., uthoritative data and remote sensing data) may be integrated for cross-validation, and crowdsourcing validation procedures can be applied to leverage volunteers for improving quality of these user-generated content.

Big Data

Although scholars and practitioners envisioned the possibilities in utilizing social media for disaster management, the computational hurdle to practically leverage social media data is currently extremely high (Huang and Xu, 2014; Elwood et al., 2012; Manovich, 2011). Social media data present challenges at least in the following four dimensions (Huang and Xu, 2014):

1. There is the huge volume of social media data. For Twitter alone, the number of Tweets reached 400 million per day in March, 2013, and that number is escalating rapidly.
2. There is the enormous velocity of real-time social media data streams. In 2014, 9100 Tweets were posted every second in Twitter.

3. There are the high variety types of social media data content. Text-based Tweets, image-based Flickr photos, or video-based YouTube posts are all telling similar stories using different media.
4. Social media data are assertive and create the trustworthiness or veracity question.

These dimensions are now widely cited as the four V's of big data (Fromm and Bloehdorn, 2014). Although the trustworthiness touches upon nontechnical challenges, the other three challenges put more demands on innovative computational solutions (Agrawal et al., 2011). To address such challenge, new computing infrastructure and geovisualization tools should be leveraged to support the exploration of social media in space and time (Roth, 2013). Section Cloud Computing to Facilitate Disaster Management presents a potential computing infrastructure - cloud computing, to address the demands on high-performance computing framework for processing social media data timely and effectively.

CLOUD COMPUTING TO FACILITATE DISASTER MANAGEMENT

Decision support systems for disaster management can only be best conducted when integrating a large amount of geospatial information in a collaborative environment and timely fashion. However, such systems pose several critical requirements to the underlying computing infrastructure. First, it must achieve high performance. The data, such as social media and forecasting data, to support effective decision-making during natural hazards, come in streams, and must be processed in a real-time fashion. Additionally, because most of these data are distributed across different agencies and companies and with different formats, resolutions, and semantics, it takes a relatively long time to identify, process, and seamlessly integrate these heterogeneous datasets. Second, it must be flexible. Tens to hundreds of computers are needed for the physical model simulations to run in a few hours to produce high-resolution results and support the associated decision-making process using multisourced data once such an event is detected. After the emergency, the computing resources should be released and reclaimed by other science, application, and education purpose in a few minutes without or with little human intervention. Third, it has to be resilient. In times of critical situation, system failures may occur because the adverse environmental conditions, such as physical damage, power outages, floods, etc. Hence, the big data storage, simulation, analysis, and transmission services must be able to operate during such adverse conditions (Pu and Kitsuregawa, 2013).

Most traditional computing infrastructure lacks the agility to keep up with these computing requirements by developing an elastic and resilient Cyberinfrasturcture (CI) for disaster management. Cloud computing, a new distributed computing paradigm, can quickly provision computing resource in an on-demand fashion. It has been widely utilized to address geoscience challenges of computing, data, and concurrent intensities (Yang et al., 2011). In fact, it can naturally serve as the underlying computing infrastructure of an operational system during a crisis in the following aspects:

- High performance: Cloud computing provides scientists with a complete new computing paradigm for accessing and utilizing the computing infrastructure. Cloud-computing services, especially Infrastructure as a Service (IaaS), a category of popular cloud services, can be easily adopted to offer the prevalent high-end computing technologies to provide more powerful computing capabilities. Many cloud providers offer a range of diverse computing resources for users' computing needs, such as Many Integrated Cores (MICs), Graphics Processing Units (GPUs), and Field Programmable Gate Arrays (FPGAs). For example, Amazon Elastic Compute Cloud (EC2) Cluster, with 17,024 CPU cores in total, a clock speed of 2.93 GHz per core, and 10G Ethernet network connection, was ranked as 102nd on the TOP 500 supercomputer lists in the November 2012. The high performance computing (HPC) capability of cloud computing can be easily leveraged to support critical scientific computing demands (Huang et al., 2013a).

- Flexibility: Hazard events often have annual or seasonal variability and are of short duration. Most events typically last a relatively short period from several hours (e.g., tornadoes) to several days (e.g., hurricanes). As a result, a real-time response system for such events would experience different computing and access requirements during different times of the year and even different hours within the day. During a disaster, the computing platform supporting an emergency response system should be able to automatically scale up enough computing resources to produce and deliver relevant and useful information for the end users, and to enable them to make smarter decisions, saving more lives and reducing assets loss. After the emergency response, the access to information can be reduced and the system can switch back to "normal mode" for reduced costs. Computing resources would be released for other science, application, and education purposes.

Applications, running on the cloud, can increase computing resources to handle spike workloads and accelerate geocomputation in a few seconds to minutes. Additional computing resources can be released in seconds once the workloads decrease. Previous studies (Huang et al., 2013a, 2013b, 2013c) demonstrated that cloud computing can help application-handling computing requirement spikes caused by massive data transfers, model runs, and data processing without expensive long-term investment for the underlying computing infrastructure. Therefore, an operational system based on cloud computing can respond to real-time natural hazards well.

• Resilience: Architectural resilience can be achieved in many ways including (1) having back-up redundant systems that automatically deploy when primary systems fail, or (2) employing multiple solutions to ensure that some minimum level of system functionality is available during massive system failures (Pu and Kitsuregawa, 2013). Cloud services provide an ideal platform to implement this resilient mechanism. Cloud computing providers offer computing and storage services that are globally distributed. For example, three major cloud providers, including Amazon, Microsoft, and Google, have multiple data centers around the world with the service. An image containing the configured application could be built in cloud services, and then a new replicated application can be easily launched on failover systems in a different cloud zone in a few minutes after a failure (Huang et al., 2013c).

CASE STUDIES

Through a variety of research studies and government practice, it has been widely demonstrated that online social technologies and social media like Facebook, Twitter, Google+, etc., can be employed to solve many problems during natural disasters. This section introduces several real applications to show how social media were used for emergency response and disaster coordination, and how cloud computing can facilitate these applications.

Tsunami in Japan

On March 11, 2011 at 2:46 pm local time, a massive magnitude 9.0 underwater earthquake occurred 70 km offshore of the eastern coast of Japan. The earthquake generated a tsunami that rapidly hit the eastern coast of Japan, and propagated across the Pacific Ocean to the west coast of the Americas. The tsunami wave hit the Fukushima power plant about 40 min after the earthquake, leading to the catastrophic failure of the cooling system. Several radioactive releases ensued as a result of an increase of pressure and temperature in the nuclear reactor buildings. Some releases were the result of both controlled and uncontrolled venting, whereas others were the result of the explosions that compromised the containment structures. The explosions were most likely caused by ignited hydrogen, generated by zirconium–water reaction occurring after the reactor core damage (Cervone and Franzese, 2014). The radioactive cloud was quickly transported around the world, reaching North America within a few days and Europe soon thereafter (Potiriadis et al., 2012; Vasilescu et al., 2008). Radioactive concentrations were recorded along the US West Coast within a week of the initial release (Cervone and Franzese, 2014).

Social media, including Twitter, quickly disseminated information about the developing disaster in Japan. Twitter data was harvested using a cloud computing solution deployed at Pennsylvania State University, which collected several million Tweets related to the Japanese disaster. The analysis of the Tweets was performed using Docker (https://www.docker.com/), an open-source virtualization software that allows quick distribution of data and computing on the cloud. First, a container was created to include all analysis software for filtering and plotting the Tweets. A second container with all the Tweets was created and continuously updated with additional data. The advantage of using Docker consists in the ability to allocate variable resources for the analysis of Tweets, varying from few tenths to over a 1000 cores. Furthermore, cloud-computing platform Amazon EC2 (Huang et al., 2010) is used to support Docker containers, and thus makes it possible to quickly deploy the analysis and meet the requirements of flexible resources during a crisis (section Cloud Computing to Facilitate Disaster Management).

Although the implementation of cloud-computing technology solves the problem of big data analysis, data has to be managed in a way that is suitable for distributed parallel processing. As discussed earlier, social media data such as Tweets, constantly flow in extremely large volume and with versatile contents. Recently, a NoSQL database has been popularly implemented as a better means to manage such data (Huang and Xu, 2014). This is because a NoSQL database can (1) store data that are not uniform and structured, and (2) support the utilization of multiple servers to improve the performance in a much easier way. In our system, a MongoDB, one of the popular NoSQL databases, therefore was set up to store all geolocated Tweets, plus all Tweets that made mention of several hashtags relative to the Japanese crisis.

Fig. 15.1 shows a trend of Tweets collected immediately before and after the Japan earthquake and resulting tsunami. Mentions of Japan in Tweets

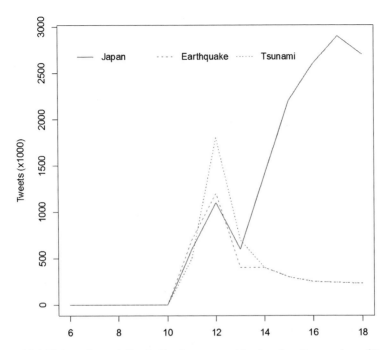

Figure 15.1 Twitter data relative to the Japanese crisis, showing the number of Tweets that include the keywords Japan, earthquake, and tsunami. In the immediate aftermath of the earthquake and resulting tsunami, Tweets including all three keywords spiked significantly, however, after March 13, the keyword Japan greatly increased as news about the crisis resonated through the social network.

peaked at nearly 3 million per day on March 16, 2011, just a few days after the event, when the widespread destruction became apparent.

Fig. 15.2 shows about 1 million geolocated Tweets collected from March 10 to March 31, 2011, within Japan. The bar graphs in the two bottom figures show the number of Tweets per longitude and latitude, to compensate for over-plotting. Taller bars indicate a higher concentration of Tweets at the corresponding longitude and latitude. The majority of the Tweets are geolocated around Tokyo and other major metropolitan areas in Japan. In the two top figures, a circle is used to indicate the 20 km restricted area around the Fukushima nuclear power plant. This exclusion zone was enacted after the nuclear leak to protect citizens from being exposed to dangerous radioactive contaminants. The data show a lack of Tweets in the exclusion zone compared to the other surrounding zones, which is to be attributed to the compliance of the residents with the mandatory evacuation order. Fig. 15.3 shows global distribution of geolocated Tweets about Japan for about 1 month after the event.

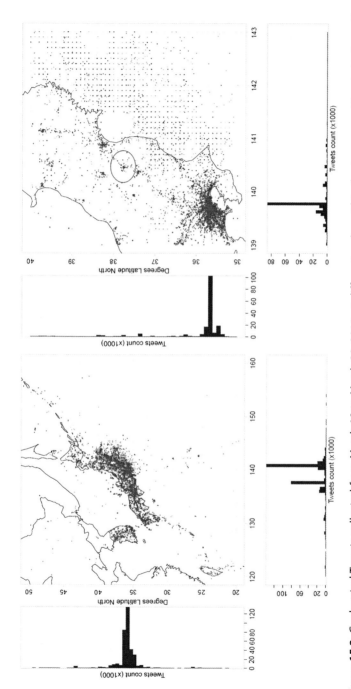

Figure 15.2 Geolocated Tweets collected from March 10 to March 30, 2011, at different scales within the Japan area. The circle shows the 20 km restricted area around the Fukushima nuclear plant. The bar graph for each axis indicates the number of Tweets along the longitude and latitude.

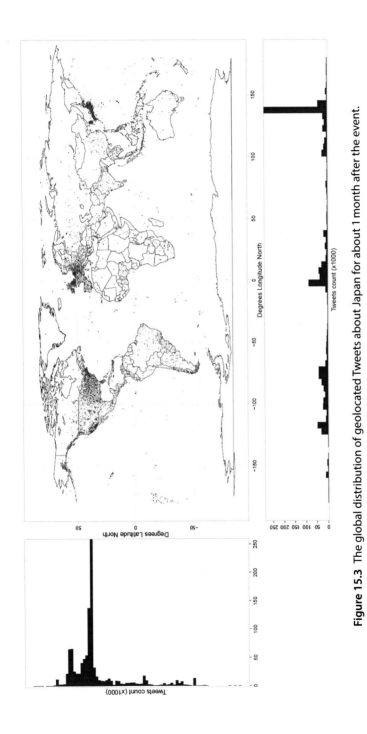

Figure 15.3 The global distribution of geolocated Tweets about Japan for about 1 month after the event.

The analysis of Tweets shows that the social intensity worldwide increased as the radioactive cloud spread over North America and Europe. The majority of the Tweets provided alarming information about the situation in Japan, and targeted both informing about the event and soliciting help for disaster response. Furthermore, although three crisis were unfolding in parallel (tsunami inundation, earthquake damage, and radioactive release), the majority of the Tweets disproportionally discussed the radioactive release over the other two. In North America, Tweets and retweets about the Japanese crisis started immediately after the initial earthquake and the resulting tsunami. The peak of Tweets was observed on March 21, which corresponds to when sensors in Alaska and along the West coast of the United States started registering a small increase in radiation dosage. One lesson learned from the analysis of Twitter data during the Japanese disaster is the good awareness of the public to the unfolding of a major crisis.

Hurricane Sandy

Hurricane Sandy, a late-season posttropical cyclone, swept through the Caribbean and up the East Coast of the United States in late October 2012. Sandy began as a tropical wave in the Caribbean on October 19. It quickly developed, becoming a tropical depression, and then a tropical storm. On October 28, President Obama signed emergency declarations for several states expected to be impacted by Sandy, allowing them to request federal aid and make additional preparations in advance of the storm. On October 29, Sandy made landfall in the United States, striking near Atlantic City, New Jersey with winds of 80 mph. It affected 24 states in the United States, including the entire eastern seaboard from Florida to Maine and west across the Appalachian Mountains to Michigan and Wisconsin, with particularly severe damage in New Jersey and New York. Within 2 days, the region was starting to recover. As of November 1, about 4.7 million people in 15 states were without electricity, down from nearly 8.5 million a day earlier. Storm surge caused subway tunnels in Lower Manhattan to close due to flooding, but some lines were able to resume limited service. Sandy's impact was felt globally as 15,000 flights around the world were cancelled. Sandy ended up causing about $20 billion in property damage and $10 billion to $30 billion more in lost business (http://www.livescience.com/24380-hurricane-sandy-status-data.html), making it the deadliest and most destructive hurricane of the 2012 Atlantic hurricane season as well as the second-costliest hurricane in US history (http://pybossa.com/).

Based on our previous work on developing a coding schema of Tweets during Hurricane Sandy (Huang and Xiao, 2015), we developed a spatial Web portal to analyze and explore the Tweets in the different disaster phases. Several open-source software and tools were used for the prototype development. The portal supports several functions from submitting a query request to visualizing or animating the query results. Users can explore the Tweets in various themes by configuring the input parameters of the query such as temporal information (time stamps when messages were posted), area of interest (AOI, also known as spatial domain), and analytical methods (visualization or charting), etc. After obtaining query results, users are able to visualize the results to get an overall view of the spatial and temporal patterns of the Tweets related to specific topics retrieved from the database (Fig. 15.4).

The aforementioned system can support geovisualizing and analyzing disaster relevant Tweets in different themes spatially and temporally for a historical event with pre-processed data. However, exploring and visualizing massive social media for real-time events requires further development. Specifically, data from different extreme natural hazard events, especially hurricane-related ones, should be examined and integrated to develop a real-time disaster management system so that it can be applied to automatically categorize the Tweets into different themes during any new disaster of different types. Such a system could help support real-time disaster management and analysis by monitoring subsequent events while Tweets are streaming, and mining useful information. As discussed in section Cloud Computing to Facilitate Disaster Management, cloud computing can then be leveraged to process multisourced and real-time social media data. In fact, a few attempts have been made to integrate real-time social media and cloud computing to support real-time emergency response. One of the examples is Esri's severe weather map (http://www.esri.com/services/disaster-response/severe-weather/latest-news-map). It harvests multisourced data, including Twitter Tweets, YouTube videos, Flickr photos, and weather reports of various events using specific (e.g., tornado, wind storm, hail storm) from NOAA and other sources. Additionally, a cloud-based mapping platform is used to display the real-time effects of the storm via social media posts (Fig. 15.5). Although the default is to show information for "tornado," registered users can log in to track and monitor the feeds of other disaster events by searching with keywords such as "fire," "snow," etc.

Earthquake in Haiti

On Tuesday, January 12, 2010 at 4:53 pm local time, a catastrophic magnitude 7.0 earthquake occurred, with an epicenter near the town of Léogâne

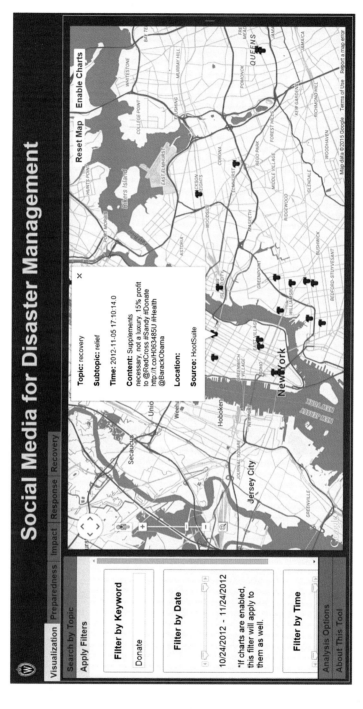

Figure 15.4 Visualizing and mining Tweets for disaster management.

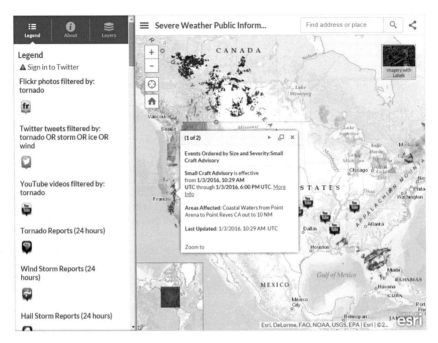

Figure 15.5 Environmental Systems Research Institute (Esri) weather map showing live storm reports, precipitation, and weather warnings along with multisourced social media content from Twitter, Flickr, and YouTube (http://www.esri.com/services/disaster-response/severe-weather/latest-news-map).

(Ouest Department), approximately 25 km (16 miles) west of Port-au-Prince, Haiti's capital (http://en.wikipedia.org/wiki/2010_Haiti_earthquake). It was the strongest quake recorded in Haiti for over 200 years. The earthquake caused devastating damage in the densely populated city of Port-au-Prince, Jacmel, and other settlements in the region. Notable landmark buildings were severely damaged or destroyed, including the Presidential Palace, the National Assembly building, the Port-au-Prince Cathedral, and the main jail. An estimated 3 million people were affected by the quake with 230,000 killed, 300,000 injured, and 1 million people left homeless (Yates and Paquette, 2011).

Many countries responded to appeals for humanitarian aid, pledging funds, and dispatching rescue and medical teams, engineers, and support personnel. Social media was successfully used as a platform to both gather and disseminate information to a global audience, and to support emergency response and disaster relief in a variety of ways.

- Public sharing of situational information and conditions through social network sites (e.g., Twitter, Flickr, and TwitPics). The collapse of traditional media elevated social media to the principal communications tool: everyone became a journalist. As soon as the earthquake struck Port-au-Prince, the capital city of Haiti, the first pictures of the devastated scenes were posted to Twitter and Facebook and were later relayed to the world by CNN. Following that, thousands of other pictures quickly spread through TwitPics and Twitter along with well wishes (Oh et al., 2010).
- Use of wikis by both nongovernmental and government relief organization to share and collaborate via the cloud in a secure arena. For example, US government agencies employed social technologies such as wikis and collaborative workspaces as the main knowledge-sharing mechanisms (Yates and Paquette, 2011).
- Crowdsourcing and identification of individuals who could translate Creole and Haitian dialects into other languages so that first responders could identify where to target aid.
- Use of social platforms for relief efforts. Twitter users spread the way to send emergency supplies or aid money to Haiti, and shared the information on how to adopt orphaned children (Oh et al., 2010).
- Remote aid via volunteers digitizing street maps and satellite imagery and uploading these results to a shared platform. Social technologies can also be leveraged to generate geographic information which is critical for emergency management. For example, a group of OpenStreetMap users around the globe rapidly produced a detailed street map of Port-au-Prince, based on digitization of satellite imagery facilitated by social networks such as Crisis Mappers Net (Li and Goodchild, 2012), which was used by crisis responders on the ground in Haiti. Before the earthquake, the OpenStreetMap map of Port-au-Prince had very limited coverage (Fig. 15.6 left), but within only 48 h, the dataset became possibly the most complete and accurate source for that area available to first responders (Fig. 15.6 right).

CONCLUSIONS

As social media applications are widely deployed in various platforms from personal computers to mobile devices, they are becoming a natural extension to human sensory system. The synthesis of social media with human

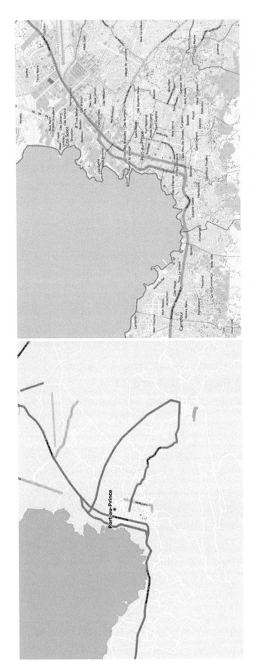

Figure 15.6 OpenStreetMap coverage before (left) and after (right) the 2010 Haiti Earthquake (http://blog.okfn.org/2010/01/15/open-street-map-community-responds-to-haiti-crisis/).

intelligence has the potential to be the intelligent sensor network that can be used to detect, monitor, and gain situational awareness during a natural hazard with unprecedented scale and capacity. However, the rate at which these crowdsourced data are being generated exceeds our ability to organize and analyze them to extract patterns critical for addressing real-world challenges, such as effective disaster management strategies. New challenges arise from an unprecedented access to massive amounts of social media data available and accumulated every day, and on how to develop and use new tools to mine knowledge from these data streams. This chapter discusses how social media can be used to assist during various stages of disaster management, and provides a literature survey of existing relevant research. However, the great potential of using social media for disaster response comes with challenges, which are identified along with a discussion of potential solutions.

Cloud computing is a promising computing infrastructure to accelerate geoscience research and applications by pooling, elastically sharing, and integrating latest computing technologies, and deploying physically distributed computing resources (Huang et al., 2013a). As discussed in the section Case Studies, cloud computing should be integrated to design and develop an elastic and resilient CI framework for archiving, processing, mining, and visualizing social media datasets disaster management. Specifically, cloud storage may be investigated and leveraged for the massive social media data archiving for immediate disaster response use and later research studies. Elastic computing power can be leveraged to handle the computing demands from mining big social media data, and massive concurrent access of an online analytical system during emergencies. Finally, several use cases are provided to demonstrate how social media and cloud computing have been employed to study several recent disasters.

REFERENCES

Agrawal, D., Das, S., El Abbadi, A., 2011. Big data and cloud computing: current state and future opportunities. In: Proceedings of the 14th International Conference on Extending Database Technology. ACM, pp. 530–533.

American Red Cross, 2012. Americans Increasingly Using Mobile Apps for Emergencies. Retrieved from: http://www.rwjf.org/en/culture-of-health/2012/09/american_red_cross.html (accessed 07.06.15.).

Ashktorab, Z., Brown, C., Nandi, M., Culotta, A., 2014. Tweedr: mining twitter to inform disaster response. In: Proceedings of the ISCRAM.

Bakshy, E., Hofman, J.M., Mason, W.A., Watts, D.J., 2011. Everyone's an influencer: quantifying influence on twitter. In: Proceedings of the Fourth ACM International Conference on Web Search and Data Mining. ACM, pp. 65–74.

Cervone, G., Franzese, P., 2014. Source term estimation for the 2011 Fukushima nuclear accident. In: Data Mining for Geoinformatics. Springer, pp. 49–64.

Cha, M., Haddadi, H., Benevenuto, F., Gummadi, P.K., 2010. Measuring user influence in twitter: the million follower fallacy. ICWSM 10, 30.

Cheong, F., Cheong, C., 2011. Social media data mining: a social network analysis of tweets during the Australian 2010–2011 floods. In: 15th Pacific Asia Conference on Information Systems (PACIS). Queensland University of Technology, pp. 1–16.

Cohen, H., 2011. 30 Social Media Definitions. Retrieved from: http://heidicohen.com/social-media-definition/ (accessed 07.06.15.).

Elwood, S., Goodchild, M.F., Sui, D.Z., 2012. "Researching Volunteered Geographic Information: Spatial Data, Geographic Research, and New Social Practice." Annals of the Association of American Geographers 102 (3), 571–590. http://dx.doi.org/10.1080/00045608.2011.595657.

FEMA, 1998. Animals in Disaster: Module Awareness and Preparedness. Accessible through: http://training.fema.gov/EMIWeb/downloads/is10comp.pdf (accessed 08.09.12.).

Fromm, H., Bloehdorn, S., 2014. Big Data—Technologies and Potential, Enterprise-Integration. Springer, pp. 107–124.

Gao, H., Barbier, G., Goolsby, R., 2011. Harnessing the crowdsourcing power of social media for disaster relief. IEEE Intelligent Systems 10–14.

Goodchild, M.F., Li, L., 2012. Assuring the quality of volunteered geographic information. Spatial Statistics 1, 110–120.

Gupta, A., Kumaraguru, P., 2012. Credibility ranking of tweets during high impact events. In: Proceedings of the 1st Workshop on Privacy and Security in Online Social Media. ACM, p. 2.

Huang, Q., Li, Z., Xia, J., Jiang, Y., Xu, C., Liu, K., Yu, M., Yang, C., 2013a. Accelerating geocomputation with cloud computing. In: Modern Accelerator Technologies for Geographic Information Science. Springer, pp. 41–51.

Huang, Q., Xiao, Y., 2015. Geographic situational awareness: mining tweets for disaster preparedness, emergency response, impact, and recovery. International Journal of Geo-Information 4, 19.

Huang, Q., Xu, C., 2014. A data-driven framework for archiving and exploring social media data. Annals of GIS 20, 265–277.

Huang, Q., Yang, C., Benedict, K., Chen, S., Rezgui, A., Xie, J., 2013b. Utilize cloud computing to support dust storm forecasting. International Journal of Digital Earth 6, 338–355.

Huang, Q., Yang, C., Liu, K., Xia, J., Xu, C., Li, J., Gui, Z., Sun, M., Li, Z., 2013c. Evaluating open-source cloud computing solutions for geosciences. Computers & Geosciences 59, 41–52.

Huang, Q., Yang, C., Nebert, D., Liu, K., Wu, H., 2010. Cloud computing for geosciences: deployment of GEOSS clearinghouse on Amazon's EC2. In: Proceedings of the ACM SIGSPATIAL International Workshop on High Performance and Distributed Geographic Information Systems. ACM, pp. 35–38.

Imran, M., Castillo, C., 2015. Towards a data-driven approach to identify crisis-related topics in social media streams. In: Proceedings of the 24th International Conference on World Wide Web Companion. International World Wide Web Conferences Steering Committee, pp. 1205–1210.

Imran, M., Elbassuoni, S., Castillo, C., Diaz, F., Meier, P., 2013. Practical extraction of disaster-relevant information from social media. In: Proceedings of the 22nd International Conference on World Wide Web Companion. International World Wide Web Conferences Steering Committee, pp. 1021–1024.

Kent, J.D., Capello Jr., H.T., 2013. Spatial patterns and demographic indicators of effective social media content during the Horsethief Canyon fire of 2012. Cartography and Geographic Information Science 40, 78–89.

Kumar, S., Barbier, G., Abbasi, M.A., Liu, H., 2011. TweetTracker: an analysis tool for humanitarian and disaster relief. ICWSM.

Leberecht, T., 2010. Twitter Grows Up in Aftermath of Haiti Earthquake. CNET News 19.

Li, L., Goodchild, M.F., 2012. The role of social networks in emergency management: a research agenda. Managing Crises and Disasters with Emerging Technologies: Advancements 245.

Livingstone, S., Helsper, E., 2007. Gradations in digital inclusion: children, young people and the digital divide. New Media & Society 9, 671–696.

Meier, P., 2012. Digital Disaster Response to Typhoon Pablo. Retrieved from http://voices.nationalgeographic.com/2012/12/19/digital-disaster-response/ (accessed 01.07.15).

Mandel, B., Culotta, A., Boulahanis, J., Stark, D., Lewis, B., Rodrigue, J., 2012. A demographic analysis of online sentiment during Hurricane Irene. In: Proceedings of the Second Workshop on Language in Social Media. Association for Computational Linguistics, pp. 27–36.

Manovich, L., 2011. Trending: The Promises and the Challenges of Big Social Data. In: Gold, M.K. (Ed.), Debates in the Digital Humanities. Cambridge, MA: The University of Minnesota Press, pp. 460–475.

Nakamura, L., 2008. Digitizing Race: Visual Cultures of the Internet. University of Minnesota Press.

Nambiar, R., Chitor, R., Joshi, A., 2014. Data Management–A Look Back and a Look Ahead, Specifying Big Data Benchmarks. Springer, pp. 11–19.

Nations, D., 2015. What Is Social Media? Retrieved from: http://webtrends.about.com/od/web20/a/social-media.htm (accessed 07.06.15.).

Neal, D.M., 1997. Reconsidering the phases of disasters. International Journal of Mass Emergencies and Disasters 15, 239–264.

Oh, O., Kwon, K.H., Rao, H.R., 2010. An exploration of social Media in extreme events: rumor theory and twitter during the Haiti earthquake 2010. ICIS 231.

Potiriadis, C., Kolovou, M., Clouvas, A., Xanthos, S., 2012. Environmental radioactivity measurements in Greece following the Fukushima Daichi nuclear accident. Radiation Protection Dosimetry 150, 441–447.

Prentice, S., Huffman, E., Alliance, B.E., 2008. Social Media's New Role in Emergency Management: Emergency Management and Robotics for Hazardous Environments. Idaho National Laboratory.

Pu, C., Kitsuregawa, M., 2013. Big Data and Disaster Management a Report from the JST/NSF Joint Workshop. Georgia Institute of Technology.

Purohit, H., Castillo, C., Diaz, F., Sheth, A., Meier, P., 2013. Emergency-Relief Coordination on Social Media: Automatically Matching Resource Requests and Offers. First Monday 19.

Roth, R.E., 2013. Interactive maps: what we know and what we need to know. Journal of Spatial Information Science 59–115.

Sakaki, T., Okazaki, M., Matsuo, Y., 2010. Earthquake shakes Twitter users: real-time event detection by social sensors. In: Proceedings of the 19th International Conference on World Wide Web. ACM, pp. 851–860.

Schadt, E.E., Linderman, M.D., Sorenson, J., Lee, L., Nolan, G.P., 2010. Computational solutions to large-scale data management and analysis. Nature Reviews Genetics 11, 647–657.

Schnebele, E., Cervone, G., 2013. Improving Remote Sensing Flood Assessment Using Volunteered Geographical Data.

Schnebele, E., Cervone, G., Kumar, S., Waters, N., 2014. Real time estimation of the Calgary floods using limited remote sensing data. Water 6, 381–398.

Schnebele, E., Oxendine, C., Cervone, G., Ferreira, C.M., Waters, N., 2015. Using nonauthoritative sources during emergencies in urban areas. Computational Approaches for Urban Environments. Springer 337–361.

Sheth, A., Henson, C., Sahoo, S.S., 2008. Semantic sensor web. In: Internet Computing, 12. IEEE, pp. 78–83.

Smith, A.B., Matthews, J.L., 2015. Quantifying uncertainty and variable sensitivity within the US billion-dollar weather and climate disaster cost estimates. Natural Hazards 1–23.

Sutton, J.N., 2010. Twittering Tennessee: Distributed Networks and Collaboration Following a Technological Disaster. ISCRAM.

U.S. Department of Education, 2010. Action Guide for Emergency Management at Institutions of Higher Education. Retrieved from: http://rems.ed.gov/docs/rems_Action-Guide.pdf (accessed 07.06.15.).

Van Dijk, J.A., 2006. Digital divide research, achievements and shortcomings. Poetics 34, 221–235.

Vasilescu, L., Khan, A., Khan, H., 2008. Disaster Management CYCLE–a theoretical approach. Management & Marketing-Craiova 43–50.

Velev, D., Zlateva, P., 2012. Use of social media in natural disaster management. International Proceedings of Economic Development and Research 39, 41–45.

Verma, S., Vieweg, S., Corvey, W.J., Palen, L., Martin, J.H., Palmer, M., Schram, A., Anderson, K.M., 2011. Natural language processing to the rescue? extracting "situational awareness" tweets during mass emergency. ICWSM, Citeseer.

Vieweg, S., Hughes, A.L., Starbird, K., Palen, L., 2010. Microblogging during two natural hazards events: what twitter may contribute to situational awareness. In: Proceedings of the SIGCHI Conference on Human Factors in Computing Systems. ACM, pp. 1079–1088.

Vieweg, S.E., 2012. Situational Awareness in Mass Emergency: A Behavioral and Linguistic Analysis of Microblogged Communications.

Warfield, C., 2008. The Disaster Management Cycle, Disaster Mitigation and Management. Accessible through: http://www.gdrc.org/uem/disasters/1-dm_cycle.html (accessed 17.10.2012.).

Witte, J.C., Mannon, S.E., 2010. The Internet and Social Inequalitites. Routledge.

Xiao, Y., Huang, Q., Wu, K., 2015. Understanding social media data for disaster management. Natural Hazards 79 (3), 1663–1679.

Yang, C., Wu, H., Huang, Q., Li, Z., Li, J., 2011. Using spatial principles to optimize distributed computing for enabling the physical science discoveries. In: Proceedings of the National Academy of Sciences 108, pp. 5498–5503.

Yates, D., Paquette, S., 2011. Emergency knowledge management and social media technologies: a case study of the 2010 Haitian earthquake. International Journal of Information Management 31, 6–13.

CHAPTER 16

Dubai Operational Forecasting System in Amazon Cloud

B. McKenna
RPS ASA, Wakefield, RI, USA

INTRODUCTION

A municipal body in the Emirate of Dubai, United Arab Emirates (UAE) required a "decision support system" to provide atmospheric and oceanographic forecasts for their coastal zones and waterway operations. In 2011, RPS Applied Science Associates (ASA) (a member of the Rural Planning Services (RPS) Group, PLC) was contracted to develop and maintain a coupled suite of atmospheric and oceanographic models alongside a data management infrastructure and Web-based presentation tool. The resulting operational METeorological and OCEANographic (MetOcean) Forecast System (MOFS) provided daily accurate and detailed 5-day forecast data.

After the initial design and calibration, the MOFS was transferred to infrastructure in Dubai. It quickly became apparent that continuous support of the required infrastructure and challenges associated with transferring data required for initialization and lateral boundary conditions across a global network led to occasional degradation in the performance of the system. Maintenance and troubleshooting of the system proved challenging as well, often hindered by poor network reliability and latency resulting in slow support response time.

As the challenges associated with a forecast system deployed over 10,000 km away became clear, solutions were sought which utilized a reliable data center with emphasis on global network performance and reliability.

After reviewing several options, including Google's Compute Engine and Rackspace's Cloud Servers, RPS ASA chose Amazon's cloud computing services, or Amazon Web Services (AWS). AWS provided "*access to the same highly scalable, reliable, fast, inexpensive infrastructure that Amazon uses to run its own global network of web sites*" (Frequently Asked Questions (FAQs)) along with a well-supported, robust Software Development Kit (SDK) to

Cloud Computing in Ocean and Atmospheric Sciences
ISBN 978-0-12-803192-6
http://dx.doi.org/10.1016/B978-0-12-803192-6.00016-5

Copyright © 2016 Elsevier Inc.
All rights reserved.

programmatically interact with the cloud infrastructure. The maturity of the Amazon SDK was an important factor in creating an automated execution environment as necessary for this project.

Our transition to Amazon AWS began in early 2013 and took approximately 4 months of effort. The resulting system automatically creates the necessary resources in the Amazon cloud, executes the suite of numerical models and delivers the model output to our clients in a rapid, reliable process.

We found that AWS did in fact enable us to provide complete functionality of the MOFS that had been developed and mitigate the network challenges being experienced. Utilizing Amazon's global cloud infrastructure, we were now able to develop and support the MOFS locally with an ability to deploy the system globally.

OPERATIONAL FORECASTING SYSTEM OVERVIEW

The operational forecasting system is a coupled suite of MetOcean models executed in the order indicated in Table 16.1.

In this one-way coupling, Weather Research and Forecasting (WRF) surface winds are used to force both Regional Ocean Model System (ROMS) and Simulating WAves Nearshore (SWAN), and ROMS hydrodynamic circulation and elevation is used to force SWAN. The output resolution desired by the client required two WRF nests, two ROMS nests, and the use of an unstructured grid for SWAN.

A global forecast from the United States' National Centers for Environmental Prediction (NCEP) is used to provide model initialization data and regional boundary conditions. The MOFS simulation is run out to 5 days once a day at 00 UTC. In the current version of the MOFS, due to many of the known observation sources being privately owned, no data assimilation is performed before the simulation.

Table 16.1 Numerical model components of the Gulf coupled MetOcean Forecast System

Component	Provides
Weather Research and Forecasting (Skamarock et al., 2005)	Atmospheric model
Regional Ocean Model System (Shchepetkin and McWilliams, 2005)	Oceanographic model
Simulating WAves Nearshore (Booij et al., 1996)	Nearshore wave model

A comprehensive round of model calibration and validation was used to establish the model domains and configurations. The client provided detailed observation data from various onshore and offshore meteorological stations as well as several Acoustic Doppler Current Profilers (ADCP) measuring current and wave data at four nearshore and two offshore stations.

The speed of execution of a forecast system is critically important as any delays could lead to outdated information or insufficient time for the user of the forecast to adequately respond to the situations predicted. To facilitate rapid execution, all models were built using a distributed memory Message Passing Interface (MPI) standard to enable communication between independent processors. The number of processors can be scaled as needed to achieve the desired performance.

Model Domains

Model domains for the MOFS were designed to represent the regional atmospheric and oceanographic circulation of the Gulf and the associated monsoonal climate representative of the Gulf region (Walters and Sjoberg, 1990). The importance of the dynamic features originating and propagating from the north-northwest region near Iran, Kuwait, and Iraq led to a need to simulate the full Gulf region. Nested domains were created over the immediate Dubai coastline to accurately resolve the bathymetry and resulting hydrodynamic circulation from the complex geographic features.

Weather Research and Forecasting Atmospheric Model

The WRF model was implemented with two domains (Fig. 16.1). The first, or parent grid, was configured to capture important regional atmospheric dynamics across the entire Gulf, as phenomena such as the Shamal winds often have a significant impact on the Dubai coastline (Perrone, 1979) and originate in the distant northwest of the Gulf. The parent grid has a horizontal resolution of 0.1 degree (approximately 10.5 km) and horizontal grid size of 94×82 grid points. There are 38 vertical levels and a time step of 30 s. This spatial and temporal resolution was found to provide a satisfactory representation of regional atmospheric dynamics as well as local atmospheric circulation features such as the land–sea breeze.

Nested within the outer domain is a second, centered over the Dubai coastline with a horizontal resolution of 0.033 degree (approximately 3.5 km) and horizontal grid size of 130×100 km. As with the parent domain, there are 38 vertical levels, but a time step of 10 s. This higher spatial and temporal

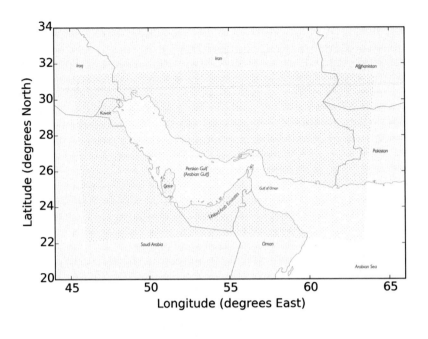

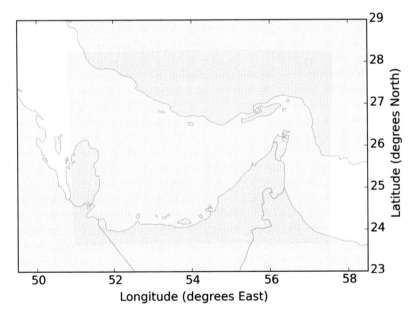

Figure 16.1 WRF model domains: (top) 10.5 km parent domain extending over entire Gulf region; (bottom) 3.5 km nested domain covering the critical coastal waterways and coastal region of Dubai, UAE.

resolution was desired by the municipality to provide detailed circulation around the complex coastline and coastal features of the shoreline.

Regional Ocean Modeling System Oceanographic Model

Due to the complex coastline along the shores of Dubai, including many man-made island features, multiple grids (Fig. 16.2) were required to accurately represent the hydrodynamic circulation along the Dubai shoreline. To represent these features with the highest precision, the client provided bathymetric Light Detection and Ranging (LiDAR) data covering nearly 1000 square kilometers with resolutions ranging from 5 m offshore to 3 m nearshore.

The first domain (Fig. 16.2 left) extends from the eastward boundary in the Gulf of Oman, through the Strait of Hormuz across the Gulf with a horizontal resolution of approximately 750 m and horizontal grid size of 300×240 km. There are 32 terrain-following vertical levels. The spatial and temporal resolution was calibrated to represent the general regional circulation using observed sea surface elevation data with the vertical levels providing detailed three-dimensional information to be used in oil-spill model applications (Knee et al., 2012).

The second ROMS domain is nested within the first domain and covers the Dubai shoreline providing a highly detailed representation of the various offshore structures (Fig. 16.2 right). The horizontal resolution is approximately 5 m and necessary to resolve the complex flow in and around the complex shoreline features and artificial islands.

Simulating WAves Nearshore Wave Model

With the Dubai coastline being a popular tourist destination, there was an interest in detailed and accurate wave forecasts that may impact public safety and impacts to the geography and infrastructure from coastal erosion and accretion. Due to the necessity for high-resolution output along the coastline and adjacent artificial islands, the SWAN model was configured using an unstructured triangular mesh (Fig. 16.3) containing 7958 triangles with 4670 vertices. The average minimum triangle edge length of the mesh among the smallest 10% of triangles, most of which were located in the nearshore Dubai coastal zone, was approximately 125 m. The maximum triangle edge length among the largest 10% of triangles in the domain, located greater than 250 km from the Dubai coast, was approximately 10 km.

Data Sources

The regional simulation of the MOFS requires a three-dimensional atmospheric and oceanographic initial condition as well as time-dependent lateral

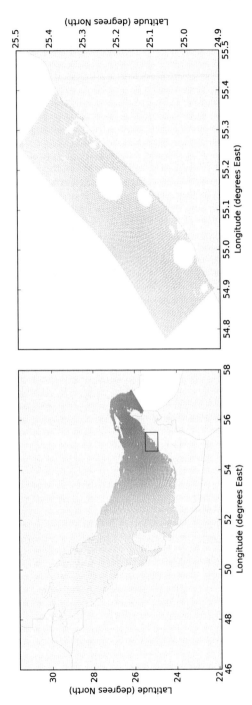

Figure 16.2 ROMS model domains. Parent domain (left) extends across the entire Gulf. High-resolution nest (right) provides detailed bathy-metric representation and circulation around the complex Dubai coastline.

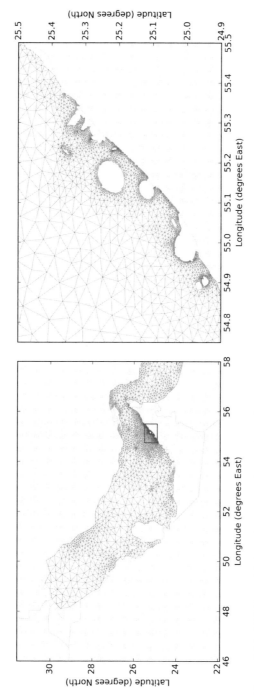

Figure 16.3 SWAN model domain—triangular mesh grid covering the entire Gulf (left) with high resolution/density in the immediate Dubai coastal waters (right).

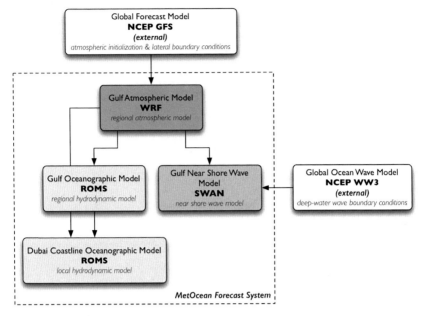

Figure 16.4 Numerical model components of the MetOcean Forecast System (enclosed in box outlined by *dotted lines*) and required input data for initialization and boundary conditions (outside box).

boundary conditions to capture features that move into the regional domain during the simulation. These initial and boundary conditions are provided by the publicly available NCEP global atmospheric and wave models and previous forecasts generated by MOFS. Fig. 16.4 shows the relationship of the MOFS components and where external data from NCEP is processed.

Weather Research and Forecasting Atmospheric Model

The parent domain of the WRF regional model requires three-dimensional initial and lateral boundary conditions provided by the NCEP Global Forecast System (GFS). The nested WRF domain, centered over Dubai, uses the parent domain for initial and lateral boundary conditions. The GFS is a global spectral, data assimilation, and forecast model system, with forecast output produced every 6 h at 00, 06, 12, and 18 UTC and extending 16 days into the future. The GFS output is available in General Regularly-distributed Information in Binary form (GRIB)-2 format and stored on NCEP's remotely accessible, United States-based servers. A script to download the 0.25 degree data (originally 0.5 degree and updated in 2015) was developed to minimize the potential of a failure from a poor network connection with several retry attempts on failure and optimizations to obtain only the necessary vertical levels and variables needed for the local forecast model to run. To expedite the execution of the MOFS and avoid unnecessary delays in beginning the numerical simulations, a fast and fault-tolerant Internet connection is essential to rapidly download the data for successful daily operation of the system.

Regional Ocean Modeling System Oceanographic Model

The ROMS oceanographic model requires three-dimensional hydrodynamic initial conditions, as well as tidal, atmospheric, and lateral hydrodynamic boundary conditions. Due to network bandwidth constraints, the MOFS uses a previous forecasted state, e.g., previous 24 h forecast output, for the ROMS initial conditions. Tidal forcing is extracted from the Oregon State University Tidal Inversion Software (OTIS) (Egbert and Erofeeva, 2002). The atmospheric forcing, including surface wind, radiative flux, air pressure, temperature, specific humidity, and precipitation is extracted from the WRF model. The hydrodynamic boundary conditions are extracted from monthly climatologies derived from the HYbrid Coordinate Ocean Model (HYCOM) (Bleck, 2002). HYCOM was run for several years with 33 vertical levels, using a higher resolution near the surface and a horizontal resolution of 0.08 degree.

Simulating WAves Nearshore Wave Model

Offshore wave boundary conditions are necessary for SWAN and are provided by deep-water wave parameters obtained from the NCEP WAVEWATCH III (WW3) global wave model. Similar to the NCEP GFS output, WW3 forecasts are produced every 6 h at 00, 06, 12, and 18 UTC and provided on a 30 arc-minute global grid. The forecast output is downloaded from the NCEP servers in the United States. Similar optimizations and fault tolerance are designed into the system as these too are essential for the daily operation of the system.

SYSTEM ARCHITECTURE

Development

Development of the MOFS began in 2011. The system's initial infrastructure was a single server with two Intel Xeon 2.80 GHz processors (8 cores total) and 16 gigabytes (GB) of RAM. The server was housed at RPS ASA facilities in Rhode Island, United States. This infrastructure was intended primarily for the development of the forecast workflow, calibration of the model parameters, and validation of the forecast output. The requirement of one forecast cycle per day also made the infrastructure a cost-effective option for determining runtime estimates. A Debian-based Linux operating system was installed along with the GNU Compiler Collection (GCC), Intel Fortran compiler, and Python 2.7 for scripts. The model binaries and supporting libraries required for the models were compiled with the Intel Fortran compiler and the GCC C/C++ compilers.

Initial runtime for a 5-day forecast simulation was nearly 16 h for the two WRF domains (6 h), two ROMS domains (6 h), and then unstructured SWAN domain (4 h). The simulation results in roughly 28 GB of raw model output. The output for delivery was reduced to the necessary vertical levels and variables then compressed to less than 2 GB. With the 00 UTC GFS and WW3 initialization data typically not available on the NCEP File Transfer Protocol (FTP) server until 04–05 UTC, a 16 h runtime meant the final forecast output would not be available for delivery to the client until 20–21 UTC, or 00–01 Dubai local time (Greenwich Sidereal Time (GST)) the following day. Given this was the middle of the night in Dubai, the forecast data would likely not be seen and utilized until the following morning around 04 UTC (08 GST), 28 h later than the original GFS/WW3 forecasts used in initialization. With predictive skill of numerical models decreasing as a function of forecast time (AMS, 2007), a delivery in delay or utilization leads to a poorer-quality forecast. It is therefore imperative we maximize performance and eliminate all potential delays so as to deliver the most recent forecast possible for the client to use in their daily operations.

Compounding the issue of a delivery time more than 24 h after initialization time, the system's reliance on the previous cycle for the ROMS initial conditions meant any system failure would require the missing cycle be run before proceeding with subsequent forecasts. With 16 h of the day consumed by a single execution of the forecast, the next forecast initialization would be delayed by the additional 16 h of rerun time. The system would thus require two full daily forecast cycles before it was caught up.

First Implementation

For the first operational release and delivery of forecast data to the client in late 2011, the infrastructure was extended from 8 to 16 cores, reducing the runtime to approximately 8 h. With the improved performance, forecasts could be delivered as early as 12 UTC (16 GST), and additional time was available to rerun any forecast cycles that may have failed.

The forecast availability time of 4 am local time (GST) was better than during development, but unfortunately, it was still on the tail end of the daily business-operating schedule in Dubai. This meant any delay in the system, such as the download of initial conditions, production of the forecasts or transfer of the output, could again push the hours of the forecast usable by the client to the following day. Despite the improvement in performance from increasing resources, there was still a risk of diminished usable forecast skill.

As it would turn out, the network connectivity between the United States and Dubai did prove to be problematic in delivering the final forecast data. At times, the transfer of the 2 GB of forecast output was taking nearly 2 h. Even if we were to further increase performance of the system, there were no guarantees we could get the output to the client earlier with the occasional network delays.

Although the first implementation of MOFS was performing as required, with few failures and the models successfully providing MetOcean forecasts for Dubai's coastal and waterway operations, network delays and occasional failures led to a desire to transfer the entire MOFS to infrastructure in Dubai where delivery times could be reduced or eliminated.

Second Implementation

In late 2012, with over a year of successful operations, the system was transferred to a server in Dubai. Before delivery, we were able to further reduce the system runtime to under 8 h by optimizing the system workflow and redesigning the download of initialization data to use multiple concurrent processes. We anticipated the network delays experienced delivering the MOFS output from the United States to Dubai, would also be present when downloading the 1 GB of the GFS and WW3 data required for the forecast system's initialization from NCEP.

Along with the transfer, the system was updated allowing it to be initialized from either the 00 UTC or 12 UTC cycles of the GFS and WW3. With 12 UTC data typically available on the NCEP servers in the United States by 16 UTC, an improved 7-h runtime meant the forecast could be available by 23–24 UTC (03–04 GST). Again, assuming the client did not use the forecast until the start of the business day (0800 GST), the client could now have a more skillful forecast period (<24 h) supporting their primary operating hours.

Moving the system to Dubai eliminated the delay we had experienced transferring the MOFS output to Dubai; however, the issue of slow or failed transfers (now of GFS/WW3 initialization data) between the United States and Dubai remained. Additionally, an unanticipated impact of the transfer of MOFS to Dubai was the high latency and low network speeds available to our team in the United States to monitor, maintain, and troubleshoot the system. Response times for support and maintenance increased significantly and development on the remote server was difficult and at times impossible.

As support and development of the system was still a requirement, a mirror copy of the system was maintained on a virtual server in the United States. This virtual server had limited capacity, only 8 virtual cores and 8 GB of random access memory (RAM). Without the complete infrastructure, it was difficult to rapidly and fully test the results of updates and modifications to the system. It was clear that supporting, testing, and developing on similar infrastructure had significant advantages.

CLOUD IMPLEMENTATION

By 2013, with the forecast system operating successfully in Dubai and several of the original challenges removed, our focus turned to optimizing the maintenance of the system and development process to continue to improve its reliability and speed of execution.

We began looking for an environment that could provide a robust global network to facilitate an improved transfer of data as well as provide low latency from the United States for our support and maintenance of the system. As seen by having the system in Dubai, an ideal environment would support a regional implementation in or near the Middle East where forecast results could be produced and delivered rapidly.

After an extensive search and analysis of physical data centers in the Middle East as well as several Infrastructure-as-a-Service (IaaS), aka "cloud-computing" providers, AWS was identified as the target infrastructure to extend the MOFS and create the robust, high-performance, global application we desired.

AWS provides the Elastic Compute Cloud (EC2), a group of services that supply on-demand computational infrastructure including storage and networking. Resource types available through AWS range from simple servers with minimal computational capacity to high-performance servers offering enhanced bandwidth and networking alongside high computational capabilities. Although there have been indications that the shared-tenancy (multiple clients sharing the underlying physical hardware) AWS environment may not be ideal for scientific computing, the high-performance computing (HPC) tier (originally termed Cluster Compute) offerings have been found to perform significantly better than the standard EC2 offerings (Jackson et al., 2010). The HPC offerings tested were found to be equivalent in system memory and storage, and better in central processing unit (CPU) speed and number of cores, than the existing infrastructure being used in the Middle East.

Amazon's per-hour cost structure of cloud resources also created an opportunity to improve the system runtime. With capacity purchased in hourly units, and assuming near-linear speedup in runtime with an increase in cores being used, an 8-h runtime using 36 cores costs the same as a 4-h runtime using 72 cores or a 2-h runtime using 144 cores. In practice, our near-linear scalability was limited to 72 cores. We believe this limit was due to the underlying Amazon infrastructure used at the time. Although the networking provided by the HPC tier offering is capable of the low latency beneficial to parallel numerical computing, Amazon required the creation of a Placement Group to ensure all networking between the requested infrastructure also supported the low latency. Further investigations are being undertaken to extend this limit.

AWS also provides several SDKs in various programming languages, which allow users to programmatically interact with the AWS environment (http://aws.amazon.com/tools). The SDKs give applications the ability to create, configure, and destroy resources in the AWS cloud as needed. The Python SDK, boto (https://boto.readthedocs.org), would allow us to utilize the Python language and ecosystem, increasingly popular with many scientists, to automate an environment capable of running the forecast system.

Design of the Cloud Implementation

To replicate the MOFS in the AWS cloud, we used Amazon's EC2, which provides various tiers of on-demand computational capacity as well as Amazon's Simple Storage Service (S3) for the persistent storage of the model domains, configurations, and forecast output.

EC2's computational resources, called Instances, provide varying amounts of CPU capacity, system memory, networking, and an operating system such as Linux. Instance types range from simple low-memory single-core instances to HPC instances with 32 high-speed cores, over 60 GB of system memory, and 10 Gigabit Ethernet connectivity to support clustered applications.

EC2's storage resources, Elastic Block Storage (EBS) Volumes, give the user persistent block-level storage devices which can be attached to (and detached from) Instances just as a hard drive can be attached or detached to a physical server. These EBS Volumes can also be archived as an EBS Snapshot, similar to a traditional disk image, and can be used to create a duplicate EBS Volume the contents of which are the same as the original Volume. The duplicate is easily created in time, as in the next day, or in space as in another region.

An Amazon Machine Image (AMI) (Amazon Linux AMI, 2015) provides a consistent operating system for the MOFS implementation in the cloud. Based on Red Hat Enterprise Linux (RHEL), this Linux AMI is created by Amazon and designed for the EC2 infrastructure. Amazon maintains a repository of additional libraries and software allowing an Instance to be easily customized for the intended application. When appropriate, we have compared the capabilities and performance provided by the repository, and compiled some libraries from source to increase overall system capabilities (such as adding Open-Source Project for a Network Data Access Protocol (OPeNDAP) support in Network Common Data Format (NetCDF)).

Our approach was to create an EBS Snapshot which contained all of the system libraries, optimized model binaries, and scripts needed to execute the complete forecast system. This Snapshot is used to rapidly create a Volume in any region in the global AWS infrastructure. The Volume is then attached to one of the HPC tier Instance types and the forecast process started.

An ideal forecast system would be capable of running simulations for any domain. Given that static domain information and model configuration files are relatively small (100 megabytes) and change independent of the underlying models (e.g., grid/domain changes), we have decided to store this data outside the system. We utilize Amazon's S3 for the storage of this static data, and our runtime scripts transfer the data at runtime allowing the MOFS to be used for various domains and model configurations. Internal bandwidth between Amazon resources is optimized making the transfer time trivial to the overall system runtime.

Cloud services are priced according to the amount of resources consumed for any part of an hour. An interesting result of this cost structure is numerical forecast applications with parallel capabilities can often improve speed of simulation for little to no extra cost. Briefly ignoring parallel overhead and assuming linear inverse scalability (i.e., doubling the CPU capacity results in halving of the runtime), the cost of running on a single node for 8 h is the same as 2 nodes for 4 h, 4 nodes for 2 h, or 8 nodes for 1 h. With the importance of speed of execution and rapid availability of forecast output for the client, cloud computing provides an opportunity to favorably customize the infrastructure.

Execution of the Cloud Implementation

Although Amazon provides an easy to use, intuitive user interface for the creation and management of AWS cloud resources, it would be both impractical and expensive for an operator to use this interface to create the infrastructure as well as run the forecast system daily as required.

Using the AWS Python SDK, boto, we are able to create an automated workflow, shown in Fig. 16.5, with the following steps (Fig. 16.5—*numbers 1–4*):

1. request a simulation be executed
2. create requested Instance and Volume
3. download necessary data and execute MOFS
4. make results available to user or application

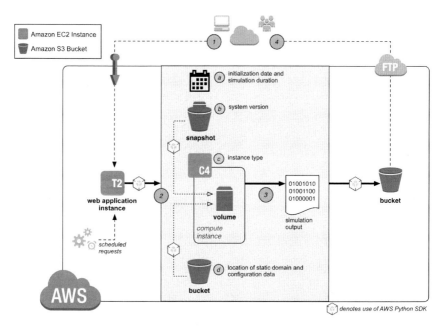

Figure 16.5 MOFS workflow. (1) request for simulation, (2) infrastructure created and models configured, (3) models executed, and (4) model output available; the inputs required for model executions are (a) initialization date and duration of simulation, (b) version of the MOFS to be executed, (c) EC2 Instance type to use (affects runtime), and (d) path to the static data needed for this regional simulation.

To provide both scheduled and on-demand execution of the cloud MOFS, we maintain a low-cost Instance containing a Python Flask (http://flask.pocoo.org) Web service application utilizing the AWS Python SDK to manage cloud resources. Requests for a forecast simulation can be scheduled on the Instance itself using standard scheduling tools (e.g., cron) or requested from an authenticated external resource via the Web service application.

The Web application requires the following to execute the forecast system (Fig. 16.5—*letters a–d*):

a. initialization date and simulation duration
b. system snapshot version
c. Instance type to execute the system on
d. location of the static domain and configuration files

EC2 provides support for automated system configuration and execution of custom commands at boot time via the *cloud-init* package (http://cloudinit.readthedocs.org/en/latest). A critical component of the cloud-based forecast system is the ability to send these commands or a script containing multiple commands alongside the SDK request for resources. The parameters of the forecast system are transferred from the web application to the compute Instance via these *cloud-init* commands.

Once the Volume containing all necessary scripts and model binaries is available to the Instance, GFS and WW3 initialization data are downloaded and preprocessed. Configuration files for WRF, ROMS, and SWAN are programmatically modified as needed for the simulation's date and duration, and then begin the execution of the forecast model suite.

Upon successful completion of all forecast components, the output is transferred to an S3 Bucket for persistent storage and external access. The cloud resources are then programmatically released and terminated so no ongoing computational charges are incurred.

Maintenance and Support of System

The transfer of the MOFS to Dubai in 2012 did result in an improvement to system reliability but ultimately led to support and maintenance difficulties associated with intermittent network speed and latency issues. The EBS Snapshot-based design of the cloud implementation provided the ability to troubleshoot, develop, test, and release incremental versions of the system in any region provided by the global AWS infrastructure.

With each update of a model version, supporting library or script, a new Snapshot is created, and the version number incremented. Previous

Snapshots of the system are retained giving us the ability to return to a previous version at any time by specifying that version in the request.

With cloud resources priced by the hour, it is now cost-effective to compare and benchmark different versions of the system against an internal suite of test cases for quality assurance and assessment of performance and identify any potential regressions associated with the updates. These executions use an entirely separate instance of cloud resources, only run for a few hours, and have no impact on the operational execution.

With the easy and cost-effective ability to create many versions of the forecast system and test as needed, care is taken to optimize each release for the cloud infrastructure. After a version has been validated for accuracy of results using our test cases, our team of scientists, developers, and engineers are able to return and look for potential optimizations of key components of the system.

RESULTS OF THE CLOUD IMPLEMENTATION

The cloud implementation of the MOFS has provided a "develop locally, deploy globally" capability for our atmospheric and oceanographic forecasting services. Previous implementations of the forecast system had limited resources, which often required in-place maintenance and troubleshooting. The ability to test, validate, and release versions independent of the operational implementation has led to significant improvements in maintenance efficiency and response time. Forecasts are being created, improving faster, and are more reliable with the cloud implementation.

ONGOING AND FUTURE SYSTEM DEVELOPMENT
Separation of Preprocessing

In the process of extending the system to use the global infrastructure, we noted impressive bandwidth speeds and quicker download times for externally hosted data being brought into the AWS infrastructure. With no cost for the transfer of incoming data, we saw an opportunity to utilize this speed alongside a low-cost instance and separate the download and preprocessing of the initialization datasets from the core execution of the models. EC2's HPC Instances are not only higher performance than lower-tier ones, but also higher cost. Operations which may be bound by network or input/output (I/O) speeds, or the execution for which is dependent on an external factor such as when it is available via FTP, can be completed by these lower-tier Instances before utilizing the high-performance ones, resulting in lower costs, shorter runtime and thus quicker forecasts.

Amazon's "low-cost instances with burstable performance", or second generation Burstable Performance Instances (T2) instances (New Low Cost EC2 Instances with Burstable Performance, 2014), operate on a CPU "budget" in which compute credits are accrued during times of zero or low utilization and can be spent to boost the computational capacity in intermittent bursts. When a burstable Instance exhausts its available credits, computation is paused until enough credits are available to continue the task.

The datasets used for initialization, the NCEP GFS and NCEP WW3, are produced four times daily at 00/06/12/18 UTC and available 4–5 h thereafter. The periodic burst required to process this data is easily matched with an appropriate T2 instance type. The first step of the forecast process, which has been slow due to network traffic in previous implementations, can be moved out of the main execution script.

An independent data acquisition process was developed to monitor the primary data sources and process that data as it becomes available instead of waiting for a request for a simulation. As both GFS and WW3 are global datasets, once preprocessed, the intermediate file, stored in S3, can be used by any regional implementation of the forecast system. With this design, we are not only getting the data into the Amazon global infrastructure as soon as it is available, but also moving it further down the forecast workflow. This has proven to be extremely useful for large datasets such as the HYCOM global ocean model, which can be as large as 160 GB per forecast cycle.

In April 2015, The US National Oceanic and Atmospheric Administration's (NOAA) announced the Big Data Project (https://data-alliance.noaa.gov), to publish NOAA's vast data offerings in the cloud and position them closer to the very cloud resources mentioned. This initiative will further enhance the ability to optimize the forecast system in the cloud.

Continuous Speed Increase

Our original benchmarks in 2013 used to select Amazon as the cloud provider showed that the original HPC offering by Amazon, the cc2.8 × large tier, was in fact a viable alternative, even an improvement on the performance of the original infrastructure. In late 2013, Amazon announced a new generation of HPC offerings, including the c3.8 × large offering, based on the latest Intel processors. The cost per hour of the new instance type was less than the cost of the previous generation. The forecast system was

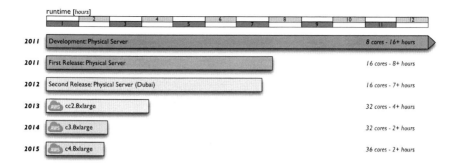

Figure 16.6 MOFS runtime as the system was improved from its development in 2011 through 2015. The AWS cloud indicates implementations in the Amazon cloud.

seamlessly transitioned to the new instance type by simply specifying that type in the SDK call to create an instance. These underlying infrastructure updates resulted in a nearly 2h reduction in forecast runtime. The new instance type was not only a lower cost per hour of utilization, but we were now running the system 1–2h less per day. Fig. 16.6 shows the reduction in runtime for a 5-day forecast simulation from the original development of our system through 2015 in which we utilize the latest Amazon HPC offering tier, c4.8 × large.

The speed of forecast production is of utmost importance for users of any forecast. The sooner the forecast is available, the quicker it can be processed, analyzed, and communicated to the stakeholders and public to better inform decisions. The ongoing improvement of speed as Amazon upgrades their infrastructure leads to an ability to deliver the same quality forecast information faster than before.

Configuration of Models

Although the EBS Snapshot-based design allows incremental version control of the core components of the system, the S3 Bucket-based design of model configurations and domains creates an environment allowing rapid replication, modification, and experimentation of model executions.

Amazon's S3 buckets are accessible by a variety of utilities similar to well-known FTP utilities. Our staff can easily access, review and even modify a configuration without needing extensive knowledge or access to the full server environment. Requesting an execution of the system can be a simple request to the web interface. The full model results, model diagnostic output, and system logs are stored in an S3 bucket to be downloaded and inspected locally by the requestor.

Due to the extremely low cost of storage in S3, our configurations can also be incrementally versioned giving us the ability to return to a previous configuration, or examine the changes and differences in detail.

CONCLUSION

With Amazon's global infrastructure providing *"access to the same highly scalable, reliable, fast, inexpensive data storage infrastructure that Amazon uses to run its own global network of web sites"* (FAQs, 2015), the MOFS is now able to consistently and rapidly create high-resolution MetOcean forecasts that can be deployed anywhere the cloud infrastructure is available.

The previous iterations, utilizing infrastructure in the United States and Dubai, resulted in a large number of potential failure points across the various networks needed in the process. Moving the system to the AWS cloud infrastructure has mitigated these challenges by placing the computational infrastructure geographically closer to the data delivery location.

In the process of moving the infrastructure, it was realized that in addition to improving on the robustness and efficiency of the system relative to forecast delivery schedules, benefits were also found which improved our development process, as well as our ability to rapidly support and maintain the forecast system for our clients. The ability to create temporary but fully functional environments has proven greatly beneficial for our teams of scientists for research and experimentation.

We have shown with cloud computing, our MOFS is no longer limited by the global network reliability or the steep entry costs to high-performance computing.

REFERENCES

Amazon Linux AMI, April 23, 2015. Amazon Linux AMI. Amazon Web Services, Inc., n.d. Web http://aws.amazon.com/amazon-linux-ami/.
AMS, 2007. Weather analysis and forecasting: an information statement of the American meteorological society. Bulletin of the American Meteorological Society 88, (n. pag. Print).
Bleck, R., 2002. An oceanic general circulation model framed in hybrid isopycnic-Cartesian coordinates. Ocean Modelling 4 (1), 55–88 CrossRef. Web. May 25, 2015.
Booij, N., Holthuijsen, L.H., Ris, R.C., 1996. The'SWAN' wave model for shallow water. Coastal Engineering Proceedings 1 (25), (n. pag. Print).
Egbert, G.D., Erofeeva, S.Y., 2002. Efficient inverse modeling of barotropic ocean tides. Journal of Atmospheric and Oceanic Technology 19 (2), 183–204 CrossRef. Web. May 25, 2015.
FAQs, April 23, 2015. Amazon S3 FAQs. Amazon Web Services, Inc., n.d. Web http://aws.amazon.com/s3/faqs/.

Jackson, K.R., et al., 2010. Performance Analysis of High Performance Computing Applications on the Amazon Web Services Cloud. IEEE, pp. 159–168. CrossRef. Web. June 5, 2015.

Knee, K., et al., 2012. Development and implementation of an operational forecasting system for the Dubai coastal zone. In: Oceans, 2012. IEEE, pp. 1–10 (Print).

New Low Cost EC2 Instances with Burstable Performance, July 1, 2014. New Low Cost EC2 Instances with Burstable Performance. Amazon Web Services, Inc., Web. June 4, 2015 https://aws.amazon.com/blogs/aws/low-cost-burstable-ec2-instances/.

Perrone, T.J., 1979. Winter Shamal in the Persian Gulf. DTIC Document (Print).

Shchepetkin, A.F., McWilliams, J.C., 2005. The regional oceanic modeling system (ROMS): a split-explicit, free-surface, topography-following-coordinate oceanic model. Ocean Modelling 9 (4), 347–404 CrossRef. Web. June 5, 2015.

Skamarock, W.C., et al., 2005. A Description of the Advanced Research WRF Version 2. DTIC Document (Print).

Walters Sr., K.R., Sjoberg, W.F., 1990. The Persian Gulf Region. A Climatological Study. DTIC Document (Print).

CHAPTER 17

Utilizing Cloud Computing to Support Scalable Atmospheric Modeling: A Case Study of Cloud-Enabled ModelE

J. Li
University of Denver, Denver, CO, USA

K. Liu
George Mason University, Fairfax, VA, USA

Q. Huang
University of Wisconsin–Madison, Madison, WI, USA

ATMOSPHERIC MODELING: AN OVERVIEW

Atmospheric processes such as radiation, convection, and aerosol movement play important roles in shaping the Earth's energy and water cycles. Modeling atmospheric processes is critical in advancing the understanding of those processes. Scientists can perform model simulations with historical climate data sets as input to predict future variations and estimate impacts of atmospheric processes on human–environment interactions. Since the last century, scientists have developed numerous atmospheric models that produce simulated data sets to quantify atmospheric parameters and to describe dynamic atmospheric processes. Examples include Energy Balance Models (e.g., Harvey and Schneider, 1985), Radiative Convective Models (e.g., Ramanathan and Coakley, 1978), and General Circulation Models (GCMs, e.g., Sellers et al., 1986).

Depending on the complexity of the model, running model simulations can be time-consuming and computationally intensive. To produce results in a timely fashion, a large number of atmospheric models have been designed to leverage parallel architecture and run on high-performance computing (HPC) facilities (e.g., Weather Research and Forecasting–Nonhydrostatic Mesoscale Model (WRF-NMM), Xie et al., 2010; Litta et al., 2011). Although traditional HPC clusters that provide parallel

Cloud Computing in Ocean and Atmospheric Sciences
ISBN 978-0-12-803192-6
http://dx.doi.org/10.1016/B978-0-12-803192-6.00017-7

Copyright © 2016 Elsevier Inc.
All rights reserved.

347

computing capabilities have been extensively used to perform model simulations, the cost of maintaining and using such facilities imposes a hardware obstacle for scientists who do not have access to such facilities to conduct model-based atmospheric research (Huang et al., 2013b). Recently, cloud computing innovations have attracted wide attention from both scientific communities and industries. Cloud computing techniques allow scientists to use elastic computing resources in the cloud on demand (Armbrust et al., 2010). Cloud computing is considered a promising solution to overcome the constraints of maintaining high-end computing resources and the desire to use computing resources on demand.

In this chapter, we describe a cloud-based computing framework that leverages on-demand cloud resources to support atmospheric modeling. The Goddard Institute for Space Studies (GISS) GCM ModelE, developed by National Aeronautics and Space Administration (NASA) (Hansen et al., 1983; Schmidt et al., 2006), was chosen as an example of atmospheric models to demonstrate the feasibility of the framework. This model produces various data sets to describe multiple atmospheric processes. The proposed framework has been tested with three cloud platforms to examine the readiness of cloud-based solutions for large-scale atmospheric modeling. Our experiments indicate that the framework facilitates atmospheric modeling by providing highly scalable, reliable, and customizable cloud computing resources.

The rest of the chapter is organized as follows: section Computing Solutions for Atmospheric Modeling provides an overview of three categories of popular computing solutions to atmospheric modeling. Section Building Cloud Infrastructure for Scenario-Based Atmospheric Modeling describes our cloud-based computing framework for atmospheric modeling. Section Case Study: ModelE presents the process of configuring and deploying ModelE onto the framework. Experiments and system demonstrations are given in the same section. Lessons learned from the implementation are discussed in section Discussion and Conclusion.

COMPUTING SOLUTIONS FOR ATMOSPHERIC MODELING

An atmospheric model typically performs complex computations to simulate the evolution of physical phenomena. Conducting atmospheric modeling is computationally intensive and requires large amounts of computing resources. Scientists have explored different types of computing solutions to support atmospheric modeling. In the following, we review typical computing solutions for atmospheric modeling. Existing solutions can be divided into three major categories: (1) centralized parallel computing with HPC clusters, (2) distributed computing such as volunteer computing resources, and (3) cloud computing.

High-Performance Parallel Computing

High-performance parallel (HPC) computing is one of the most popular computing solutions to address the computational challenges of running complex models (Menemenlis et al., 2005; Huang and Yang, 2011). This computing paradigm uses the computing resources of an HPC cluster to run tasks in a parallel manner. An HPC cluster generally comprises a head node and multiple computing nodes. The computing nodes can run independently and communicate through a computer network. Middleware is installed and configured on all nodes to monitor and support communication between the head node and computing nodes. The head node is responsible for (1) scheduling and dispatching tasks to computing nodes, (2) activating the computing tasks by configuring the middleware, and (3) collecting results from the computing nodes. Several open-source middleware solutions can be used to deploy an HPC system (e.g., Message Passing Interface over CHameleon (MPICH)).

However, building an HPC system is costly and time-consuming. A large financial investment and several weeks or even months are required to purchase the servers and configure the hardware and software while building up a middle-scale HPC system (Huang et al., 2013a). As a result, only a few organizations can afford or have access to these expensive computing facilities. Such systems are also difficult and expensive to maintain and operate. Besides, an HPC system is not feasible for different types of parallel applications. Because of the complex nature of parallel applications, the deployment of such an application relies on the customization of hardware and software environments tailored to the application. Once an HPC cluster is set up for a specific application deployment environment, limited customization can be made to both hardware and software environments. The cluster cannot be easily used to run another parallel application for two reasons. First, because traditional HPC systems do not provide an option for resource consumers to customize the underlying physical machines, a computing pool is only designed to run a specific type of computing task, such as Message Passing Interface (MPI)-based tasks and graphics processing unit (GPU)-based tasks (Li et al., 2013). An HPC system can only run one type of computing task. Second, different applications have different requirements for hardware and software. System administers may need to reconfigure the system environments.

Volunteer Computing

Recently, volunteer computing has become a popular computing solution to climate modeling. The idea of volunteer computing came from the popularity of the Internet and the increasing power of personal computing devices. Volunteers around world can contribute their computing resources (e.g., personal computers) and storage and help scientists to run the applications at home (Anderson and Fedak, 2006). Compared to HPC clusters, computing resources from volunteers are free. A volunteer computing project starts when scientists set up a project server which can assign and distribute tasks to volunteers and collect results from the volunteers' computers. Volunteers participating in the project run the applications using their idle computing resources.

One example is the implementation of the weather@home project (Massey et al., 2014). This project is designed for volunteers to participate in large-scale weather forecasting. Each volunteer can download a weather simulation task and run the task on a personal computer. The results of the task will be uploaded to the weather simulation center. Kondo et al. (2009) evaluated the volunteer computing paradigm and reached a conclusion that volunteer computing provides large benefits to scientists. The major challenge of using volunteer computing is ensuring volunteers can complete tasks successfully and on time. There is a high possibility that volunteers may terminate the assigned tasks Darch and Carusi (2010). Therefore, volunteer computing is not suitable for time-critical tasks that should be completed within a tight period.

Cloud Computing

In recent years, there has been an explosion of interest in using cloud computing to access computing resources (e.g., networks, servers, storage, applications, and services) via the Internet (Armbrust et al., 2010; Huang et al., 2010). Both commercial platforms (e.g., Amazon Web Services (AWS) cloud, Microsoft Azure) and open-source solutions (e.g., Openstack, Eucalyptus, and Nebula) have been utilized by cloud consumers. Cloud computing provides "Infrastructure as a Service" (IaaS), "Platform as a Service" (PaaS), "Software as a Service" (SaaS), and "Data as a Service" (DaaS) for end users in a "pay-as-you-go" mode (Buyya et al., 2009; Armbrust et al., 2010). Consumers purchase cloud resources on demand. Cloud computing solutions yield a cost-effective approach to the issue of lacking computing resources and remove the need for advanced facility management.

Cloud computing solutions have the following characteristics. First, cloud computing resources are highly customizable and configurable. The virtualization technique allows users to configure virtual computing instances of their choice. Once an instance is created and properly configured with the target applications or models, a virtual image can be produced and reused to create other instances within the same environment. In the process of launching multiple instances from the newly created image, users can choose different hardware configurations (e.g., different number of CPU cores). Second, cloud computing can save costs in load-balancing and auto-scaling capabilities, which can scale up or down the virtual machines to meet the computing needs of different applications. Scientists can also manually adjust the number of cloud instances and the capacities of cloud storage. Third, cloud computing platforms such as Amazon Elastic Compute Cloud (EC2) are featured with on-demand usages and low system maintenance. Users are not responsible for maintaining facilities after terminating cloud instances. Although cloud computing resources are more expensive than volunteer computing resources, using cloud computing resources is more cost-effective compared to using traditional HPC systems. Because most cloud computing providers guarantee service availability more than 99% of the time, cloud resources are more reliable than volunteer ones. Therefore, we designed a cloud-based modeling environment for atmospheric models.

BUILDING CLOUD INFRASTRUCTURE FOR SCENARIO-BASED ATMOSPHERIC MODELING

An Overview of the Architecture

We have developed a cloud-based computing framework to support atmospheric modeling. As discussed in section Atmospheric Modeling: An Overview, to evaluate the performance of a model, scientists run the model multiple times and each time with a different set of model configuration parameters such as spatial domains or spatial and temporal resolutions. The results from model runs are used to evaluate the accuracy of the model. The amount of computing resources needed by running model simulations varies with the number of model runs. Scalable computing solutions are necessary to provide a variable number of computing instances to meet the dynamic demands of multiple model runs. Besides, depending on the configuration of a model run, the duration of a model run and the volume of

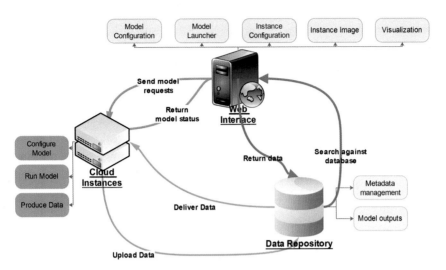

Figure 17.1 The cloud-based model simulation framework.

model outputs change drastically (e.g., 10-year vs. 1-year simulation). Providing computing resources on demand is both energy efficient and usage efficient.

In our framework (Fig. 17.1), cloud computing resources are utilized to perform model runs and store model outputs. Fig. 17.1 shows the architecture of the framework. The framework consists of three core components: a Web portal, a data repository, and a set of cloud instances. The configuration interface is installed on a Web server for configuring the model and computing facilities. Cloud instances perform atmospheric model runs. A cloud-based data repository is used to archive model outputs. The framework has three unique design features. First, the system provides a Web interface to interact with cloud instances which hides the complex cloud configurations. Users without cloud experience can configure, build, and run models. Second, the system is highly configurable and scalable for different simulation purposes. Both computing instances and data storage can be adjusted for different models. Third, as all components of the system are deployed in the cloud, maintenance costs have been reduced. For example, local Web servers are not needed to deploy the framework.

Web Portal

The portal is the entry point to interact with all server-side operations of the framework and provides two types of functions: system configuration and analysis manipulation. The portal includes five modules, including model configuration, model launcher, instance configuration, instance image, and data visualization. The first two modules are designed for scientists to configure models. The model configuration module provides an interface for users to configure a set of model parameters. The parameters are used to customize a precompiled atmospheric model. The configuration of model parameters will be recorded in a model launcher file. One module run is associated with one launcher file. Launcher files are created through the model launcher module with parameters specified from the model configuration module. Besides the model parameters, the launcher file also stores Linux command lines to start a model run and store the status of the model run. The model launcher files will be shipped to cloud instances.

Both the instance configuration and instance image are designed for configuring cloud instances. The instance configuration provides a set of parameters related to hardware configurations of cloud instances. Examples of hardware parameters are CPU speed, memory, and storage capacity. The instance image stores a copy of an operating system with a preconfigured model image. Considering the variations of cloud platforms, we only focus on the platforms that support Linux systems as most atmospheric models are designed to run on Linux systems. Scientists can create, modify, and delete the system image through this module. Upon the configuration, the configuration interface will initiate the process of creating cloud instances with the instance image. The cloud instance runs one or a set of model runs. This setup is useful when the user wants to create a limited number of instances for cost or other reasons. Launcher files will be dispatched to the cloud instances during the creation process. The data visualization module provides a set of multidimensional visualization functions through which users can view, manipulate, and analyze model outputs. More information for the visualization module can be found in Sun et al. (2012) and Li et al. (2013). The portal is hosted on a Web server. Besides a set of interfaces for configuring models and instances, the Web server also stores instance images.

Cloud Instances

Every cloud instance is an individual computing unit to perform one model run. The hardware configuration of the cloud instance is determined by users' selection through the configuration interface. When creating an instance, the instance image stored on the Web server is used. Both the model and hardware configuration can vary with users' selections. Each cloud instance includes one or multiple model launcher files. Each file is associated with one model run using a different set of model parameters.

Within the framework, the cloud instances can be created from a public cloud platform (e.g., Amazon EC2 and Microsoft Azure), or from a private cloud platform built upon open-source cloud solutions (Eucalyptus and OpenStack).

Although using a public cloud, users can configure the same number of instances as the number of model runs with each instance supporting one model run. Users can also configure a large number of model runs with fewer instances as each instance can perform multiple runs. For example, a user configures 10 instances and creates 100 model runs. Because the 10 instances can have different sets of hardware configurations, each instance may be assigned different number of launcher files depending on the hardware configuration. In this way, every model run is dispatched to one instance with hardware configuration tailored to the model run. When an instance starts, the instance will automatically run model simulations based on the information provided by launcher files one by one. During the running process, every instance will update its system status and model progress regularly. Once the simulation is finished, the results will be uploaded to cloud-based data storage and the instance will be released for other applications.

Cloud-Based Data Repository

To design the data repository, we consider the needs of the elasticity and the scalability of cloud storage solutions. First, depending on the configurations of model runs (e.g., 300 runs versus 3 runs), the volume of data sets produced by atmospheric modeling can vary significantly. Upon the configuration of model runs, the storage capability of the data repository will be adjusted according to the number of model runs and the estimated volume of outputs for each run. This process usually requires scaling up or down the data storage. The storage capacity should be adjusted at the modeling stage. Second, the volume of data sets stored in the data repository can change during the evaluation and manipulation. Scientists may visualize data to produce immediate data sets or delete model outputs upon evaluation. The capacity of data storage is also adjustable at the data analysis stage. Besides the scalability and the elasticity, cloud storage solutions allow scientists to store those data sets at distributed network locations to facilitate fast data access.

We designed a cloud-based data management module which includes a metadata system to store the metadata information for model outputs and a file system to store the model outputs. The metadata system stores model configurations, spatiotemporal information for model outputs, parameters

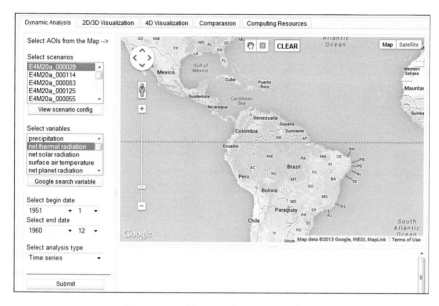

Figure 17.2 The visualization interface.

produced from model runs, paths pointing to data sets on the storage server, and other relevant information related to the model outputs. A set of query operations are provided to perform queries on metadata and retrieve associated data sets. With the data visualization interfaces provided by the Web portal, users can query against the metadata system to access the raw model outputs. The database is integrated with the visualization module seamlessly. Fig. 17.2 shows the interface of the visualization module supported by the data management system.

CASE STUDY: MODELE

ModelE: An Overview

To evaluate the feasibility of the framework, we selected ModelE (Hansen et al., 1983; Del Genio and Yao, 1993). ModelE is the latest GISS model incorporating numerous improvements in basic physics, stratospheric circulation, and forcing fields (Schmidt et al., 2006). Numerous projects have been conducted using the GISS ModelE (e.g., Koch and Hansen, 2005). ModelE simulates more than 300 variables at a global scale. Model outputs are provided in Network Common Data Format (NetCDF). The spatial resolution of model outputs is 4 degrees by 5 degrees in latitude and longitude.

We selected ModelE as an example for three reasons. First, because ModelE supports both parallel and serial processing, it can be deployed and executed with any instance with at least a CPU core. The low hardware requirement allows the model to be portable to different cloud environments. Scientists can configure different types of cloud instances to perform model runs. Second, the model supports both short-term and long-term simulations. Scientists can configure the model to simulate a short temporal scope to evaluate the performance of different instances and identify the most suitable hardware configurations to perform long-term model runs to better utilize cloud resources. Third, ModelE comes with input data sets and a model configuration file so limited configuration is needed to create and customize cloud instance images. This model is a good yet simple example of testing all functions of our framework.

Customizing ModelE to the Cloud Framework

To build a standard model image, we first installed the serial version of ModelE on a Linux cloud instance and then created a system image using the instance. The serial version was used for testing for three reasons. First, the performance of cloud resources is generally not comparable to HPC systems when used to run parallel applications. Second, the serial version can run with one instance, and can be used to identify the appropriate hardware configurations for running a particular model. Third, the serial version has lower hardware requirements and requires fewer configuration efforts. We chose a Community ENTerprise Operating System (CentOS) system as the operating system to build the model. The operating system was updated to meet the configuration of the model. Additional development tools and data packages have been installed as well. The complete list of configuration requirements of the model can be found at the ModelE official Web site. The configuration process may not apply to all operating systems. Thus, building an image is necessary to avoid any issues of building and configuring the model for multiple computing environments.

The basic image is stored on the Web server and will be used to create cloud instances. The image includes a basic operating system, all necessary packages for running the model, and a shared script for monitoring the progress of the model. The shared script takes the configuration of the model sent from user requests as the input and modifies the configuration of the model on the instance (see Script 17.1). Upon the modification, the script launches the model scenario and records the time cost of running the model scenario.

Script 17.1 An example of the script file for scenario configuration

```
#time:19491202
echo "Begin to run the model at " >logModel19491202.txt
cd decks
make setup RUN=E4M20one
cd ..
sleep 30 &
wait
date
sed -i 's/YEARE=1961,MONTHE=1,DATEE=1,HOURE=0/YEARE=1949,
MONTHE=12,DATEE=2,HOURE=0/g'
decks/E4M20one/I
date >>logModel19491202.txt
echo "----real start----" >>logModel19491202.txt
date >>logModel19491202.txt
time1=$(date +%s)
./exec/runE_new E4M20one -np 1
date >>logModel19491202.txt
echo "----real end----" >>logModel19491202.txt
time2=$(date +%s)
echo "Model finished in: " $(($time2-$time1))
>>logModel19491202.txt
```

Experiment 1: Identifying Appropriate Computing Configurations

To evaluate the performance of cloud computing solutions, we have conducted the tests using both public and private clouds: Amazon EC2 and NASA's Nebula. In this experiment, we aim to explore the performance of different computing facilities in supporting large-scale climate simulations. For each cloud, we configured and set up one-core, two-core, four-core, and eight-core machines (Table 17.1) to examine the roles of multicore computing units to improve the simulation. To compare the performance on clouds versus the local computing facilities, we also conducted the tests using a local server with eight cores. The configurations of computing facilities are recorded in Table 17.1. All instances or servers used in the experiments are 64-bit Linux systems with different numbers of CPU cores, CPU speed, and memory capacities. For each computing facility, we set up the simulation periods as 1 day, 2 days and 7 days with an executable script. As only monthly data sets are produced from the model, the experiments did not produce any physical model outputs.

Table 17.1 Configurations of cloud instances used for the experiments

Cloud environment	Core	Speed (MHz)	Memory (GB)
Amazon	1	2.3	0.5
	2	2.3	7
	4	2.3	7
	8	2.3	7
Local server	8	2.4	23
Nebula	1	2.9	2
	2	2.9	4
	4	2.9	8
	8	2.9	16

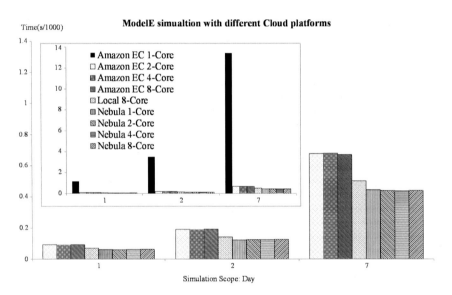

Figure 17.3 Execution time for ModelE simulation using different computing facilities.

Fig. 17.3 shows the execution time for simulation scenarios. The test results shows that the basic one-core Amazon instance with limited memory is insufficient to run a ModelE simulation and thus was excluded from the test scenario (The test results are included as an inset figure to the main performance figure). In Fig. 17.3, the total execution time increases as the simulation period increases, in general. This is because a scenario with a longer simulation time takes more computing time as would be expected. The overall performance of Nebula is best among the three groups of computing facilities mainly because the CPU speed of Nebula facilities is the fastest compared to the other facilities. The worst execution time is found when using the Amazon EC2 one-core machine with around 0.5 GB memory. The inadequate memory leads to the extremely low performance. However, when running the simulations using instances equipped with more CPU cores and better memory capacities, the improvement is not significant. For example, the performances of Amazon instances with two, four, or eight cores are similar. The performances of Nebula instances with 2, 4, 8, or 16 GB are similar as well. Because the serial processing version is used, the variations in CPU cores do not significantly influence the speed.

Besides the configuration of instances, we also noticed that virtualization of cloud-computing solutions may affect the performance. Although the CPU speed of the local server is slightly faster than that of Amazon instances (2.4 vs. 2.3 MHz), the time costs are much less (e.g., about 30% faster for 7-day simulation). By contrast, although the CPU speed of the Nebula instances is significantly faster than that of the local server (2.9 vs 2.4 MHz), the performances are similar (e.g., about 5% faster for 7-day simulation). To overcome the negative impacts of virtualization, scientists may configure instances with better computing capabilities than physical machines. To summarize, a cloud instance with a one-core CPU at a speed of 2.3 MHz and 4 GB memory seems to be sufficient to run the serial version of the model.

Experiment 2: Utilizing Cloud Computing to Support Massive Model Runs

Working with atmospheric scientists, we have conducted a simulation experiment with 300 model runs to produce 10-year simulation results using ModelE. Each model run was configured with a different set of climate parameters. There are three purposes of running this large number of model runs with cloud computing resources. First, the simulation examines the sensitivity of climate parameters through adjusting input parameters. Running models repetitively is an important way to evaluate the accuracy of atmospheric modeling. Second, it evaluates the robustness and the stability of cloud computing resources in supporting massive long-term simulations. Third, as the outputs from a large number of model runs are huge, it would be useful to evaluate the readiness of cloud storage solutions for managing large volumes of data.

Eucalyptus-Based Cloud Computing to Support Massive Model Runs

A 10-year simulation takes approximately 100 h to complete on a single computer with a 2.88 GHz CPU speed. We used the Eucalyptus-based

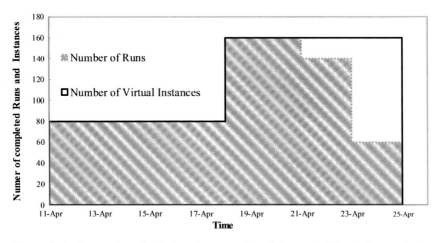

Figure 17.4 The number of virtual instances and model runs at different time periods.

cloud platform to spin off 80 and 160 instances. A virtual computing cluster is created based on these instances. The middleware Condor is used to schedule the tasks and collect results. At any time, each virtual instance was able to perform one model run. Fig. 17.4 shows the number of virtual instances to support model runs at different periods. On April 11, 2012, 80 instances were used to run the 300 tasks. Around April 16, the first 80 runs finished. Around April 18, another 80 instances were allocated to the cluster to speed up the simulation process. Since then, 160 instances were available for the model runs. On April 21, 2012, the 80 model runs starting from April 16 were completed. On that day, 80 instances were rearranged to run the remaining 60 tasks. Since that time, the total model running tasks by the system were 140. The 80 runs starting from April 18 were finished around April 23. With the support of the virtualized resources, the 300 runs were successfully completed within 14 days, rather than 1387 days had the 300 runs been executed by only one computer.

Experimental Result Analysis

Once a model run was completed, the model outputs were uploaded to the data repository for analysis. The total volume of data sets from the 300 runs is about 750 GB. Fig. 17.5 shows examples of the visualization functions. Fig. 17.5A shows the two-dimensional display of four climate parameters (i.e., *net solar radiation, net thermal radiation, net radiation of the planet, and surface air temperature*) of the same time step. This function allows users to view multiple atmospheric parameters at the same time step to evaluate the

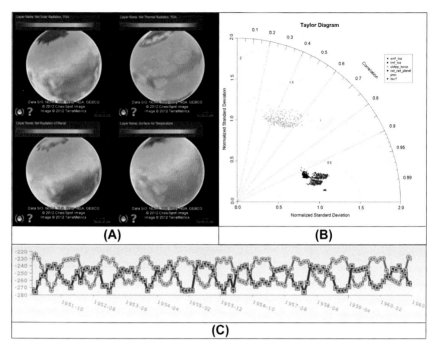

Figure 17.5 Examples of visualization: (A) basic comparative data view, (B) Taylor diagram, and (C) time series plots of data from two AOIs.

relationships among multiple climate parameters. Fig. 17.5B is an example of the Taylor diagram designed to quantify the correlation between the aggregated results of a variable from model simulations and the aggregated results of the same variable from observations. According to Fig. 17.5B, the accuracy of different parameters varies. Precipitation (denoted as yellow dots) is the most accurate. Fig. 17.5C shows the comparison of the same climate parameters from two Areas of Interest (AOIs) over years. This function is useful to examine the latitudinal component in determining the variations of atmospheric parameters.

DISCUSSION AND CONCLUSION

Cloud computing techniques have gained popularity in atmospheric applications. This chapter describes a cloud-based computing framework to support atmospheric modeling. We first discussed different computing paradigms for atmospheric modeling with an emphasis on applicability of each paradigm to support atmospheric modeling. We then described the design and the implementation of a cloud-based computing framework in detail. In this framework, cloud computing resources are used to support model runs and store model related data sets. To demonstrate the usability of the framework, we introduced the customization and experiments of running ModelE with the framework. According to the implementation and the experiments, we have found that cloud computing solutions provide potentials and benefits for atmospheric modeling that can be of great use to the atmospheric community.

First, compared to the traditional HPC systems, a cloud--based framework requires much less effort in hardware maintenance but provides a high degree of customization of cloud instances. As the hardware and the software requirements vary significantly with different atmospheric models, both computing facilities and operating environment should be configurable. With cloud platforms, a virtual image with model configured can be created as an instance template. Scientists can configure both instances and operating systems of their choice. In Experiment 1, when we change the configuration of the computing facilities, the performance of the model does not always improve linearly with the increase of computing power. This helps scientists configure instances tailored to particular models. Because the instances can be destroyed when finishing a project, no ongoing maintenance effort is needed.

Second, the load balancing and the autoscalability mechanisms of cloud computing can provide important cost benefits to run atmospheric modeling. Previous studies have shown that the computing performance of cloud resources is not comparable to HPC clusters (Huang et al., 2013b). According to our experiments, cloud computing facilities provide good performance for serial processing models which require fewer computational resources. Because of the scalability, scientists can launch a variable number of instances to run model simulations. Scientists can also decide how many resources are needed in the specific time range. If more resources are needed, scientists can configure more virtual machines whereas fewer instances can be created when fewer resources are needed. Besides launching a variable number of computing instances on demand, scientists can configure the storage capacity to store metadata and raw data sets. The volume is adjustable for model simulations.

However, cloud computing solutions may not be applicable for all atmospheric models, in particular for the models only run with parallel processing clusters due to the following reasons. First, because cloud instances are created with virtualization, which introduces additional computational and communication overhead, the performance can be much worse when running parallel processing version of models. Second, computing uses private or public networks to connect different virtual machines and to enable communications between instances. Although such a network connection has better network connection speed than the network speed between volunteer computing facilities, the speed does not meet the requirements of the high bandwidth for the paralleling processing models. If the modeling process needs to interact frequently and rapidly, cloud computing facilities are not applicable unless special optimizations have been applied. Third, although cloud computing is much cheaper than supercomputing, it is not as cost-effective as volunteer computing. In particular, because the cost of parallel processing is not evaluated in our studies, the cost of running long-term simulations is unknown. As the costs of cloud resources are determined by the instance types, storages, bandwidth, and other factors, a cost-evaluation mechanism is needed to apply cloud computing for various atmospheric models (Gui et al., 2014).

REFERENCES

Anderson, D., Fedak, G., May 16–19, 2006. The computational and storage potential of volunteer computing. In: Sixth IEEE International Symposium on Cluster Computing and the Grid, 2006. CCGRID 06, Singapore.

Armbrust, M., Fox, A., Griffith, R., Joseph, A.D., Katz, R., Konwinski, A., Lee, G., Patterson, D., Rabkin, A., Stoica, I., Zaharia, M., 2010. A view of cloud computing. Communications of the ACM 53 (4), 50–58.

Buyya, R., Yeo, C.S., Venugopal, S., Broberg, J., Brandic, I., 2009. Cloud computing and emerging IT platforms: vision, hype, and reality for delivering computing as the 5th utility. Future Generation Computer Systems 25 (6), 599–616.

Darch, P., Carusi, A., 2010. Retaining volunteers in volunteer computing projects. Philosophical Transactions of the Royal Society of London A: Mathematical, Physical and Engineering Sciences 368 (1926), 4177–4192.

Del Genio, D., Yao, S., 1993. Efficient cumulus parameterization for long-term climate studies: the GISS scheme. The Representation of Cumulus Convection in Numerical Models, Meteorological Monographs 46, 181–184.

Gui, Z., Yang, C., Xia, J., Huang, Q., Liu, K., Li, Z., Yu, M., Sun, M., Zhou, N., Jin, B., 2014. A service brokering and recommendation mechanism for better selecting cloud-services. PloS One 9 (8), e105297.

Hansen, J., Russell, G., Rind, D., Stone, P., Lacis, A., Lebedeff, S., Ruedy, R., Travis, L., 1983. Efficient three-dimensional global models for climate studies: models I and II. Monthly Weather Review 111 (4), 609–662.

Harvey, L.D., Schneider, S.H., 1985. Transient climate response to external forcing on 100–104 year time scales part 1: experiments with globally averaged, coupled, atmosphere and ocean energy balance models. Journal of Geophysical Research: Atmospheres 90 (D1), 2191–2205.

Huang, Q., Yang, C., 2011. Optimizing grid computing configuration and scheduling for geospatial analysis: an example with interpolating DEM. Computers & Geosciences 37 (2), 165–176.

Huang, Q., Yang, C., Nebert, D., Liu, K., Wu, H., November 2–5, 2010. Cloud computing for geosciences: deployment of GEOSS clearinghouse on Amazon's EC2. In: Proceedings of the ACM SIGSPATIAL International Workshop on High Performance and Distributed Geographic Information Systems, pp. 35–38 San Jose.

Huang, Q., Yang, C., Benedict, K., Chen, S., Huang, Q., Yang, C., 2013a. Utilize cloud computing to support dust storm forecasting. International Journal of Digital Earth 6 (4), 338–355.

Huang, Q., Yang, C., Benedict, K., Rezgui, A., Xie, J., Xia, J., Chen, S., 2013b. Using adaptively coupled models and high-performance computing for enabling the computability of dust storm forecasting. International Journal of Geographical Information Science 27 (4), 765–784.

Koch, D., Hansen, J., 2005. Distant origins of Arctic black carbon: a Goddard Institute for Space Studies ModelE experiment. Journal of Geophysical Research 110 (D4), Atmospheres (1984–2012).

Kondo, D., Javadi, B., Malecot, P., Cappello, F., Anderson, D.P., May 23–29, 2009. Cost-benefit analysis of cloud computing versus desktop grids. In: Parallel & Distributed Processing, 2009. IPDPS 2009, Rome, Italy, pp. 1–12.

Li, J., Jiang, Y., Yang, C., Huang, Q., Rice, M., 2013. Visualizing 3D/4D environmental data using many-core graphics processing units (GPUs) and multi-core central processing units (CPUs). Computers & Geosciences 59, 78–89.

Litta, A.J., Sumam, M.I., Mohanty, U.C., 2011. A comparative study of convective parameterization schemes in WRF-NMM model. International Journal of Computer Applications 33 (6), 32–39.

Massey, N., Jones, R., Otto, F.E.L., Aina, T., Wilson, S., Murphy, J.M., Hassel, D., Yamazaki, Y., Allen, M.R., 2014. Weather@home—development and validation of a very large ensemble modelling system for probabilistic event attribution. Quarterly Journal of the Royal Meteorological Society. http://dx.doi.org/10.1002/qj.2455.

Menemenlis, D., Hill, C., Adcroft, A., Campin, J.M., Cheng, B., Ciotti, B., Fukumori, I., Heimbach, P., Henze, C., Köhl, A., Lee, T., Stammer, D., Taft, J., Zhang, J., 2005. NASA supercomputer improves prospects for ocean climate research. EOS, Transactions American Geophysical Union 86 (9), 89–96.

Ramanathan, V., Coakley, J.A., 1978. Climate modeling through radiative–convective models. Reviews of Geophysics 16 (4), 465–489.

Schmidt, G.A., Ruedy, R., Hansen, J.E., Aleinov, I., Bell, N., Bauer, M., Yao, M.S., 2006. Present-day atmospheric simulations using GISS ModelE: comparison to in situ, satellite, and reanalysis data. Journal of Climate 19 (2), 153–192.

Sellers, P.J., Mintz, Y.C.S.Y., Sud, Y.E.A., Dalcher, A., 1986. A simple biosphere model (SiB) for use within general circulation models. Journal of the Atmospheric Sciences 43 (6), 505–531.

Sun, M., Li, J., Yang, C., Schmidt, G.A., Bambacus, M., Cahalan, R., Huang, Q., Xu, C., Nobble, E., Li, Z., 2012. A web-based geovisual analytical system for climate studies. Future Internet 4 (4), 1069–1085.

Xie, J., Yang, C., Zhou, B., Huang, Q., 2010. High-performance computing for the simulation of dust storms. Computers, Environment and Urban Systems 34 (4), 278–290.

CHAPTER 18

ERMA® to the Cloud

K. Sheets
NOAA, National Weather Service, Bohemia, NY, USA

A. Merten
NOAA, National Ocean Service, Seattle, WA, USA

R. Wright
NOAA, National Ocean Service, Silver Spring, MD, USA

INTRODUCTION

In 2011, the US White House and Congress challenged the executive departments to be cloud first with Information Technology. National Oceanic and
Atmospheric Administration's (NOAA) Environmental Response Management Applications (ERMA®) was an early candidate for moving an application from local hardware resources to the cloud. ERMA is an online-mapping
tool designed for environmental disaster response (e.g., oil spills, hurricane
impacts) that meets the varying demands for information technology (IT)
through the lifetime of a disaster and the response to the disaster (Maps and
Spatial Data, 2015). At NOAA, this temporal continuum is typically viewed
as having several phases: (1) the emergency response phase (hours to weeks),
(2) injury assessment phase (weeks to years), and (3) restoration phase (months
to decades). The emergency phase, in which fast response is paramount and
the scramble for information is high and often accompanied by high public
interest, generates the heaviest demand on IT infrastructure. During this first
phase, bandwidth is needed to support distributing information or interactive
information for constituents and stakeholders while also supporting the addition of large volumes of data to the databases used to create the map layers
being accessed through the Web-enabled user interface/experience. The second phase, injury assessment—can begin by overlapping with emergency
response, but it often then lasts much longer. In the last phase restoration,
there is a greater need for computing resource for analysis, but less bandwidth
necessary for high volumes of Website visits, as not as many people are

Cloud Computing in Ocean and Atmospheric Sciences
ISBN 978-0-12-803192-6
http://dx.doi.org/10.1016/B978-0-12-803192-6.00018-9

Copyright © 2016 Elsevier Inc.
All rights reserved.

365

following the case with the same intensity and frequency as they do during the emergency response phase. Much of the data needed for restoration analyses is loaded into ERMA during the response phase of an incident. This model of having varying IT performance needs at different phases within the response continuum caused NOAA managers to designate ERMA as an ideal early candidate for the cloud. The acquisition of computing resources on an as-needed basis rather than having to buy and maintain physical servers built to handle maximum loading would maximize resource availability only when needed to accommodate supporting an incident of national significance and potentially save resources and costs at other times. The major cloud service providers include options for failover across geographic locations as well as machine to machine, which is an added benefit when natural disasters can take down critical infrastructure such as power grids in an affected region.

ERMA® is an online mapping tool that integrates both static and real-time data, such as Environmental Sensitivity Index (ESI) maps, ship locations, weather, and ocean currents, in a centralized, easy-to-use format for environmental responders, decision makers, and the public. ERMA enables a user to quickly and securely upload, manipulate, export, and display spatial data in a Geographic Information System (GIS) map. Common datasets used for an oil spill incident can be trajectory models, field collections of the shoreline, photos, overflight information, and resource location and deployment status. Developed by NOAA and the University of New Hampshire, through the Coastal Response Research Center, with funding from the US Environmental Protection Agency, US Coast Guard, and US Department of the Interior, ERMA provides environmental resource managers with the data necessary to make informed decisions for environmental response (2015). ERMA was declared the US Government's Common Operation Picture (COP) for the Deepwater Horizon Oil Spill disaster by that incident's National Incident Commander. ERMA is still in use for visualizing and accessing data collected for the response and the complex injury assessment (Fig. 18.1). ERMA was a primary location for the public to access data and information about the spill and aided in government transparency. On the first day that ERMA went live publicly, it received over 3.4 million hits, requiring replicated databases and load-balancing coding that dispersed the direct load on the infrastructure. It was this experience that drove the ERMA team toward exploring cloud infrastructure as potentially the most efficient and effective way of managing information during a disaster. Since the Deepwater spill, ERMA has been used as a COP during the response to Hurricane Sandy (Fig. 18.2), several smaller spills and drills, and for situational awareness in the Arctic.

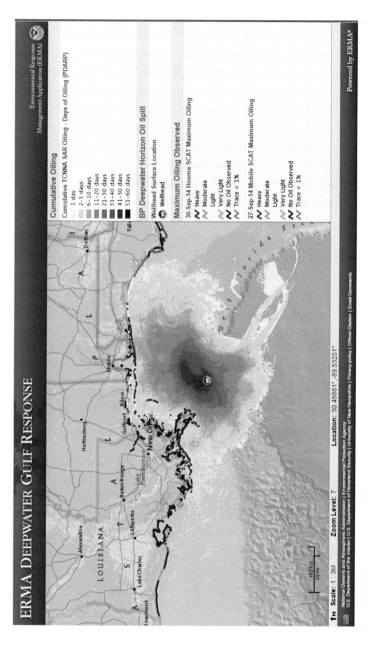

Figure 18.1 ERMA Deepwater Gulf Response contains a wide array of publicly available data related to the 2010 Deepwater Horizon oil spill in the Gulf of Mexico. This image displays the maximum observed shoreline oiling and the total days of observed oil in the Gulf of Mexico.

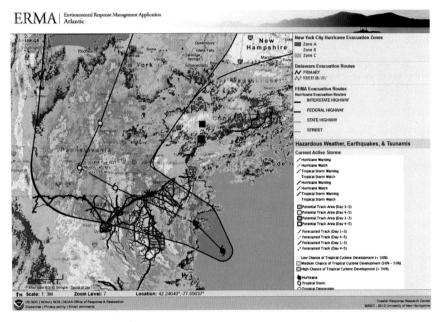

Figure 18.2 Atlantic ERMA shows the track of Sandy as it approached landfall in New Jersey. The map shows evacuation zones, real-time road closures, and weather warnings as were used as a reference during the response following landfall of the storm.

THE PROCESS OF MOVING TO THE CLOUD

As NOAA's first operational cloud services contract, it was incumbent that NOAA establish solid, comprehensive requirements. The requirements ended up falling into four categories: type of cloud service, computing infrastructure requirements, IT security, and contracting.

When considering a cloud implementation, the first question to answer is whether one needs Infrastructure as a Service (IaaS), Platform as a Service (PaaS), or Software as a Service (SaaS). In the case of ERMA, an alternative to owning and maintaining hardware was the primary appeal of a cloud solution. This translates to IaaS, meaning hardware is provided in the cloud and the government would provide the operating system and anything built on top of the operating system. PaaS would have relied on the cloud service provider (CSP) to provide the operating system as well as the computing resources. SaaS would have meant turning ERMA over to the cloud service provider to be reconstructed and future development would be done by the CSP and their contract partners. This shift of control of software and security adds time and money to the software development life cycle.

Documenting the computing resource requirements was more of a challenge than originally anticipated. Cloud virtual machines and sizes do not directly correlate to more traditional hardware such as blade servers housed in racks. The requirement for the equivalent computing power was included in the performance work statement, but the lack of a crosswalk between existing computing resources and cloud-based computing resources created a serious challenge in estimating cost. One of the key concepts of cloud computing is only buying what you use, which works well when running a large computer model, getting outputs and then turning off the machines until the next large model run. In the case of an operational Web-based application, there is never total downtime. It is necessary to keep an instance running for the Web application to function even if it is in "waiting mode." Given the application can never be down, this introduced the question of how to properly scale up resources to transition to a larger virtual machine without any downtime. What is the cloud equivalent of adding more cores to the environment? These considerations ultimately led to a more holistic redesign of ERMA software configuration than originally anticipated for the migration.

SECURITY CONSIDERATIONS

The Federal Information Security Management Act (FISMA) establishes the level of documentation, assessment, and monitoring required for information technology systems. The government-wide Federal Risk and Authorization Management Program (FedRAMP) provides cloud service providers levels of certification to be considered by Federal organizations. The certification level depends on the system security needs for given FISMA categories (low, moderate, or high) and for what level of services (IaaS, PaaS, or SaaS). This was both timely and helpful as IaaS CSPs were being approved as ERMA submitted the performance work statement for cloud services. However, this only certified the infrastructure; it was NOAA's responsibility to certify the application. Any prime and subcontractors with access to either or both systems also required information security documentation, which is discussed in more detail in the following. All in all, IT security requirements were largely tailored for traditional, physical on-site hardware, so documenting a virtual system according to the physical system requirements was uncharted territory.

CONTRACTING, PROCUREMENT, AND PLANNING

Given the recent cloud-first focus of the Federal government, NOAA expected many of its offices and projects would be interested in establishing cloud service contracts. However, much like the IT security standards, the NOAA IT contracting process is geared toward traditional, physical systems or professional services contracts. These factors meant that additional staff from the office of the Chief Information Officer (CIO) and Acquisitions and Grants Office (AGO) were interested in contributing to the acquisition process for cloud services so this initial process could be leveraged for future projects. For the most part, these added perspectives were helpful, though they did prolong the development of documentation and required the ERMA project managers to focus ancillary people back to the specific technical needs and status of ERMA.

Once a contract for cloud infrastructure services was awarded, the CSP and ERMA teams, including government information technology security officers (ITSOs), met regularly to plan the migration of ERMA to the cloud. At this early phase in the migration, documenting to the level required to satisfy NOAA IT security requirements was a major focus of our work. The IaaS was with Amazon, but was acquired through a subcontractor of our prime contractor. Amazon has two tiers of security documentation, a redacted version they share with partners and a broader version they share directly to the Federal government. Once this documentation was provided, the NOAA ITSO also required the prime and subcontractor to provide artifacts showing that their systems and interaction with the Amazon cloud did not introduce new risks to ERMA. The efforts to sort out the documentation for the ERMA team to begin working on the cloud took approximately 5 months from contract award.

SYSTEM DESIGN

Although the ERMA team had experience with using private cloud infrastructure, no one had implemented an operational system, which required 24 h/7 days a week support with geographic redundancy. The requirement that there be no downtime raised questions about some of the key design elements to ERMA so that the system could be switched from one cloud farm to another with no impact to the end user. For this reason, as ERMA was being transitioned to the cloud, the project as a whole required reconstruction to make the most of a cloud infrastructure operating environment.

The precloud ERMA architecture comprised several independent Web sites, each supported by its own database. The cloud rearchitecture consolidated these regional databases into a single national database. The new database architecture and workflow added complexity to testing the migration project. The project was no longer simply a move from one environment to another, but a change in the coding behind some of the key functionality within ERMA. These significant changes required an exhaustive test plan with an "all hands on deck" approach for the team using ERMA. Every use case, as well as load tests, had to be fully tested before confidence could be gained that ERMA was ready to be operational on the cloud.

The team experimented with the types of virtual machines available on the Amazon cloud. Unfortunately, running ERMA on the largest cloud instance possible was slower than the dedicated cores allocated to ERMA on the physical hardware in the legacy systems. This forced the team to investigate alternative designs and use of the cloud resources. Because ERMA is also required to retain backups of the system, for case histories and legal record retention, the questions of the types of storage and how quickly historical data could be produced were explored. Once a satisfactory new design was settled upon, hundreds of hours of translating code and copying files between systems remained before the system could be up and ready for testing.

PROJECT MANAGEMENT

Despite having to address additional IT security requirements and a redesign of the system, NOAA management had expected a quicker transition and dramatic cost savings with moving to the cloud. The migration of ERMA to the cloud had been interpreted by decision makers to be an instant cost-saving option for NOAA. The savings would come from reduced hardware costs, because they are leased, and the user only having to pay for the computing needed on-demand, instead of buying, in advance, anticipated computing/storage space, in the event of a large disaster that may or may not ever occur. These "selling points" were an oversimplification of the overall migration process, so time had to be taken to educate the full team on why cloud migration was challenging, time-consuming, and likely more expensive, at least with the initial investment. Given the complexities of this migration, management required continuous reporting of complex challenges in nontechnical terms, and explanations of why those challenges affected schedules, budgets, and viable mitigation strategies. The varied levels of experience with cloud computing, emerging requirements and standards, and high visibility, caused this project to undergo multiple rebaselines to adhere as much as possible to traditional project management principles.

Rebaselining multiple times was inevitable even though project management best practice of allocating 20% margins in both budget and schedule was followed when the project was initiated. Migrating ERMA from physical traditional hardware to a cloud infrastructure was a pioneering maneuver in which there were very few or no examples to follow, including agency policies. Being first has many unknowns, so consider adding more time and budget to your planning. For example, create a budget line item for unknowns with up to 50% of initial estimates for unexpected expenses. In the case of ERMA, this would have funded contracts with FISMA security documentation specialists, additional computing power for development and experimentation, etc. When scheduling, add at least 10–15% more time than developers estimate for all tasks and then add standard margins to the increased estimate. All of this equates to adding 50% margin in both time and budget.

A waterfall-system life-cycle design was followed for this major migration project. In hindsight, a more iterative or agile approach may have made it a smoother transition. There was about a year of back-end code redevelopment followed by 2 months of testing cycles of one to two weeks duration before ERMA was ready to go live on the cloud. Once ERMA was deployed and analytics were run, the development team was able to "right size" the cloud computing resources and eventually draw down some costs. In the time since establishing requirements for the cloud service provision request for proposal and now (July 2015), cloud service providers have generated tables to relate physical hardware computing power (e.g., number of cores and amount of RAM) to virtual cloud-computing instances. The equivalency tables revealed initial instances stood up for ERMA had less computing power than the physical systems previously powering ERMA. System loads were monitored for periods after going live to determine which systems within the cloud infrastructure were either overtaxed or underutilized. Several iterative adjustments were made to the systems to ensure they were all performing well with each other, so as not to have a bottleneck of competing processes resulting in a subpar, or slow, application user experience. Changes were made on both the live production site as well as the failover site to ensure that if the failover site needed to be activated, it was configured with the proper resources.

LESSONS LEARNED

The ERMA migration to the cloud experience yielded the following lessons learned for implementing new technology, especially in a virtual environment:

1. Changing the infrastructure of a major IT application or system is not a trivial task.
2. Know your system requirements and stick to them.
3. Ensure the entire team, from program managers to IT administrative staff to end users, is engaged and onboard beginning with the design phase through deployment.
4. Include cloud-knowledgeable IT security staff from the onset.
5. Budget for significant overlap of legacy and new systems to allow for full performance comparison and failover.
6. Keep your stakeholders informed.
7. Cost savings with the cloud are not instantaneous, but a long-term reality. In fact, system computing resource costs will likely increase the first several years.
8. The cloud is reliable with little to no downtime.

Also, before accepting a mandate to migrate a project to cloud, consider if a cloud environment is well suited for the project. Some key things to keep in mind for this decision are:

1. Is your system dependent on computing resources 24/7?
   If yes, then cost savings for the project may not be as dramatic as marketing materials suggest because the project will not be able to release resources and therefore must pay for every type of machine being used, even if you are not fully using the available computing power. Adjusting the size of the cloud machine becomes the most effective way to adjust and scale costs.
2. Does the system depend on high availability or high data integrity (see FISMA categories, 2015)?
   If yes, as of July 2015, a FISMA "high" approved cloud service provider does not exist, so cloud migration is not an option for a FISMA high system.
   Cloud migration is a viable option for FISMA "moderate" or "low" systems.
3. Does the system require "local" storage?
   It is difficult to interface local storage with cloud architecture, so it is advised to add in schedule slack to account for custom coding and

design documentation for interfacing different security systems. If you chose to continue with implementing a cloud-computing solution, it is best to account for migrating and gaining security certification for the entire system to be cloud based (computing and storage instances).

Cloud computing is touted as the obvious solution for reducing computing costs when compared to traditional on-site hardware maintenance. The additional option of virtual machines, even in a private farm, is overlooked. Virtual machines, whether privately operated or as part of a commercial cloud, are the future for computing systems. Efforts should also be made to align private virtual machine farms, sometimes referred to as a private cloud, to the same technology being used in commercial cloud farms. This common design will allow hybrid solutions to be developed when development and other low or moderate security systems can leverage commercial clouds while having a clear path to migrate to private virtual machine environments which are the preferred solutions for systems with high security ratings. As more government systems become operational on the cloud many of the risks and issues encountered with ERMA will likely reduce or disappear as policy and governance catches up with the technology. The Cloud is a viable end-to-end solution for low or moderate FISMA systems.

REFERENCES

ERMA, May 26, 2015. Web Application: Deepwater Gulf Response Environmental Response Management Application. National Oceanic and Atmospheric Administration. Retrieved from: http://gomex.erma.noaa.gov/.
FISMA, 2015. Standards for Security Categorization of Federal Information and Information Systems. National Institute of Standards and Technology, Gaithersburg, MD. Retrieved from: http://csrc.nist.gov/publications/fips/fips199/FIPS-PUB-199-final.pdf.
Maps and Spatial Data, May 26, 2015. Retrieved from: http://response.restoration.noaa.gov/erma/.

CHAPTER 19

A Distributed, RESTful Data Service in the Cloud in a Federal Environment—A Cautionary Tale

R. Mendelssohn, B. Simons
NOAA/NMFS/SWFSC, Santa Cruz, CA, USA

INTRODUCTION

In the 1960s–1970s, companies like Advanced Data Processing (ADP) and Electronic Data Systems (EDS) rose on the premise that they would simplify information technology (IT) for firms by running and maintaining the computers and software needed by the firms, and the firms would access the information remotely. Computers and software to perform the same processing were also available to companies to run in-house. Some companies chose to use these services, some companies on evaluation chose to run in-house using vendor hardware, and some companies chose to use the services but then purchased their own equipment instead.

Today, "the Cloud" plays a similar role, promising to ease the burden of IT infrastructure and computational costs by storing data and running programs and applications on equipment housed elsewhere. Since the 1960s–1970s, technology has greatly improved in its ability to interact with and retrieve information remotely and the same or similar technologies are available to be implemented locally. For example, Apache has a suite of software projects aimed at distributed computing and databases (http://projects-old.apache.org/indexes/quick.html). There are also solutions such as OpenStack (https://www.openstack.org) and OpenCompute (http://www.opencompute.org), as well as many flavors of Linux that come with many of features of "Cloud computing" built in. Ubuntu (http://www.ubuntu.com/download/cloud), among others, has a distribution specifically designed for building an in-house enterprise cloud.

Therefore, despite all the intensive cloud promotion, the situation is very much like it was back in the 1960s–1970s, and it is a classical economic problem: whether it is preferable to lease or purchase infrastructure needed

Cloud Computing in Ocean and Atmospheric Sciences
ISBN 978-0-12-803192-6
http://dx.doi.org/10.1016/B978-0-12-803192-6.00019-0
Copyright © 2016 Elsevier Inc.
All rights reserved.
375

to do business. Viewed in this manner, the pros and cons of putting services in the cloud can be examined in a rational way, rather than just rushing to implement the latest fad.

This is particular true in a Federal (or other government) environment, because of limitations due to budget cycles, fixed budget amounts, one-time resources, contract restrictions, inability to borrow funds, and the fact that "success" can lead to greater costs but without the concomitant greater income. As we will see, these factors can greatly change the relative benefits of the cloud versus in-house services.

We examine these issues in the context of the Environmental Research Division's Data Access Program (ERDDAP), a data Web service and middleware. A porting of ERDDAP to Amazon Elastic Compute Cloud (EC2) that allows multiple copies to be spun up as needed is described in Meisinger et al. (2009). As part of the Federal GeoCloud project, an ERDDAP instance was run on Amazon EC2 also in 2009, and costs and performance were analyzed. In the end, providing in-house services appeared the better option.

ENVIRONMENTAL RESEARCH DIVISION'S DATA ACCESS PROGRAM

ERDDAP is a Representational State Transfer (RESTful) Web data service designed by Bob Simons, which when installed provides not only the services for machine-to-machine interaction but also a Web page built on top of the services for human interaction. ERDDAP acts a middleware between the user and various local and remote data (files, databases etc.) and data services, as well as distributed and federated access through services (Fig. 19.1). In a distributed system, the actual data are stored in different locations but accessed through common services. In a federated service, multiple instances of the service can appear to the user as if it is just one system–this can be used for load balancing at a single location as well as for unifying a distributed system.

ERDDAP allows users to subset data so that they get only the data desired, and it can return the data in a variety of formats such as Gridded data formats (http://coastwatch.pfel.noaa.gov/erddap/griddap/documentation.html#fileType) and Table data formats (http://coastwatch.pfel.noaa.gov/erddap/tabledap/documentation.html#fileType), regardless of the original format of the data. Thus, ERDDAP acts as a data service, a data service federator, and a data service middleware, as it can ingest data from a

ERDDAP
Easier access to scientific data

Brought to you by NOAA NMFS SWFSC ERD

ERDDAP > List of All Datasets

Or, Do a Full Text Search for Datasets: ⦿ Search

Or, Search for Datasets by Category:
cdm_data_type, institution, ioos_category, keywords,
long_name, standard_name, variableName

Or, Search for Datasets with Advanced Search ⦿

Pick a Dataset

943 matching datasets, listed in alphabetical order.

Grid DAP Data	Sub-set	Table DAP Data	Make A Graph	W M S	Source Data Files	Title	Sum-mary	FGDC, ISO, Metadata	Back-ground Info	RSS	E mail	Institution	Datas
		data				* The List of All Active Datasets in this ERDDAP *	⦿	M	background			NOAA NMFS SWFSC	⦿ allDatasets
data		data	graph	M		AMSRE Model Output, obs4MIPs NASA-JPL, Global, 1 Degree, Monthly	⦿	F I M	background ⊘	RSS	✉	Remote Sensing	jplAmsreSstMon
		data	graph		files	AN EXPERIMENTAL DATASET: Underway Sea Surface Temperature and Salinity Aboard the Oleander	⦿	M	background ⊘	RSS	✉	NOAA OAR AOML	nodcPJJU
data		data	graph	M		Aquarius Sea Surface Salinity, Version 2, DEPRECATED, Global, 2011-2014, 3-Month	⦿	F I M	background ⊘	RSS	✉	NASA/GSFC OBPG	jplAquariusSSS3M
data		data	graph	M		Aquarius Sea Surface Salinity, Version 2, DEPRECATED, Global, 2011-2014, 7-Day	⦿	F I M	background ⊘	RSS	✉	NASA/GSFC OBPG	jplAquariusSSS7D
data		data	graph	M		Aquarius Sea Surface Salinity, Version 2, DEPRECATED, Global, 2011-2014, Daily	⦿	F I M	background ⊘	RSS	✉	NASA/GSFC OBPG	jplAquariusSSSDa
data		data	graph	M		Aquarius Sea Surface Salinity, Version 2, DEPRECATED, Global, 2011-2014, Monthly	⦿	F I M	background ⊘	RSS	✉	NASA/GSFC OBPG	jplAquariusSSSMo
data		data	graph	M		Aquarius Sea Surface Salinity, Version 4, Global, 2011-2015, 3-Month	⦿	F I M	background ⊘	RSS	✉	NASA/GSFC OBPG	jplAquariusSSS3M
data		data	graph	M		Aquarius Sea Surface Salinity, Version 4, Global, 2011-2015, 7-Day	⦿	F I M	background ⊘	RSS	✉	NASA/GSFC OBPG	jplAquariusSSS7D
data		data	graph	M		Aquarius Sea Surface Salinity, Version 4, Global, 2011-2015, Daily	⦿	F I M	background ⊘	RSS	✉	NASA/GSFC OBPG	jplAquariusSSSDa
data		data	graph	M		Aquarius Sea Surface Salinity, Version 4, Global, 2011-2015, Monthly	⦿	F I M	background ⊘	RSS	✉	NASA/GSFC OBPG	jplAquariusSSSMo
		data	graph			Argo Float Data from the APDRC DAPPER Server	⦿	M	background ⊘	RSS	✉	NOAA PMEL	apdrcArgoAll
data		data	graph	M		AVISO Model Output, obs4MIPs NASA-JPL, Global, 1 Degree, Monthly	⦿	F I M	background ⊘	RSS	✉	Centre National	jplAvisoSshMon
	set	data	graph			CalCOFI Cruises	⦿	F I M	background ⊘	RSS	✉	NOAA SWFSC	erdCalCOFIcruise
	set	data	graph			CalCOFI Egg Counts	⦿	F I M	background ⊘	RSS	✉	NOAA SWFSC	erdCalCOFIeggcr
	set	data	graph			CalCOFI Egg Counts Positive Tows	⦿	F I M	background ⊘	RSS	✉	NOAA SWFSC	erdCalCOFIeggcr
	set	data	graph			CalCOFI Egg Stages	⦿	F I M	background ⊘	RSS	✉	NOAA SWFSC	erdCalCOFIeggst
	set	data	graph			CalCOFI Fish Counts	⦿	F I M	background ⊘	RSS	✉	NOAA SWFSC	erdCalCOFIfishcn
	set	data	graph			CalCOFI Fish Counts Positive Tows	⦿	F I M	background ⊘	RSS	✉	NOAA SWFSC	erdCalCOFIfishcn
	set	data	graph			CalCOFI Fish Sizes	⦿	F I M	background ⊘	RSS	✉	NOAA SWFSC	erdCalCOFIfishsiz
	set	data	graph			CalCOFI Larvae Counts Positive Tows	⦿	F I M	background ⊘	RSS	✉	NOAA SWFSC	erdCalCOFIlrvcnt
	set	data	graph			CalCOFI Larvae Counts, Scientific Names A to AM	⦿	F I M	background ⊘	RSS	✉	NOAA SWFSC	erdCalCOFIlrvcnt
	set	data	graph			CalCOFI Larvae Counts, Scientific Names AN to AR	⦿	F I M	background ⊘	RSS	✉	NOAA SWFSC	erdCalCOFIlrvcnt
	set	data	graph			CalCOFI Larvae Counts, Scientific Names AS to BA	⦿	F I M	background ⊘	RSS	✉	NOAA SWFSC	erdCalCOFIlrvcnt
	set	data	graph			CalCOFI Larvae Counts, Scientific Names BCE to BZ	⦿	F I M	background ⊘	RSS	✉	NOAA SWFSC	erdCalCOFIlrvcnt
	set	data	graph			CalCOFI Larvae Counts, Scientific Names C to CE	⦿	F I M	background ⊘	RSS	✉	NOAA SWFSC	erdCalCOFIlrvcnt
	set	data	graph			CalCOFI Larvae Counts, Scientific Names CD to CH	⦿	F I M	background ⊘	RSS	✉	NOAA SWFSC	erdCalCOFIlrvcnt

Figure 19.1 A partial display of some of the 943 datasets served by the Environmental Research Division (ERD) ERDDAP server. ERDDAP is a Web service, and this is a built-in Web page that comes with the service, though the Web page is not needed to use the service. All services in ERDDAP are defined by a universal resource locator (URL), and anything that can send a URL and receive a file can use the services.

number of other data services. This helps with the problem that many scientists spend a lot of their time reformatting data and can be reluctant to explore new data sources because of their fears about the effort needed to use new data formats. To the extent possible, ERDDAP standardizes dates and coordinate variables (latitudes, longitudes, etc.). ERDDAP provides powerful search capabilities, Federal Geospatial Data Committee (FGDC), and International Organization for Standardization (ISO) metadata, and on-the-fly visualization of the data, and everything is a service completely defined by an URL.

ERDDAP, because of its flexibility, is ideally suited to provide services for a distributed and federated data system, providing a consistent interface and base URL regardless of the type or location of the data. When a data request is made to ERDDAP for a dataset that is served by a remote but federated ERDDAP server, the "local" ERDDAP server seamlessly transfers the user's request to the remote ERDDAP, and the response is returned to the user from the remote ERDDAP. This is in contrast to when ERDDAP accesses a dataset from a remote location and returns the subset. In the federated case, the request is passed to the remote ERDDAP server and the data are returned from there, whereas in the latter case the data must be passed by the remote service to ERDDAP, reformatted, and sent to the user. The federated system removes one possible bottleneck in the delivery of the data.

Presently, ERDDAP is mostly used as a single-server application. However, as mentioned, there have been two efforts to put ERDDAP on the cloud, as well as a discussion of ERDDAP in the context of "Heavy Loads, Grids, Clusters, Federations, and Cloud Computing" (http://coastwatch. pfeg.noaa.gov/erddap/download/grids.html#cloudComputing). In the first effort, described in Meisinger et al. (2009), ERDDAP was divided into three different types of processes: (1) the ERDDAP utility that provides the Web interface and the transformation engine; (2) the ERDDAP crawler, which regularly queries the data sources for updates of data and metadata; and (3) the Memcache component that is the shared state efficiently distributed between all instances. The prototype was designed to allow for adding and removing instances of each component flexibly as needed.

In the second effort, which was part of the Federal GeoCloud project, the aim was to see if there was indeed better bandwidth and response for the same or better costs by putting the service as well as some of the data in the cloud, and with the added possibility of spinning up more copies of

ERDDAP as needed. From these experiences we have identified the following places where bottlenecks can arise in an ERDDAP application:

- **A remote data source's bandwidth**—Unless a remote data source has a very high bandwidth Internet connection, ERDDAP's responses will be constrained by how fast ERDDAP can get data from the data source.
- **ERDDAP's server's bandwidth**—Unless the server hosting an ERDDAP instance has a very high bandwidth Internet connection, ERDDAP's responses will be constrained by how fast ERDDAP can get data from the data sources and how fast ERDDAP can return data to the clients. The only solution is to get a faster Internet connection.
- **Memory**—If there are many simultaneous requests, ERDDAP can run out of memory and temporarily refuse new requests. (ERDDAP has a couple of mechanisms to avoid this and to minimize the consequences if it does happen.)
- **Hard drive bandwidth**—Accessing data stored on the server's hard drive is vastly faster than accessing remote data. Even so, if the ERDDAP server has a very high bandwidth connection, it is possible that accessing data on the hard drive will be a bottleneck.
- **Too many files in a cache directory**—ERDDAP caches all images, but only caches the data for certain types of data requests. It is possible for the cache directory for a dataset to have temporarily a large number of files. This will slow down requests to see if a file is in the cache. This is likely a rare bottleneck, and the parameter in setup.xml lets you set how long a file can be in the cache before it is deleted. Setting a smaller number would minimize this problem.
- **CPU**—Making graphs is the only thing that takes significant CPU time (roughly 1 s per graph). So if there were many simultaneous unique requests for graphs (a single Web Map Service often makes six simultaneous requests), there could be a CPU limitation. On a multicore server, it would take a lot of requests before this became a problem.

Relatively speaking, hard drives and the Internet have significant latency problems, and the Internet has a bandwidth problem. Looking at the sequence of events it takes to request and receive data from an ERDDAP server, the time for the request to get to ERDDAP is trivial. What matters most is the time it takes for ERDDAP to fulfill the request once it has it in hand, which is constrained by the hard drive (for nonmetadata requests), the CPU, and memory (notably image requests), as well as the time for the response to be sent back to the user.

If we put this in the context of the cloud, there might be performance gains from faster bandwidth, the ability to store multiple copies of the data, and spinning up multiple copies of the ERDDAP service. However, multiple copies of ERDDAP will rarely provide a performance gain, all else being equal, unless there are multiple copies of the data in fast storage at the same cloud location, and similarly if most the data are distributed and not at the location of the cloud service, the faster bandwidth will not have much effect on the overall system performance, as the speed of the remote data provider will be the limiting factor.

This is a nontrivial point, because cost comparisons are often done with the cheaper, slower storage, which are not suitable for this type of data service, whereas the appropriate comparison is with the more expensive fast storage.

Costs

The costs discussed here are for Amazon EC2 (https://aws.amazon.com/ec2/) as of March 2009 when we did the Federal GeoCloud project. Although the costs of data storage on the cloud have dropped significantly since then, cost of local storage has also dropped significantly (the same $38K that bought an enterprise level 192 terabyte (TB) redundant array of independent disks (RAID) can now buy a much faster and improved 288TB RAID), so the relative story has not changed.

As discussed above, a data service like ERDDAP needs fast data access, and our experience is that all of the less expensive cloud-storage options, such as Amazon Simple Storage Service (S3), were much too slow to be of any practical use. Access to data in Amazon S3 is also cumbersome, as the block access needed to efficiently subset a field is not provided unless something akin to the FUSE-based file system backed by Amazon S3 (s3fs) (https://github.com/s3fs-fuse/s3fs-fuse) is used; s3fs allows S3 files to be accessed as if they were stored in a regular file system for a Unix-based system, but this capability comes with a great performance cost.

At that time Amazon was charging $150/month per TB for the faster storage. At that price, 192TB of storage for a year would cost $345,600 compared to the $39,000 the equivalent RAID cost us. Moreover, the RAID has next business day (NBD) coverage of parts for 3 years, and these enterprise-level RAIDs normally last 5 years or more. Even just taking the 3-year comparison, even assuming our costs for Amazon are off by an order of magnitude, there is a real cost for the storage. And as we noted, to really be able to overcome a data access bottleneck by being in the cloud, we would need multiple copies of the data, with the concomitant increase in costs.

Amazon also charges for input/output (I/O); in March 2009 when we did the project it was $0.10/GB for data transfers into and out of the system. As discussed in the following, even if the costs are equivalent over a 5-year period, there are other issues with using cloud services. No matter what the requests to ERDDAP, there will be a data transfer out of the system, and if the request is for a subset of a remote dataset accessed through a service, there will be a data transfer into the system. Also, the data are constantly being updated, so there will be an I/O cost to the updates.

Many of the datasets we presently serve at ERD are very high-resolution satellite data, on the order or a kilometer resolution or better, so it is very easy to have very large requests. For example, for a dataset at 0.01 degree a request for one 2×2 degree square would have 400 series multiplied by the number of time periods requested.

At a very modest 50TB/day of combined data update, remote data access and data delivery, we are looking at I/O charges of $1825/year. But for a larger and more successful operation, say with combined I/O of 500TB/day, we are looking at $18,250 a year, and so on. Moreover, most importantly, this charge is open-ended, a point we will return to the following.

We want to emphasize that we are not saying that Amazon or other cloud services are overpriced. Besides better bandwidth, they provide automatic duplication, power backup, and server and drive maintenance, which may offset other costs of the services, and are important considerations in looking at the trade-offs in using a cloud-based service. As we said in the introduction, there are trade-offs between buying equipment (in-house services) and renting (cloud-based services) and the trade-offs are not as one-sided as much of the cloud promotion would have you believe.

WHY A FEDERAL (OR OTHER GOVERNMENTAL) SETTING MATTERS

We have seen that for a Web data service, cloud computing has some potential benefits, but it also has a number of drawbacks including cost, risks, and whether in practical terms it solves the real bottlenecks in the service. Even if we assume for the sake of argument that the fixed costs for a cloud solution and for a local solution are about the same over say a 5-year period, operating in a Federal environment changes the equation. This is because:
- Increasingly how funds are being obtained
- Constraints on contracts
- Higher usage does not imply higher revenue
- URL restrictions

How Funds Are Being Obtained

Increasingly funds for Web data services in the Federal government are given in blocks, in what amounts to an internal grant application system. Even if the costs to store the data locally and the costs to store the data on the cloud service are the same over a 5-year period (which it is not), the RAID can be bought in the fiscal year that the funds are obtained, and will last for the 5-year period. To pay for the equivalent cloud storage, a contract must be given to the service provider, and this leads to the second problem.

Constraints on Contracts

Increasingly contracts in the Federal government are limited to a 1-year period of performance, or at best 2 years. Any large IT-related contract is a very difficult process, and a long-term and large IT-related contract is an extremely difficult process. The upshot of this is given, say, a $40K grant, a RAID that will last for 5 years can be bought fairly readily, but it will be almost impossible to contract out the same funds for cloud storage over a 5-year period because contracts cannot be of that length, and funds cannot be carried over between fiscal years. Thus there are real, practical impediments for paying for cloud services.

In addition, what happens if we store the data in the cloud for several years, and then do not get promised funding for the next year? How is the data service to be continued? If you cannot afford the cloud storage, you most likely do not have the funds to purchase local storage, and if you have been purchasing local storage for backup, why pay the double cost of local storage plus cloud storage?

More Is Not More

As we have discussed, I/O costs for most cloud operations are not fixed but are based on usage. For a private company, if the service is a success, and usage is greater than initially budgeted for, presumably that means a larger income to pay for the increased usage, or at least the ability to borrow funds to tide over to the next budgeting cycle. Basically, more usage means more income. This is not the case in the Federal government. Even assuming a contract has been worked out, and even assuming that there is a mechanism for adding funds to that contract, because the services are not free of charge, more is not more, increased usage does not generate increased income. If usage is much higher than that budgeted for, the funds can run out before the end of the contract with no way to pay for an extension. This could mean the shutting down of the service for an extended period of time. Or conversely, if extra funds are allocated for the service as a possible buffer, and demand is less than what was expected, once the period of performance of the contract ends, the unspent funds are lost.

In contrast, the cost of the local storage and servers is a fixed cost, and the cost of Internet access is generally a fixed cost that is not usage dependent. On the cloud, bandwidth costs are per GB. The more you use, the more you pay. Locally, bandwidth costs are fixed. At any given moment, if the bandwidth is totally utilized, each additional user causes all users to get less bandwidth. It is nice that the cloud scales bandwidth as needed, but it is impossible to budget for cloud-bandwidth costs. If you set limits for cloud-bandwidth costs (e.g., $X can be spent in a given month), then the service will just stop when the limit is reached. With local bandwidth, budgets can be drawn with certainty to have enough funds to cover the activities required by the data service.

Lest this be thought of just a "scare" against cloud services, the ERD-DAP server at ERD increasingly is serving very high-resolution satellite data, less than a kilometer in resolution in some cases, daily in temporal extent. We have never been able to forecast when a new or past user will start to access these datasets, or how frequently, as research projects and research and management needs vary considerably from year to year. Thus, the amount of access year over year has continually surprised us, and it would be unlikely that we could accurately price a 1-year cloud service contract. The possibility of having to shut down the service is real, and is something people often do not think about.

But, but

But can you not just store the data locally as a backup to the cloud service and switch to local services? First, we wish that we had those kinds of resources. But the reality is that for the amount of storage required for ERD's data services and the funds we are able to obtain, it is an either/or situation.

There is a further technical issue, and that has to do with URLs. As of this writing, National Oceanic and Atmospheric Administration (NOAA) has not allowed NOAA URLs to be on anything but the NOAA network, and similarly no non-NOAA URL can be on the NOAA network. So the only way to run the service on the cloud is either to use a non-NOAA URL, which raises issues of branding of the service and official clearance, or to use a cloud service as a back end that the user does not see. However, in most cases this back-end approach means that the local Internet speed becomes the main bottleneck, and it is unclear then that there are advantages being gained by the use of the cloud service. Moreover, this would mean that if it were necessary to switch the ERDDAP service from being cloud-based to being locally hosted, the base URL of the ERDDAP service would be different, breaking any scripts, programs, or Web pages using the URLs.

CONCLUSION

Certainly for a large-scale, one-time (or infrequent) computation problem, the ability to pay for the computation power of a cluster only for as long as needed, an ability which is provided by the cloud, and not to invest in rarely needed computation resources, would be a very enticing proposition. Private businesses also may face different costs and benefits than those faced by agencies of the Federal government, which is the reason the chapter has the title it does. A major point is that even if the costs in-house and in the cloud are the same over a 5-year period for a large, fast, distributed 24/7 Web data service, limitations due to how Federal funding works, and Federal rules for contracting and for use of funds can make it almost impossible to take advantage of a cloud-based service.

The main thing is to demystify "the cloud." Cloud services are just large, centralized providers of the hardware (servers, storage, and networking devices) and software (clusters and distributed file systems), hardware, and software that are available for purchase by a Federal agency or private business. The question then does come down to when is it advantageous to buy and when is it advantageous to rent.

ACKNOWLEDGMENTS

The views or opinions expressed herein are those of the author(s) and do not necessarily reflect those of NOAA or the Department of Commerce. Any mention of trade names does not imply endorsement by NOAA.

REFERENCE

Meisinger, M., Farcas, C., Farcas, E., Alexander, C., Arrott, M., de la Beaujardiere, J., Hubbard, P., Mendelssohn, R., Signell, R., 2009. Serving ocean model data on the cloud. In: OCEANS 2009, MTS/IEEE Biloxi – Marine Technology for Our Future: Global and Local Challenges, October 26–29, 2009, pp. 1–10.

CHAPTER 20

Conclusion and the Road Ahead

M. Yuan

Geospatial Information Sciences, School of Economic, Political, and Policy Sciences,
University of Texas at Dallas, Richardson, TX, USA

Cloud computing is complex and evolving. It took the US National
Institute of Standards and Technology (NIST) more than 3 years and 16
revisions to reach a definition of cloud computing in late 2011 (Mell and
Grance, 2011). According to NIST, the transformative computing paradigm
possesses five essential characteristics, three service models, and four deployment models, all of which are discussed in this edited collection of *Cloud
Computing in Ocean and Atmospheric Sciences*.

With advances in environmental sensing and data acquisition, demands for
computation and management continue to grow exponentially, but meanwhile
the demands fluctuate with occurrence and development of the events of interest. The five essential characteristics of cloud computing nicely address the challenge with on-demand self-service, broad network access, resource pooling,
rapid elasticity, and measured service. In contrast to other computing paradigms
that rely on institutional or regional network clusters, cloud computing deploys
its resources and capabilities in three service models of Software as a Service
(SaaS), Platform as a Service (PaaS), and Infrastructure as a Service (IaaS).

This edited book provides numerous implementations of these service
models that are tailored to search, disseminate, analyze, and visualize data
for ocean and atmospheric applications. Organizations work to deploy
existing environmental models and tools in the cloud, such as Unidata on
Thematic Real-Time Environmental Distributed Data Services
(THREDDS) and Advanced Weather Interactive Processing System II
(AWIPS II), the European Centre for Medium-Range Weather Forecasts
on high-resolution forecast (HRES) and ensemble forecast (ENS),
National Aeronautics and Space Administration (NASA)'s Modern-Era
Retrospective Analysis for Research and Applications (MERRA) data and
analytics service, and United Kingdom Met Office on four-dimensional
(4D) visualization of high-resolution numerical weather prediction data.
Collaborative research projects bring community efforts to develop the

Cloud Computing in Ocean and Atmospheric Sciences
ISBN 978-0-12-803192-6
http://dx.doi.org/10.1016/B978-0-12-803192-6.00020-7

Copyright © 2016 Elsevier Inc.
All rights reserved.

National Flood Interoperability Experiment, Atmospheric Analysis Cyberinfrastructure, a Polar Cyberinfrastructure Portal, PolarHub, the General Circulation Model (GCM) ModelE, and Global Earth Observation System of Systems interoperability for Weather, Ocean and Water (GEOWOW). Cloud computing offers four deployment models to make available the computing resources and capabilities in private cloud, community cloud, public cloud, or hybrid cloud according to organizational needs and computational considerations. Examples of these deployment models are also discussed in this collection.

Preferences for cloud-based solutions prevail as people grow accustomed to doing things on any device anywhere anytime. This is particularly desirable for ocean and atmospheric sciences as weather and climate is ubiquitous, and so is the need for information. This edited book provides several cloud-based solutions to disseminate data rapidly and make information accessible across platforms. With IaaS, cloud computing offers applications and virtual machines that can packaged into a Docker Container, for example, to achieve portability and flexibility for running on a public cloud or a private cloud. PaaS provides application development engines that can easily leverage data and libraries in the cloud to create Web or mobile applications. SaaS offers tools or utilities that can be consumed directly remotely. The late comer to the cloud-service family, InfoaaS, has processed information ready for consumption.

Ocean and Atmospheric communities in academia, industry, and government agencies have experimented with all the service models with success. LiveOceanServer, for example, ports the capabilities of the regional ocean modeling system (ROMS) to the Microsoft Azure cloud with an application programming interface (API) to allow data access and with middleware to translate forecast data into consumable information on the client side. NASA's MERRA Analytics Service and the associated Climate Data Services API work together to converge data management and data analytics that traditionally are handled separately. Various open source projects provide computing frameworks to handle big data in distributed and paralleled processing. Geographic information system (GIS) Tools for Hadoop, for example, interface MapReduce and Amibari/Oozie to enable geoprocessing tools to manage and analyze massive data in the cloud. The convergence harmonizes data flows and analytical tools, and consequently analytical processes can be rerun as needed for collaboration and reproducibility. The grand vision of e-Science and open-science is being realized with cloud-based solutions.

Many chapters in the book elaborate on the key challenges to cloud computing. Technological challenges on Internet connectivity and bandwidth, while grand and critical, may be the simplest to address. Organizational challenges could be difficult to overcome. Some organizations may not be able to accommodate pay-as-you-go models, for example, according to their existing procurement and budgeting processes. Many organizations are concerned with security and privacy issues that result from having information resources in the cloud and handled off site. Technological and organizational challenges hamper the Cloud First Initiative in the US Federal Government (Figliola and Fischer, 2015). In a 2010 White House report, Vivek Kundra, the then US Chief Information Officer, issued a 25-point implementation plan to reform federal information technology (IT) management and directed a shift to a cloud-first policy. In 2011, he outlined a strategy to migrate 25% (or $20 billion) of the Federal Government's Information Technology spending to cloud-computing solutions (Kundra, 2011). In the past 4 years, the migration has, however, been slow. A 2015 report by the Congressional Research Service identified five challenges in federal implementations of cloud services: meeting federal security requirements, overcoming cultural barriers with agencies, meeting new network infrastructure requirements, having appropriate expertise for acquisition processes, and funding for implementation. The report also concluded that budget concerns and data-center consolidation are two main drivers to cloud adoption by federal agencies.

From a long-term perspective, cloud services are expected to lower the overall IT cost through savings on maintenance, personnel, space, utilities, and new system acquisitions. Meanwhile, the Federal Data Center Initiative launched in 2010 with a plan to close 1100 of the total federal data centers in 2015 with an expected saving of $3 billion by 2015. As a result, federal agencies turn to cloud solutions for data storage and management. Overall, the outlook for cloud computing is promising. Both technological and organizational challenges are being addressed and will be eventually resolved. Large-scale Internet investments by government agencies (such as the US Lambda Rail (http://www.nlr.net/) and National Broadband Plan (https://www.fcc.gov/national-broadband-plan)) and by industry (such as Google Fiber (https://fiber.google.com/about/)) can significantly improve connectivity and bandwidth issues. Research in cloud computing has been developing bandwidth-elastic or bandwidth-aware solutions to alleviate bandwidth constraints (Popa et al., 2013; Lin et al., 2014). Ocean and atmospheric science communities have already made significant progress in building cloud-based cyberinfrastructure for data dissemination, data analytics, model runs, ensembles, and scientific visualization. This progress will continue making ocean–atmospheric data more accessible, facilitating data integration and model validation, promoting collaboration in research and education, democratizing ocean and atmospheric research, and broadening science participation.

Cloud computing is transforming how we access information, share content, and communicate. In the road ahead, three key research directions in cloud computing can continue to benefit ocean and atmospheric sciences: *Everything as a Service (XaaS)*, *Big Data mobile computing*, and *fog computing*. The idea of Everything as a Service originated from SaaS, IaaS, and PaaS (Robinson, 2008). Later, XaaS expands to Desktop as a Service, Healthcare as a Service, Marketing as a Service, and Privacy as a Service (Duan et al., 2015). It conveys the idea of providing data, functions, and utilities as services via the cloud. The chapter on Climate Analytics as a Service is very much on the forefront of XaaS. It is easy to envision other possibilities, like Flood Forecast as a Service, Wildfire Monitoring as a Service, or Eddy Tracking as a Service in the near future. Essentially, users are able to access computing and information resources on demand over the Internet, instead of being constrained by hardware and software available on premises. Cloud computing has already been shown to benefit small start-up businesses on application developments. Likewise, cloud computing can enable research teams in small institutions access to advanced computing facilities that usually are only housed in large research centers. This trend democratizes science participations and promotes XaaS innovations.

Cloud-supported Big Data computing and analytics will be critical to advancing cloud computing for ocean and atmospheric sciences. For ocean and atmospheric sciences, the main Big Data challenges arise from the spatiotemporal scales, the magnitude of data, and the difficulties associated with validation of long-term predictions based on models (Kambatla et al., 2014). Constant flows of sensor data pose intensive demands on data processing, management, storage, dissemination, analytics, and visualization. Transmitting diverse, massive amounts of ocean and atmospheric data over the Internet may not be feasible. Technological advances to enable Big Data computing and analytics in the cloud can alleviate the demands. New cloud solutions are being developed to provide building blocks for stream and complex event processing (Assunção et al., 2015). This book includes three pioneer projects that push cloud-supported Big Data computing and analytics for atmospheric research: Unidata efforts to cloud-enabled AWIPS II,

Environmental Systems Research Institute (ESRI) efforts to interface Hadoop and MapReduce for climate data analysis in the cloud, and the United Kingdom's National Weather Service efforts to use cloud services for a 4D browser visualization of environmental data. Overall, only a few tools are available for Big Data analytics in the cloud (Hashem et al., 2015). Although the computational complexity is daunting for cloud-supported Big Data computing and analytics, the now common expectation of information being accessible via mobile devices pushes for mobile cloud computing to enable computational intensive mobile applications with support from services of computational clouds (Fernando et al., 2013). New research is being developed with application partitioning and offloading to incorporate cloud-computing features in mobile applications (Khan et al., 2014). Innovations in cloud-supported mobile Big Data computing can be of great use to realize ubiquitous ocean and atmospheric research and operations.

The third promising development is fog computing that extends cloud computing and services to the edge of the network. Furthermore, fog computing pushes for applications and services that do not fit well in the cloud paradigm, including applications in which knowing where the computation and storage take place is important (e.g., videoconferencing), geo-distributed applications (e.g., sensor networks), fast mobile applications (e.g., peer-to-peer vehicle communications), and large-scale distributed control systems (e.g., smart grids) (Bonomi et al., 2014). Fog computing is also the key to Internet of Things (IoT) technologies by connecting the cloud and smart devices. With high-speed Internet connections to the cloud and physical proximity to smart devices, fog services enable real-time applications, location-based services, and mobility support (Stojmenovic, 2014). Smart weather sensors are already available to automatically adjust landscape-watering schedules according to actual weather conditions, for example. Fog computing provides the intermediate computing hubs (or mini clouds) to process data close to the sensors and pull necessary resources from and communicate the results to the designated cloud services. The fog also allows sensors or devices to become the virtualization platforms themselves, and hence they can provide computing and storage capacities as needed (Vaquero and Rodero-Merino, 2014). Such capabilities suit well the distributed nature of environmental observation networks and can reduce demands on data transmissions.

Advances in information and computing technologies have revolutionized practices in many sciences. Ocean and atmospheric sciences are among those disciplines that embrace innovations in information technologies and computing to help scientific advances and facilitate collaboration. Cloud computing introduced a new paradigm that changes the way we think about computing resources. Cloud computing accelerates e-Science and open-science and furthermore broadens science participation through cloud services. Ocean and atmospheric science communities are making good progress in migrating data, tools, and models to the cloud. A number of common challenges exist on connectivity, bandwidth, and procurement. New research developments have proposed technological solutions to overcome the Internet limitations. In the United States, cloud-first and associated policies are setting the stage for possible changes to procurements. Cloud computing is making transformative impacts in information management and technologies at the national scale. New technological trends in XaaS, cloud-supported Big Data mobile computing, and fog computing bring new possibilities to overcome technological challenges in cloud computing to make computing resources and scientific findings even more accessible and better facilitate scientific collaboration and discovery. This book gives an overview of the key developments of cloud computing in ocean and atmospheric sciences. Research and developments highlighted in the chapters have established a solid foundation for the next phase of bigger than cloud computing that spreads across XaaS, is able to analyze Big Data via mobile devices, and establishes fog services on smart mobile sensors for real-time analytics of the environment.

REFERENCES

Assunção, M.D., Calheiros, R.N., Bianchi, S., Netto, M.A., Buyya, R., 2015. Big Data computing and clouds: trends and future directions. Journal of Parallel and Distributed Computing 79, 3–15.

Bonomi, F., Milito, R., Natarajan, P., Zhu, J., 2014. Fog computing: a platform for internet of things and analytics. In: Big Data and Internet of Things: A Roadmap for Smart Environments. Springer International Publishing, pp. 169–186.

Duan, Y., Cao, Y., Sun, X., June 2015. Various "aaS" of everything as a service. In: 16th IEEE/ACIS International Conference on Software Engineering, Artificial Intelligence, Networking and Parallel/Distributed Computing (SNPD), 2015. IEEE, pp. 1–6.

Figliola, P.M., Fischer, E.A., 2015. Overview and Issues for Implementation of the Federal Cloud Computing Initiative: Implications for Federal Information Technology Reform Management. Congressional Research Service, Library of Congress. http://www.fas.org/sgp/crs/misc/R42887.pdf (accessed 16.10.15.).

Fernando, N., Loke, S.W., Rahayu, W., 2013. Mobile cloud computing: A survey. Future Generation Computer Systems 29 (1), 84–106.

Hashem, I.A.T., Yaqoob, I., Anuar, N.B., Mokhtar, S., Gani, A., Khan, S.U., 2015. The rise of "big data" on cloud computing: review and open research issues. Information Systems 47, 98–115.

Kambatla, K., Kollias, G., Kumar, V., Grama, A., 2014. Trends in big data analytics. Journal of Parallel and Distributed Computing 74 (7), 2561–2573.

Khan, A.R., Othman, M., Madani, S.A., Khan, S.U., 2014. A survey of mobile cloud computing application models. Communications Surveys & Tutorials, IEEE 16 (1), 393–413.

Kundra, V., 2011. Federal Cloud Computing Strategy. http://www.mail.governmenttrainingcourses.net/pdfs/Federal-Cloud-Computing-Strategy1.pdf (accessed 16.10.15.).

Lin, W., Liang, C., Wang, J.Z., Buyya, R., 2014. Bandwidth-aware divisible task scheduling for cloud computing. Software: Practice and Experience 44 (2), 163–174.

Mell, P., Grance, T., 2011. The NIST Definition of Cloud Computing. http://faculty.winthrop.edu/domanm/csci411/Handouts/NIST.pdf (accessed 13.10.15.).

Popa, L., Yalagandula, P., Banerjee, S., Mogul, J.C., Turner, Y., Santos, J.R., August 2013. Elasticswitch: practical work-conserving bandwidth guarantees for cloud computing. In: ACM SIGCOMM Computer Communication Review, vol. 43(4), pp. 351–362 ACM.

Robison, S., 2008. The Next Wave: Everything as a Service. Executive Viewpoint. www.hp.com.

Stojmenovic, I., November 2014. Fog computing: a cloud to the ground support for smart things and machine-to-machine networks. In: Telecommunication Networks and Applications Conference (ATNAC), 2014 Australasian. IEEE, pp. 117–122.

Vaquero, L.M., Rodero-Merino, L., 2014. Finding your way in the fog: towards a comprehensive definition of fog computing. ACM SIGCOMM Computer Communication Review 44 (5), 27–32.

INDEX

Note: Page numbers followed by "f" indicate figures, "t" indicate tables.

Printed in the United States
By Bookmasters